P9-DHD-857

The Honors Class

The Honors Class

Hilbert's Problems and Their Solvers

Benjamin H. *Yandell*

A K Peters
Natick, Massachusetts

Editorial, Sales, and Customer Service Office

A K Peters, Ltd.
63 South Avenue
Natick, MA 01760
www.akpeters.com

Library of Congress Cataloging-in-Publication data

Yandell, Ben.
The honors class : Hilbert's problems and their solvers / Benjamin H. Yandell.
p. cm.
Includes bibliographical references and index.
ISBN 1-56881-141-1
1. Mathematics–History–20th century. 2. Mathematicians–Biography. I. Title: Hilbert's problems and their solvers. II. Title.

QA26 .Y36 2001
510′.9′04–dc21 2001036795

Printed in Canada
06 05 04 03 02 10 9 8 7 6 5 4 3 2 1

$$\lim \frac{(\text{Editing}_{\text{Janet Nippell}} - \text{Coauthorship})}{\text{Time}} \rightarrow 0$$

Thank you.

Table of Contents

Acknowledgments

Many people read parts of this book in manuscript, or in a few cases, the manuscript as a whole. They have helped improve it in various ways—most often by pointing out errors, sometimes by suggesting new material or anecdotes to add. I would like to thank Juan Carlos Alvarez, Tom Apostol, Karin Artin, Michael Artin, Tom Artin, Nina Baym, Andrei Bolibruch, Paul Cohen, Martin Davis, John W. Dawson Jr., Eva Dehn, Helmut Dehn, H. M. Edwards, Benedict Freedman, Andrew Gleason, James Glimm, Yulij Ilyashenko, Irving Kaplansky, Viatcheslav Kharlamov, Stephen Kleiman, Peter Lax, Trueman MacHenry, Yuri Manin, Yuri Matiyasevich, Masayoshi Nagata, Virginia Narehood, Maria Dehn Peters, Hilary Putnam, Constance Reid, Nancy A. Schrauf, James Serrin, Jim Stasheff, John Stillwell, and Oleg Viro.

I want to especially thank Martin Davis for service beyond the call of duty and Benedict Freedman (one of few mathematicians to have written a best-selling novel) for early, coherent criticism of bad habits in my prose style. Any errors that remain are my fault.

Thanks to Mary Jane Hayes at the Institute for Advanced Study for her help.

Thanks to reference librarians all over the country. These are very nice people.

Thanks to many people for the use of photographs.

Thanks to Nancy A. Schrauf and Sabine Gahlmann for helping me with sources in German. Nancy made the translations from German quoted in the text, unless the source identifies a translator—particularly those from Siegel, Braun, Schneider, and about Schneider.

My friend Doug Rabin redrew the figures.

Thanks to Ariel Jaffee, Kathryn Maier, and Klaus Peters at A K Peters.

Introduction:
The Origin of the Coordinates

Introduction

At the Second International Congress of Mathematicians in Paris in 1900, David Hilbert delivered a talk, "Mathematical Problems." He spoke of the "deep significance of certain problems for the advance of mathematical science," saying:

> If we would obtain an idea of the probable development of mathematical knowledge in the immediate future, we must let the unsettled questions pass before our minds and look over the problems which the science of today sets and whose solution we expect from the future. To such a review of problems the present day, lying at the meeting of the centuries, seems to me well adapted. For the close of a great epoch not only invites us to look back into the past but also directs our thoughts to the unknown future.

A great problem must be clear, he said, because "what is clear and easily comprehended attracts, the complicated repels us. . . . [It] should be difficult in order to entice us, yet not completely inaccessible, lest it mock our efforts." He talked about how the efforts to solve famous historical problems had enriched mathematics and in conclusion presented a list of unsolved problems, now known as "Hilbert's problems."

Solving one of Hilbert's problems has been the romantic dream of many a mathematician. Hermann Weyl wrote that by solving one, a mathematician "passed on to the honors class of the mathematical community." In the last hundred years, solutions and significant partial results have come from all over the world. Hilbert's list is a thing of beauty, and aided by their romantic and historical appeal, these well-chosen problems have been an organizing force in mathematics.

Hilbert quotes an "old French mathematician": "A mathematical theory is not to be considered complete until you have made it so clear that you can explain it to the first man whom you meet on the street." Mathematics has come a long way from the time when that goal seemed reasonable, expanding and fragmenting to the point that it is inconceivable that one person could survey the whole and make a similar list of problems. In the worst case such a list would be the work of a series of committees with no common members. The International Mathematical Union did better than this in 2000, when it published *Mathematics: Frontiers and Perspectives* ("It is inspired by the famous list of problems that Hilbert proposed") but still credited four editors, had thirty signed articles, and was 459 pages long. Some of Hilbert's problems are indeed technical, but others are reassuringly down-to-earth: the irrationality and the transcendence of certain numbers, the solvability of diophantine equations, the question of how common prime numbers are as we count higher and higher. These are not easy to solve, but it is possible to understand what they are about.

Hilbert's emphasis on problems grounded mathematics in curiosity. One of his own contributions was a simplified proof that π is a transcendental number (not the solution to any polynomial equation with whole numbers as coefficients, "transcending" the realm of algebra). Pi arrived early on the human scene, out of the relationships between the diameter, circumference, and area of a circle; *e*, another transcendental number, pops out of calculus. The seventh problem asks if $2^{\sqrt{2}}$ is transcendental. The transcendental numbers are superabundant but hard to snare, and transcendence, though simply defined, is devilishly difficult to prove. This is the kind of hauntingly simple, mysterious question that has always intrigued students of mathematics.

Hilbert's Göttingen is still remembered as a Camelot for mathematics and physics. All roads led to Göttingen from 1900 to 1933, just as the mathematical community there had designated a point on the town square "the origin of the coordinates." The mathematics of quantum mechanics reigns in what is called "Hilbert space." This flow of mathematicians through Göttingen did not diminish the attractiveness of Hilbert's problems.

I'll follow the progress of the problems. Who worked on them? Who solved them? How were they solved? Which have not been solved? And what developed in twentieth century mathematics that Hilbert left out? The people and ideas involved in the solutions to Hilbert's problems provide a connect-the-dots for the mathematical culture of the first half of the twentieth century and the first strokes toward a picture of the second half.

Advice: How to Read This Book

> "That I have been able to accomplish anything in mathematics is really due to the fact that I have always found it so difficult."
>
> –*David Hilbert*[1]

My mathematical readers know how to read this book. They will read it the way they read mathematics books. If they don't understand something (a state they are used to), they will keep reading in the hope that they will understand the next thing. They will skip sections that don't interest them. So this is to my nonmathematical readers. If you are reading for the story, keep reading if you don't understand something. Skip a bit if you want—the biographical narrative will pick up again. Pretend you are reading *Moby Dick* and you've come to another section on whaling. However, entertain the possibility that some of the most vivid material might be in the passages explaining the mathematics.

This book presents the story of all of Hilbert's problems. A few sections will probably be of interest only to people reading for the mathematics. These sections tend to be short, don't have much biographical story, and appear later. If you are reading for story, don't miss the final three chapters, on Poincaré, on Kolmogorov, and "We Come to Our Census."

David Hilbert.

The Origin of the Coordinates

David Hilbert and Göttingen

To understand something of Hilbert's problems and their solvers we must start with the "origin of the coordinates," as Göttingen mathematicians jokingly called a point on the town square from which four churches could be seen. That origin is David Hilbert and Göttingen. What follows is a sketch. Readers who want more background, or who simply want to read a beautiful book, should go to Constance Reid's biography, *Hilbert*.

The son of Otto and Maria Hilbert, David Hilbert was born on January 23, 1862, near Königsberg, East Prussia, to a family of professionals at a time when high middle-class culture was flourishing. His father was a judge. One uncle was a lawyer, another the director of an academic high school, or gymnasium. His paternal grandfather was a judge as well, distinguished enough to be called *Geheimrat*, a title carrying almost no duties and great honor that Hilbert would one day also possess. The German states were uniting and preparing to enter the period of nationalist expansion and prosperity that would end in 1914 when Hilbert was fifty-two. This was also a time of cultural expansion, a time rich for mathematics.

Königsberg had a distinguished mathematical tradition. Its observatory had been established by Friedrich Wilhelm Bessel (1784–1846), for whom the Bessel functions so useful to physics are named. Carl Jacobi (1804–1851) taught at the university. Jacobi's replacement, Friedrich Richelot (1808–1875), had discovered the work of Karl Weierstrass (1815–1897), then teaching at a gymnasium in a small town. Richelot journeyed there to award Weierstrass an honorary doctorate from Königsberg. Weierstrass moved on to the university in Berlin and during Hilbert's youth was the most distinguished mathematician in Germany. He was at the center of an effort to introduce "rigor," by which mathematicians generally mean a water-tight consistency among the parts of a subject, into analysis, that is, calculus and its extensions. Invented independently by Isaac Newton (1642–1727) and Gottfried Wilhelm Leibniz (1646–1716), calculus had been intuitive and pragmatic prior to that time. Phrases like "infinitely small" (but apparently big enough to matter) were used freely. Mathematicians of the mid-nineteenth century were disturbed by this. Weierstrass, building on work by Augustin-Louis Cauchy (1789–1857) and others, took concepts like "limit" or "continuous" and defined them rigorously for the first time. In a way that is common in mathematics and would be a hallmark of Hilbert's contribution as well, understanding the difficulties of exact statement opened up new territory for concrete, constructive work. Weierstrass introduced the epsilons and deltas that are used to define the basic concepts of calculus. They did not spring, full blown, from the heads of Newton and Leibniz.

The philosophical tradition of Immanuel Kant (1724–1804) had an almost palpable presence in Königsberg. Every year on Kant's birthday, the crypt next to the Königsberg cathedral was opened and fresh wreaths of laurel placed on his bust. Kant was concerned with how we know things and had argued that our knowledge of certain mathematical truths was *a priori*, there from the beginning, not learned from experience. He argued that the basic concepts of arithmetic were *a priori*, as were the basic concepts of geometry.

Mathematicians would, as a practical matter, come to side with Kant on arithmetic. But on geometry they had already shown by the time of Hilbert's birth that the concepts manifested in Euclid's axioms are based on experience, because we are in fact free to make logically consistent geometries that differ from Euclid's. The discovery and elaboration of non-euclidean geometries, as they gradually became more clearly and widely understood in the 1820s through 1870s, wrought a revolution in mathematics. Hilbert was born into a time when "intuition" could no longer be fully trusted. Every assumption had to be reconsidered. Hilbert was perfectly suited to this task.

During Hilbert's childhood another boy came to Königsberg, Hermann Minkowski (1864–1909), whose life would be linked with his, though the two did not meet until their university years. Successful merchants, the Minkowskis fled persecution in Russia in 1872, starting anew in Königsberg, exporting white linen rags. Hilbert and Minkowski illustrate two distinct types of mathematical talent. Hilbert was not thought to be exceptional as a child. He remembered himself as slow. It was said that he had trouble understanding new ideas. Hilbert had to work something out for himself before he could understand it. He had difficulty with the memorization involved in studying languages at the gymnasium. In contrast, Minkowski was a child prodigy. His mathematical ability was obvious. He had a prodigious memory, read Shakespeare and Schiller, and memorized most of Goethe. He understood new ideas rapidly, and in his home such speed must have seemed almost normal. Minkowski's brother Oskar later established, in 1889, that there was a sugar-regulating substance in the pancreas (subsequently isolated and called insulin). Hilbert progressed through school in an orderly fashion and graduated with distinction. As mathematics came more into the curriculum, his excellence, originality, and industry became recognized. Minkowski raced through gymnasium, rapidly surpassed his teachers in mathematics, and graduated in five-and-a-half years instead of the usual eight. Though two years younger, he enrolled at the university before Hilbert.

The gymnasiums provided a broad, liberal education for the sons of the merchant and professional classes throughout the German-speaking world, which included much of central Europe. (Königsberg, for instance, some 350 miles northeast of Berlin, is now part of Russia, renamed Kaliningrad, with Poland in between.) Appreciation of music and literature was high and science flourishing. For those so inclined, mathematics had an appeal as obvious as poetry. *Gymnasium* in German, as well as English, connotes a building-up of the brain through exercise, and mathematics was regarded as an important form of mental exercise. Teachers were held in high esteem, and as noted in the example of Weierstrass, some of the foremost scientists and thinkers of Germany spent part or even all of their careers at the gymnasiums.

Upon graduation Hilbert enrolled at the University of Königsberg. There was only one professor of mathematics at Königsberg, Heinrich Weber (1842–1913). When he moved on, Ferdinand Lindemann (1852–1939) took his place. Both were distinguished and useful to Hilbert, but neither inspired him. One of the delightful aspects of university life in Germany in this period, however, was its freedom, which many used to drink beer, carouse, and fight duels. For a self-directed, serious student it allowed free, wholehearted study, and students commonly traveled to other universities to work directly under the professors that seemed most vital or were preeminent in their specialties.

In his second semester Hilbert took himself to Heidelberg, where he heard lectures by Lazarus Fuchs (1833–1902) on linear differential equations. The next step, perhaps, would have been to go to Berlin, but Hilbert, not to prove the most peripatetic of mathematicians (eventually all of mathematics would come to him), decided to return to Königsberg. Hilbert and Minkowski hadn't met. Minkowski was at Berlin but was also soon to return, having won, though only eighteen, a significant prize from the Paris Academy. Hilbert's father advised him that trying for a friendship with "such a famous man" would be "impertinence."[1] David, steady and sure of himself, nevertheless became friends with this interesting fellow student. Daily or almost daily contact with an equal was of crucial importance to Hilbert throughout his career.

In 1884 Adolf Hurwitz (1859–1919) came to Königsberg. Twenty-four years old, he was unpretentious, sweet-tempered, excited about mathematics, and already started on a substantial career. Hilbert and Minkowski established a close friendship with this new teacher. Hilbert said of Hurwitz, "His wise and gay eyes testified as to his spirit." The three met daily at "precisely five" to walk "to the apple tree" and discuss mathematics.[2] All his life, Hilbert would do mathematics as he walked and talked. During this period Hilbert truly became a mathematician: "On unending walks we engrossed ourselves in the actual problems of the mathematics of the time; exchanged our newly acquired understandings, our thoughts and scientific plans; and formed a friendship for life."

Under the full professor, Lindemann, Hilbert completed his degree and wrote a thesis of some originality on algebraic invariants. At his promotion he chose to defend two propositions. The first was about electromagnetism. The second was: "That the objections to Kant's theory of the *a priori* nature of arithmetical judgments are unfounded."[3] The thesis on algebraic invariants was classic, detailed, pure mathematics. The question about electromagnetism acknowledged the connection of mathematics to science, while the question about Kant is philosophical. Taken together they foreshadow the breadth of Hilbert's career.

After his degree Hilbert took the next obvious steps. He set out to write a second thesis for his *Habilitation*, which would certify him to become a privatdozent. The career ladder in German universities at this time was *Privatdozent* (allowed to teach but paid directly by students who came to lectures), *Extraordinarius* (paid by the university), and then *Ordinarius* (full professor, of which there were very few). Ever practical, Hilbert also took and passed the examination to teach at gymnasium.

At Hurwitz's urging, Hilbert went to Leipzig to study with Felix Klein (1849–1925). This is Klein of Klein's bottle, a well-known topological form, though his mathematical work went far beyond that classic illustration. Arguably the most prominent mathematician in Germany, Klein recognized Hilbert's talent. He was getting ready to move to Göttingen. During this period, Hilbert scribbled inside a notebook:

> Over this gloomy November day
> Lies a glow, all shimmery
> Which Göttingen casts over us
> Like a youthful memory[4]

At Klein's suggestion, Hilbert traveled to Paris to meet the French mathematicians. He paid calls and attended lectures, meeting Henri Poincaré (1854–1912), Camille Jordan (1838–1922), Charles Hermite (1822–1901), and others, while corresponding with Klein, telling him of his progress and passing on gossip and his own evaluations. Hilbert always knew how to get along with his fellow mathematicians. Back in Germany, he visited Klein in Göttingen and passed through Berlin, returning to Königsberg "content and full of joy."[5] He got ready to start lecturing. Hilbert was almost twenty-five, in a field where one's greatest work is often done before thirty. He had done nothing that set him apart from other solid but essentially ordinary mathematicians. There is no indication that the young Hilbert was disturbed or impatient. He decided to use the lectures he would give to teach himself, never repeating lectures, so that he would have to learn a new area of mathematics to prepare for every course he taught. At the same time he resolved with Hurwitz to undertake "a systematic exploration"[6] of mathematics during their daily walk to the apple tree. Hilbert settled into a happy period of work. In early 1888, after a year and a half of lectures, he set out on another round of visits, to twenty-one mathematicians.

Up to this time, most of Hilbert's publications had been in the area of algebraic invariant theory, a striking example of nineteenth century computation-based mathematics. A basic element of algebra is the polynomial, with any chosen number of variables raised to any chosen powers. A simple example is $x^2 + 2xy + 3y^2$. A basic operation is to change variables. Viewed spatially (in the case of two variables), the type of change studied in invariant theory is rotating coordinate x and y axes drawn on the plane, not necessarily keeping them perpendicular, and also possibly expanding or contracting the scale on each axis. Starting with Carl Friedrich Gauss (1777–1855), mathematicians observed that certain combinations of constants remained essentially unchanged by such rotations and scale changes. For the binary quadratic form, $ax^2 + bxy + cy^2$, one is $b^2 - 4ac$. This particular invariant was called "the discriminant" by Gauss. Other mathematicians investigated more complicated invariants. Arthur Cayley (1821–1895) and James Sylvester (1814–1897) in England and then Paul Gordan (1837–1912) and Alfred Clebsch (1833–1872) in Germany elaborated the theory greatly, as did many others.

Some branches of mathematics are selected for their usefulness in understanding the physical world. Nobel laureate in physics Eugene Wigner expressed his own astonishment that marks on paper or the blackboard could so deeply penetrate the secrets of nature. Physics selects and has played a large role in driving what mathematicians find interesting in many areas of analysis, as it does in probability theory. Interest in number theory, on the other hand, resides deep in human nature. Facts or patterns involving numbers have prompted curiosity from the time of the earliest writings. A search for beauty and surprise is the key motivation here, though today there are applications of number theory, most strikingly in computers and encryption. Geometry would be a hybrid—useful from the beginning, but most people have also been struck by its logical solidity and beauty.

Where does algebraic invariant theory fit? It was a window that allowed mathematicians to find hidden structure in the sea of polynomials. The general notion of the invariance of a mathematical object under a given group of transformations or variable changes is of fundamental technical importance for modern mathematics and physics. The apparatus of vectors and matrices as a way of economically dealing with transformations in linear algebra had not been fully worked out in the nineteenth century, and we can think of invariant theory as being loosely a part of that development.

When Hilbert arrived at the subject, the outstanding problem in invariant theory was Gordan's problem, which conjectured that there was a finite "basis" from which all algebraic invariants of a given polynomial form could be constructed by applying a specified set of additions and multiplications. Relying on brute-force calculation, Gordan had proved his theorem for the special case of binary forms in 1868. To apply the same technique to more difficult cases was hopelessly cumbersome, and in the ensuing twenty years many mathematicians had tried without success. Gordan, in Erlangen, was near the beginning of Hilbert's itinerary for the 1888 round of visits. Gordan had a forceful personality and was passionate about invariant theory. Like Hilbert, he liked to walk as he did calculations, stopping to refuel at beer gardens. Hilbert left "enchanted" with Gordan's problem. As he continued his peregrination from Klein to Fuchs to Helmholtz to Weierstrass to Kronecker, he thought about it. Before he left Klein, Hilbert had discovered a radically simpler proof of the binary case. Back in Königsberg he continued to work on the larger problem and by September 6, 1888, had solved it in an unexpected way.

Hilbert must have reasoned that the brute-force approach would have already worked if it were going to. He stepped back and looked at Gordan's problem. What would be the consequences if the conjecture weren't true? Through a relatively simple chain of reasoning he obtained a contradiction. If the assumption that Gordan's conjecture wasn't true led to a contradiction, then it must be true. Hilbert didn't find a basis, as everyone had been trying to do. He merely proved that if we accept Aristotle's law of the excluded middle—any statement is either true or its negation is true—then such a basis had to exist, whether we could produce it or not.

At first this result was greeted with disbelief. Gordan said, "*Das ist nicht Mathematik. Das ist Theology*." Cayley at first failed to grasp the proof. Lindemann thought the proof *unheimlich* ("uncomfortable, sinister, weird").[7] Only Klein got it right away: "Wholly simple and, therefore, logically compelling." Within the next five years organized opposition disappeared, and this was the result that initially made Hilbert's reputation.

Such proofs by contradiction are common today. Hilbert was hardly the first to use this kind of argument. However, previous uses had not dealt with a subject of such obvious calculational complexity. Since a pure existence proof doesn't produce a specific example that can be checked, one had to trust the logical consistency of the growing body of mathematics to trust the proof. Mathematics had been expanding its reach dramatically, moving more toward complex, abstract questions. Many mathematicians, Leopold Kronecker (1823–1891) in Berlin, in particular, were bothered by this headlong leap into the infinite, accessible only by inference, not finite construction. Georg Cantor (1845–1918), teaching at Halle in 1888, had invented set theory in the 1870s and was writing about infinite numbers of different sizes and even doing arithmetic with them. But Kronecker would admit only numbers or other mathematical objects that were finitely "constructible." Kronecker was important for Hilbert's own work in algebra. It took considerable openness of mind for Hilbert to make the leap needed to solve Gordan's problem.

Hilbert continued to work on invariant theory and lecture as a privatdozent, as the world of mathematics digested his proof. He published two more papers and in 1890 wrote up all his work in a unified treatment. Although he would continue to defend "existence proofs" of the type he had supplied for Gordan's problem, he also recognized the virtues of displaying a specific answer. He began work with algebraic number fields and proved what is called Hilbert's *Nullstellensatz*, a technical, though fundamental, result about algebraic surfaces. In 1892, using this new result and his existence theorem as a pushing-off point, he was able to give a method for actually constructing a finite basis for any group of invariants.[8] He returned to Kronecker, the now-dead finitist, for a crucial insight. If there was still any resistance to Hilbert's existence proof, it evaporated with his description of an actual construction.

At this point the German system of professorships struck the hour and began to shift like a musical clock. Kronecker's death had freed a chair in Berlin. Doors opened and mathematicians popped out of the universities and began moving around. Hurwitz landed in Zurich as a full professor. Minkowski stayed in Bonn, where he had gone after Königsberg, but became extraordinarius, and Hilbert, now also extraordinarius, filled Hurwitz's spot in Königsberg.

Hilbert was now in a position to marry. He was a good dancer and a flirt and had been enjoying a lively social life. However, one young woman in particular, Käthe Jerosch, became the focus of his attentions. He and Käthe were married in October 1892. By the next August their only child, Franz, was born. This was a most fortunate marriage and seemed so from its inception. In the words of one of Hilbert's pupils: "She was a full human being in her own right, strong and clear, and always stood on the same footing with her husband, kindly and forthright, always original."[9] Congratulating the new husband, Minkowski wrote that he expected "another great discovery."[10] Early in the next year Hilbert found a new, simplified proof of the transcendence of both e and π (following Hermite and Lindemann), one that has become standard today.

Hilbert announced that he would do only number theory. Lindemann soon had an offer from Munich. Hilbert was appointed full professor, vacating his extraordinarius. Minkowski, who had also entered the field of number theory, filled Hilbert's old spot. Before this last switch of chairs had been completed, and before Hilbert had actually done much number theory, the German Mathematical Society assigned Hilbert and Minkowski the job of reporting on the status of the subject. Hilbert took algebraic number theory, Minkowski rational number theory. Hilbert plunged in. He liked to go in and organize a whole branch of mathematics. Minkowski was sidetracked by his high critical standards for writing and the press of other interests. The two enjoyed being back together, but soon Weber left Göttingen, creating a vacancy there and leaving Klein effectively in charge. In 1895 Hilbert was summoned as an ordinarius, and Minkowski took Hilbert's position at Königsberg.

A university town, Göttingen lies in the center of Germany, close to what was once the border between East and West. Much of the *Altstadt* is intact, even some of the old city wall. Half-timbered houses with red-tiled roofs line crooked, narrow streets. Just beyond the wall, the newer houses are made of handsome sulfur-yellow brick. The rolling countryside close at hand made the walking Hilbert loved easy, and Göttingen has not been too much overtaken by development, even today. The motto on the *Rathaus* reads: "Away from Göttingen there is no life."[11]

Gauss had placed number theory at the summit of mathematics, and now Hilbert arrived at the university of Gauss a hundred years later to work on number theory. Klein had embarked on a program to make Göttingen the center of the mathematical world and had already succeeded far enough so that the *Bulletin of the American Mathematical Society* listed the courses to be given there. Students attended from the United States and from all parts of Europe. Klein's lectures were famous for their organization. He wrote on the blackboard as he spoke, never erasing, and left it covered with a flawless summary of the material at the end.

At first Hilbert and Käthe experienced some discomfort in this milieu. Klein, though polite, was very formal. Mathematical society in Göttingen was precisely structured. Klein's wife, a granddaughter of the philosopher Hegel, did not welcome all and sundry to her house for expansive social gatherings, as Käthe Hilbert soon would. Further, Hilbert's lecture strategy was to display the way he actually thought about mathematics to his students. His lectures were full of the pauses, backtracking, error, and general confusion that occur in the process of thought. He lacked the high manner of a German professor of this period. The students soon realized that surprising insights and fresh ideas came out as Hilbert talked. He acted as if he were not aware there was a social hierarchy and began taking long walks with students and privatdozents after lectures. Käthe threw open the house, so Göttingen began to seem warmer. Hilbert's mathematical life thrived on conversation. He and Käthe built one of the yellow brick houses and included a covered walkway in back so Hilbert could do mathematics on foot on inclement days. They got a dog, "the first of a long line of terriers, all to be named Peter."[12]

Hilbert threw himself into his report on algebraic number theory. He felt the subject was less understood than it should be, full of unconnected results that were grouped, if at all, on the basis of chronology. Normally he hated to learn mathematics by reading, but in this case he read the literature. The result, Hilbert's *Zahlbericht*, would be a classic. Minkowski and Hurwitz proofread and edited the manuscript so thoroughly that even the careful Hilbert became impatient. When the report finally appeared it was greeted with enthusiasm. In addition to summary and recapitulation, there were original remarks. It would prove typical of Hilbert's "summaries" that they could be mined by future mathematicians. Once the logical structure was elucidated and the hidden connections made explicit, the stage was set. He himself proceeded to work on broadening the law of reciprocity—a favorite of Gauss's. The result, unfolding through a series of papers, again set the stage for a new area of mathematics, class fields. Hilbert tended to finish with subjects and wash his hands of them. Now he announced he would lecture on geometry.

Hilbert's work on geometry had two distinct parts. In his obituary essay on Hilbert's career, Hermann Weyl (1885–1955) describes the first:

> The Greeks had conceived of geometry as a deductive science which proceeds by purely logical processes once the few axioms have been established. Both Euclid and Hilbert carry out this program. How-

> ever, Euclid's list of axioms was still far from being complete; Hilbert's list is complete and there are no gaps in the deduction.[13]

One area that needed clarification and amplification should be evident to students. In first-year geometry, the person offering a proof and the person understanding it make constant visual reference to the drawing that appears at the top. These glances make no official appearance in the formal proof (and generally weren't acknowledged in Euclid either). Over time, erroneous proofs had been offered in which the error was hidden in an incorrect drawing or interpretation of a drawing. Axioms that capture visual judgments such as "betweenness," called axioms of order, were needed. Hilbert was not the first to recognize this. Mathematicians like Moritz Pasch (1843–1930) and Giuseppe Peano (1858–1932) had set out to translate what we understand as geometry into the notation of symbolic logic, without pictures. This was too bloodless for Hilbert. In his lectures he took a middle path. He set up everything on a rigorous axiomatic basis. However, he did draw pictures and use words like "points" and "lines." Logically the pictures were only an aid, and anything that satisfied the axioms could be understood as a "line," "point," or "circle." There was nothing magical about the words, and one could just as well use "tables, chairs, and beer mugs"[14] (Hilbert's example). Weyl writes of a model for the axioms of geometry built out of electric currents, with all the results of geometry true of these electric currents. Hilbert emphasized that the pattern of deduction is logically separate from our intuitions. This part of Hilbert's work on geometry is a masterful synthesis and display, and a preparation for the second part of his work.

Again in the words of Weyl:

> It is one thing to build up geometry on sure foundations, another to inquire into the logical structure of the edifice thus erected. If I am not mistaken, Hilbert is the first who moves freely on this higher "metageometric" level: systematically he studies the mutual independence of his axioms and settles the question of independence from certain limited groups of axioms for some of the most fundamental geometric theorems. His method is the *construction of models*: the model is shown to disagree with one and to satisfy all other axioms; hence the one cannot be a consequence of the others. . . . The general ideas appear to us almost banal today, so thoroughgoing has been their influence on our mathematical thinking. Hilbert stated them in clear and unmistakable language, and embodied them in a work that is like a crystal: an unbreakable whole with many facets.[15]

A key idea in the construction of models is relative consistency. Hilbert starts with real number arithmetic. He assumes it is consistent, that is, free of the possibility of contradictory deductions. Using René Descartes's (1596–1650) analytic geometry, he then exhibits a model of euclidean geometry. A point is a pair of real numbers. A line is a set of pairs of numbers that satisfy the equation for a line; a circle . . . and so on. Euclid's axioms are all true statements about these "lines" and "points," that is, they are true statements about these sets and pairs of real numbers. Euclidean geometry is reduced to a fraction of all the true statements about real numbers. We conclude that if real number arithmetic is consistent, Euclid's geometry is consistent.

What about consistent non-euclidean geometries, in which different versions of the postulate on parallel lines hold true? This method constructs a model for a given non-euclidean geometry using pieces of and the context of euclidean geometry, thereby achieving the result that a non-euclidean geometry is consistent as long as euclidean geometry is—which in turn goes back to the consistency of arithmetic. Euclidean models for non-euclidean geometries existed before Hilbert. What Hilbert did was investigate a wide range of variations of many different axioms of euclidean geometry. As Weyl says: "In the construction of his models Hilbert displays an amazing wealth of invention."[16] Hilbert published his lectures as *The Foundations of Geometry* in 1899. (It has gone through many editions, has been improved, and had topics added. Most people, including, I suspect, Weyl, have in mind one of these later editions.)

Hilbert then started a new examination of what is known as Dirichlet's principle, a useful but unproven assertion of the existence of solutions to certain differential equations that commonly arise in physics. Indeed, Weierstrass had proved that, in its standard form, Dirichlet's principle was false. Mathematicians had tried to prove it in some reasonably general form but failed. Hilbert accomplished this by putting certain limits on the boundary values involved, limits that did not exclude any situations of physical interest, but excluded Weierstrass's counterexamples. Problem solved. He moved on to lecture on the calculus of variations in general. In these last two topics Hilbert entered the realm of analysis, the part of mathematics most closely associated with physics; he would increasingly involve himself with physics.

The turn of the century was coming, and Hilbert received an invitation to give one of the major talks at the Second International Congress of Mathematicians in Paris. He had done his major work in algebra, number theory, and geometry—and though barely begun, was already doing work in analysis, the fourth major branch of mathematics. He was working creatively on foundational issues (models, relative consistency in his work on geometry). After some thought and discussion with Minkowski and Hurwitz, he decided to lecture on the significance of individual problems and to give a list of problems that he thought would be the most fruitful for mathematics in the new century. He only had time for ten problems when he delivered the talk, but there were twenty-three in the published version. Near the end of his introduction Hilbert made a statement that lies at the core of his beliefs, showing us the engine of his desire and, more poignantly, giving us a benchmark:

> This conviction of the solvability of every mathematical problem is a powerful incentive to the worker. We hear within us the perpetual call: There is the problem. Seek its solution. You can find it by pure reason, for in mathematics there is no *ignorabimus*.

Hilbert continued to do important work on integral equations, on what came to be known as Hilbert space, and on Waring's problem, which he solved in 1908. He made contributions to solving some of his own problems. His first lecture in Königsberg had been delivered to one student; now his lectures drew hundreds, and students from all over the world flocked to Göttingen. Germany continued to be prosperous and peaceful. Exceptional students were fostered. We will meet this world in our account of the problems.

Hilbert received a "call" to a chair in Berlin and shrewdly used his growing prominence. He remained in Göttingen but managed to have a new professorship authorized, which was filled by Minkowski. Klein continued to busily organize the Göttingen mathematics department as well as scientific education in the rest of Germany. Once, when Klein had filled an entire blackboard with numbers about German middle schools, Minkowski said, "Doesn't it seem to you, *Herr Geheimrat*, that there is an unusually high proportion of primes among those figures?"[17] Minkowski and Hilbert loved mathematics, and they loved it together. Their two families were also close. Each day, Minkowski made time to be alone with each of his children to talk. Life in the household was rich, as it had been in the Minkowski family home in Königsberg. Minkowski also made time for Hilbert's son Franz.

Hilbert's idiosyncrasies became more pronounced. Once or twice when teaching a large calculus class he became so tangled in the details that he gave up and walked out. Yet that class is also well known for its insights. When bicycles came into fashion, Hilbert acquired one and took mathematical bicycle rides. At home he liked to work in his garden. A blackboard was mounted on a neighbor's wall. He would work, pull weeds, write on the board, talk to visitors, jump on the bicycle and ride it in a figure-eight around his two rose beds, then return to the board.

Franz was developing mental illness and became the source of a growing grief. Hilbert himself suffered a mental breakdown in 1908 but, after some months at a rest home in the Harz Mountains, was able to return with full vigor for the fall. Then in early 1909, when Hilbert was eager to explain to his friend his new solution to Waring's problem, Minkowski fell ill with appendicitis. It proceeded rapidly and inevitably. Minkowski was entirely lucid. He saw his family at the hospital. Hilbert said later, "He spoke his regrets upon his fate, since he still could have accomplished much; but he decided that it would be good to correct the proof-sheets so that his latest electrodynamical works could be more easily read and better understood." In a letter to Hurwitz, Hilbert wrote, "The doctors themselves stood around his bed with tears in their eyes."[18] Hilbert announced the death to the university. A former student said, "I was in class when Hilbert told us about Minkowski's death, and Hilbert wept. Because of the great position of a professor in those days and the distance between him and the students it was almost more of a shock for us to see Hilbert weep than to hear that Minkowski was dead."

Hilbert worked increasingly on one of his own problems, the axiomatization of physics. When the First World War began, students and sons of professors enlisted with patriotic fervor. "Hilbert thought the war was stupid, and said so,"[19] Reid says. One day the professors were assembled. In the past at some of these assemblies extra food had been distributed. This time the purpose was to announce the news that unrestricted submarine warfare would begin. Hilbert said to his neighbor, "And I thought we would get swine! But, you see, the German people are like that. They don't want swine. They want unrestricted submarine warfare."[20] Late in the war the French mathematician Gaston Darboux (1842–1917) died. Hilbert prepared a memorial to appear in a German publication. A mob of students appeared at his house. Hilbert threatened to quit if he didn't get an apology. He got one, and the memorial was published.

In evaluating mathematicians, although he did have a bias in favor of Göttingen, Hilbert was fair. Before the war he was presented with a dissertation from Jakob Grommer, a student who had attended a Talmudic school in Eastern Europe and did not have a gymnasium certificate. Grommer had planned to be a rabbi, but when his betrothed, the daughter of the rabbi he was to replace, saw that his hands and feet were deformed she wouldn't marry him. Grommer received his doctoral degree. Hilbert didn't care whether a mathematician was Jewish, or about outward aspect. He said, "If students without the gymnasium diploma will always write such dissertations as Grommer's, it will be necessary to make a law forbidding the taking of the examination for the diploma."[21] Emmy Noether (1882–1935) provides another example of Hilbert doing what was necessary for a great mathematician to have a place at Göttingen.

Hilbert had some success during the war with the axiomatization of physics and made a contribution to general relativity. However, he recognized the limitations of his contribution. He said: "Every boy in the streets of Göttingen understands more about four-dimensional geometry than Einstein. Yet, in spite of that, Einstein did the work and not the mathematicians."[22] Toward the end of the war Hilbert became increasingly involved with the foundations of mathematics. There was a new challenge to what he took to be the full scope of mathematics from a group called "intuitionists," led by L. E. J. Brouwer (1881–1966). The intuitionists, as Kronecker had, quarreled with the free investigation and use of infinite structures. Intuitionists don't believe a mathematical object exists unless you can construct it starting from primary objects, like lines, points, and numbers, that can be clearly known through intuition.

Immediately after the First World War German mathematicians were not included in international conferences, and the German economy was destroyed by hyperinflation. But soon students from other countries began to return. Max Born made the physics department a world center. Born's first two assistants were Wolfgang Pauli and Werner Heisenberg. In 1922 Hilbert turned sixty. His creative

David Hilbert and Hermann Weyl in Göttingen (photo by Richard Courant, courtesy of Natascha Artin Brunswick).

life as a mathematician was now focused on furthering his program in logic. In the fall of 1925 he was diagnosed as having pernicious anemia, then a rapidly fatal disease. A cure was discovered almost simultaneously, and some of the rare, raw liver extract needed was obtained in time, though irreversible damage was probably done. Hilbert continued to work and lecture until he retired in 1930, then continued to lecture. The world of Göttingen was about to collapse. Klein and Hilbert had built a sunny place where, in Weyl's words, "the passionate scientific life"[23] could be lived. It had survived the First World War and the chaos of the

Käthe Hilbert (photo by Elisabeth Reidemeister, courtesy of Natascha Artin Brunswick).

David Hilbert (photo by Richard Courant, courtesy of Gertrud Moser).

1920s in Germany. It would not survive Hitler. By late summer of 1933, Göttingen was practically empty. Hilbert said, "When I was young, I resolved never to repeat what I heard the old people say—how beautiful the old days were, how ugly the present. I would never say that when I was old. But, now, I must." Sitting at a banquet next to the new Nazi minister of education, he was asked how mathematics in Göttingen was faring now that the "Jewish influence" was gone. Hilbert said, "Mathematics in Göttingen? There is really none anymore."[24]

After the winter term of 1933–34, Hilbert did not return to the university. With each year he had fewer visitors. His memory was failing, but he remained intermittently sharp on his research in logic. If memory is not your strong point, you can lose some of it. In 1937, at age seventy-five, he told a reporter, "Memory only confuses thought—I have completely abolished it for a long time."[25] Käthe was losing her sight. In 1942 Hilbert slipped and broke his arm. On February 14, 1943, he died. Käthe died two years later. Rumors of Hilbert's death could not be confirmed outside Germany until after the war.

Hilbert made a statement to Harald Bohr (1887–1951) that sums up his career:

> That I have been able to accomplish anything in mathematics is really due to the fact that I have always found it so difficult. When I read, or when I am told about something, it nearly always seems so difficult, and practically impossible to understand, and then I cannot help wondering if it might not be simpler. And on several occasions it has turned out that it really was more simple![26]

David Hilbert (photo by Elisabeth Reidemeister, courtesy of Natascha Artin Brunswick).

Hilbert had the poet John Keats's "negative capability." He allowed and sought both abstraction and construction as he expanded and streamlined mathematics.

* * *

Hilbert's twenty-three problems appear with his speech, "Mathematical Problems," in an appendix at the back of this book. How many of them have been solved? What does it mean for a Hilbert problem to be solved?

Hilbert's problems have the characteristics of any good founding document. Each one is a short essay on its subject, not overly specific, and yet Hilbert makes his intent remarkably clear. He leaves room for change and adjustment. Hilbert's goal was to foster the pursuit of mathematics. He helps us find the vital center of each problem by using italics—often in more than one place. So the question of how many of the problems have been solved becomes part of a more ambiguous question: How many have been posed? Sometimes Hilbert suggests a plan for investigation of an area. Or he may be pursuing an intuitive feeling, and the problem can be paraphrased: Why don't you look in that direction? However, most of his problems have been identified—in something approaching a consensus—with a single, clearly stated mathematical question. That consensus, when it exists, is what I have taken to be the problem.

When can a mathematical publication be taken as a solution to one of Hilbert's problems? Sometimes it is clear. The question addressed in the paper is by consensus an accurate statement of the problem, and the mathematical answer is precise, rigorous, without flaw, and answers every bit of the question posed. But how do we treat partial solutions? How do we treat partial solutions that taken together solve the problem? What if they solve most of the problem? What if Hilbert's conjecture as stated is false but can be reformulated using more precise, modern language so that it is true or might be true? The ambiguities of what a solution to a Hilbert problem is apply directly to the question of which mathematicians should be written about and in what detail. Here I add two practical criteria: I wrote more about mathematicians I found interesting and about whom I was able to find information. Mathematicians, forgive me my omissions. I have tried to acknowledge sources and lacunae.

Hilbert introduced his problems by type—foundations, number theory, etc.—and I follow his order. Within each section, however, I treat the problems in the historical order in which they were solved. I have moved the tenth and thirteenth problems out of Hilbert's order to the sections on foundations and analysis, respectively, because the character of their solutions clearly places them there.

The Foundation Problems

« *1, 2, 10* »

Set Theory, Anyone?

Cantor

What is called the foundation of mathematics deals with questions that interest philosophers as well. The metaphor is the foundation of a house. The questions are of too general a character to lie comfortably in any of the classical categories of mathematics: algebra, number theory, geometry, or analysis. Symbolic logic, investigations of independence and consistency in systems of axioms (i.e., whether or not an axiom is already implied by the other axioms and whether its addition leads to contradictions), theories of computability, and the less immediately practical aspects of set theory lie in the foundation. Hilbert's first two problems are foundation problems, as is the tenth problem, though it was presented as a question in number theory. None of these problems was solved in a way that Hilbert envisioned, yet he helped set the stage for the solutions. Explicit contradictions had developed in Cantor's set theory, and Hilbert placed the foundation problems first on his list.

One of the great achievements of nineteenth century mathematics was understanding what are called "the real numbers": any number that can be expressed in decimal notation, whether the chain of digits stops, goes off to the right forever in a regular pattern, or goes in a pattern that does not repeat. The natural numbers, 1, 2, 3, 4, 5, . . ., are real numbers, as are fractions. Square roots of positive numbers, cube roots, etc., are real numbers. Negative numbers, irrational numbers, transcendental numbers—all are real numbers. Any number we arrive at in algebra or calculus that doesn't involve $\sqrt{-1}$ is a real number. (The square root of negative one, or i, is an "imaginary" number.) What is included as a real number was not easily arrived at. The connotations of the names given to numbers—"natural" for the positive counting numbers, or positive integers; "rational," with its associations of ratio and reason—reflect this history. The square root of two and π are both "irrational" (not writable as a ratio of whole numbers), π so irrational that it is also "transcendental." An irrational square root was called a "surd," meaning deaf, silent (expressing the attitude toward it). The word "absurd" was first used in English in 1557, according to the *Oxford English Dictionary*, for the purpose of pronouncing the number 8 – 12 (or –4) absurd.

The path that led to acceptance of the different kinds of numbers as equally "real" came through algebra and the development of algebraic notation. The convention of symbols one uses to write a problem has a tremendous effect on how easy it is to solve. Algebra problems used to be written out in words. Take, for instance, Girolamo Cardano's (1501–1576) version of the solution of $x^3 + mx = n$ (as we express the problem), from William Dunham's *Journey Through Genius*:

> Cube one-third the coefficient of *x*; add to it the square of one-half the constant of the equation; and take the square root of the whole. You will duplicate [repeat] this, and to one of the two you add one-half the number you have already squared and from the other you subtract one-half the same. . . . Then, subtracting the cube root of the first from the cube root of the second, the remainder which is left is the value of *x*.[1]

What we recognize as algebraic notation began to appear in the 1500s and 1600s. For example, an equation we might write as

$$5BA^2 - 2CA + A^3 = D$$

the French mathematician François Vieta (1540–1603) would have written as

B5 in A quad – C plano 2 in A + A cub aequatur D solido.

This was a major advance. Vieta had the idea to consistently use letters instead of words for the constants and variables—vowels for variables and consonants for constants. (Some of the Greeks had used letters for numbers, but only sporadically.) We generally use the famous *x*, as in "solve for *x*" (and *y* and *z*), for our variables. We use the other letters of the alphabet for constants and, in general, don't bother to capitalize.

Each symbol we use had an inventor, or at least a first appearance on a surviving piece of paper. According to the venerable and always bold eleventh edition of *The Encyclopedia Britannica* (1910–1911), the = sign instead of the word "equals" came from an Englishman, Robert Recorde, in the 1550s. A German named Michael Stifel supplied the + sign for addition, the – sign for subtraction, and the √ sign for square root. (The + and – signs were possibly first used by German shippers, who wrote them on chests that were heavy or light.) An Englishman named William

Oughtred gave us the × sign for multiplication. Mathematician Thomas Harriot (who was sent as a geographer by Sir Walter Raleigh on an expedition to Virginia) started using uncapitalized letters.[2]

The new notation led to an explosion in problem-solving. Now insight could often be replaced by manipulation according to rules, and insight into the real difficulties of a problem was easier to come by. It is difficult to imagine Newton inventing calculus or classical mechanics without some of this new notation. The way of writing in algebra contains a way of thinking. If we write $x^2 = 2$ and become more and more familiar with solving problems with the notation, eventually philosophical scruples are cast aside and we say $\sqrt{2}$ is a number. If we end up with $x^2 = -1$ through algebraic manipulation and thereby are led repeatedly to $x = \sqrt{-1}$, we eventually say, "Oh, all right—let's talk about these numbers and call them imaginary. What can it hurt?" Soon people are designing things like the grid that delivers electric power to homes and businesses using lots of imaginary numbers. Imaginary numbers occur in the possible solutions of almost any equation, including most of the equations used in physics or engineering.

Understanding the real numbers led to the creation of set theory. In the nineteenth century, the real numbers came to be associated with the points on a line and each point on the line to be associated with a number. This was naturally expanded so that each complex number was associated with a point on the complex plane (one axis being pure real numbers, the other pure imaginary numbers). As work in analysis led to and required a study of the topology of a line or of the complex plane, their systems of neighborhoods and closeness, restrictions limiting efficient reference to infinite sets of points or numbers were standing in the way. Bernhard Riemann (1826–1866) and Poincaré crossed the landscape in seven-league boots. Yet their mathematics often had a gestural quality. The flexible and powerful language of set theory is a large part of what they were missing.

Today, mathematicians talk about sets that are infinite—like the set of all whole numbers, the set of prime numbers, or the set of all fractions between 0 and 1—as objects that can be studied and reasoned about, like other objects in mathematics. Before Cantor, mathematicians could not refer to an infinite as an entity. There was a lack of consistency in this. A line (which people had in fact been referring to as a discrete entity since before Euclid) went on forever, and even a line segment was made up of an infinitude of "points." Once one allows oneself to talk freely about sets of points on a line segment, one is led to more extensive observations comparing the "size" of infinite sets of points and of completed infinites as having measurable quantity. Cantor expanded mathematics. The first Hilbert problem is a question he raised, and much of the rest of twentieth century mathematics has been conducted in his language. Though originally led to talk about infinite sets by a problem in analysis, Cantor yearned for and embraced the infinite as a matter of character as well.

Georg Cantor was born in St. Petersburg, Russia, in 1845, seventeen years before Hilbert. He was the eldest son of Maria Anna (Böhm) and Georg Woldemar

Cantor. His father's wholesale company did business in Hamburg, Copenhagen, London, New York, Rio de Janiero, and Bahia. The family background was one of civic participation, commercial success, musical accomplishment, and philosophical interests. Cantor went to Berlin to attend the university, studying under Weierstrass. He also attended lectures by Ernst Eduard Kummer (1810–1893) and Kronecker, obtained his degree, and went to the university in Halle as a privatdozent. There he rose and eventually became a full professor. Halle is between Göttingen and Leipzig, but closer to Leipzig. It was not at the center of the mathematical world. Cantor hoped to move to a more important university but never did, due to the controversy that surrounded his work and possibly the bouts of mental illness (probably manic depression) that afflicted him later in life. After Cantor's death, in an attempt to prove his mathematics "tainted" or "decadent," the Nazis made much of whether he was Jewish. He was raised an evangelical Lutheran. His wife, Vally Gutman, was from a Jewish family in Berlin. Cantor's personal life has prompted serious scholarship in English by Joseph Dauben and by I. Grattan-Guinness.

Cantor possessed some of his family's musical ability and in later years expressed regret that he hadn't become a musician. In school he did drawings that showed artistic talent. He was particularly fond of Shakespeare and had strong philosophical and religious interests. In a letter written at the time of his confirmation, Cantor's father urged him to study widely in science, language, and literature. Cantor did this, though gradually he became increasingly involved with mathematics. His first important work was under Weierstrass, who was testing the boundaries of analysis. For example, when did an infinite series converge? What is a precise definition of convergence? An infinite series of numbers converges if the numbers, as one proceeds through the list, move consistently closer to a specific number that is said to be its limit. Any wandering the numbers do from the limit value gets progressively smaller so that, in the long run, any wandering is negligible. Other questions would be what exactly is a continuous function, or what kinds of hybrid, pathological functions exist that could undermine the truth of a given theorem? What is the most general statement of this or that theorem?

Cantor worked on trigonometric series. One important type, the Fourier series, follows a specific recipe created by Joseph Fourier (1768–1830). We can think of a trigonometric series as a way of approximating a finite segment of any reasonably smooth curve or approximating any function that repeats itself forever. Even functions with discontinuities or jumps can be approximated, like:

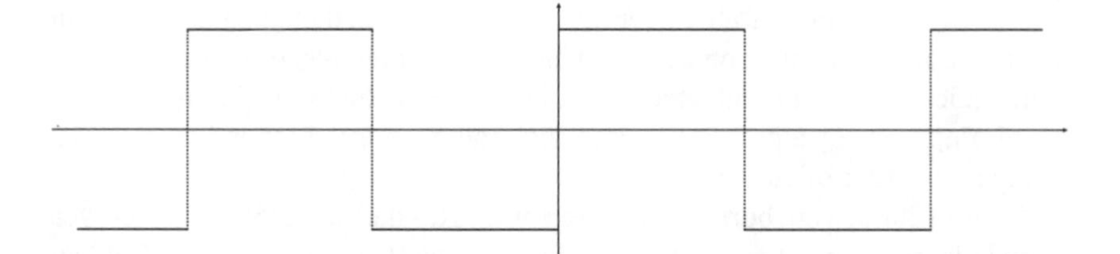

The square edges of the square wave, as it is called, seem very different from the smooth flow of the sine or cosine waves that are the bases of trigonometric series. But we can see how the sharp jumps of the square wave are approximated by superimposed, smooth sine waves in the following pictures:

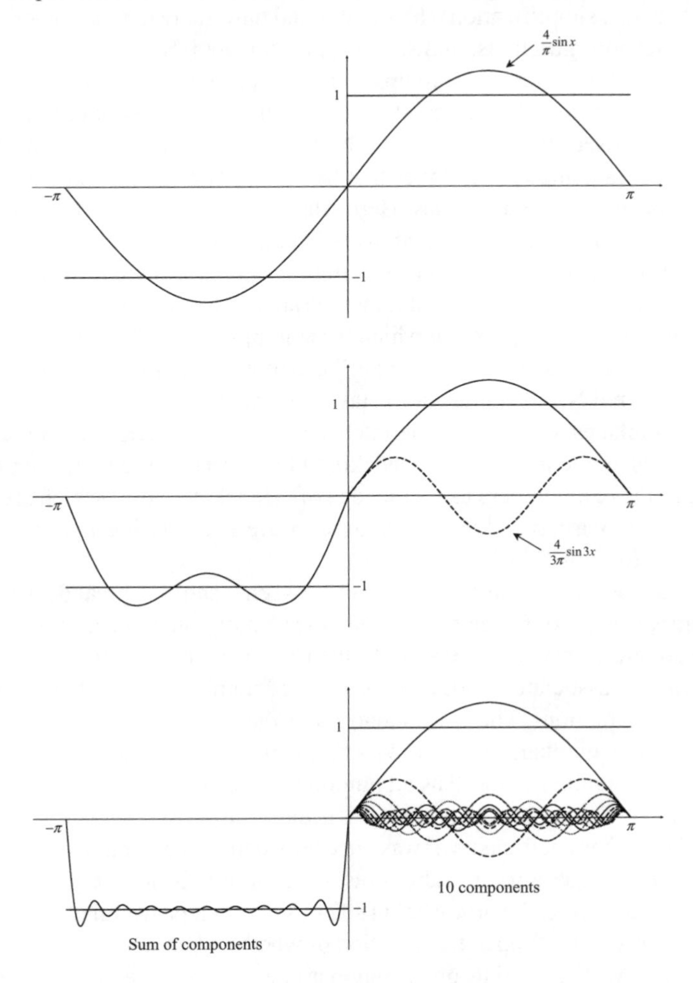

A sine wave approximation gets closer and closer to a square wave as higher frequencies are added.

A trigonometric series decomposes functions that might appear quite unsmooth or unmusical into an underlying initial vibration (specified by the initial interval of the function—or its period of repetition) and its subsequent harmonics. This mathematical construct has a physical reality. A computer signal, for instance, is made up of square waves. If you take a fast computer and look at the data with an oscilloscope that can't process high frequencies, the wave form displayed on the screen has rounded edges similar to the first pictures of the sine waves, as if the higher terms of the Fourier series were left off.

Cantor proved that if a function could be expressed as a trigonometric series, there was only one way to do it (the theorems he proved are more technically complicated than this simplification. He investigated how many jumps a function could have, in what configurations, and still be approximated. Surprisingly, some functions with an infinite number of jumps could be approximated. It depends on how the jumps are distributed. They can't be evenly dispersed; they have to clump up.

Cantor crossed his Rubicon when he talked about these jumps as being in "sets," that is, as completed infinites that had integrity as mathematical objects in the same way a line segment does. Here was born the modern notion of a set as something that can be described and investigated as an object in and of itself. Today it is difficult to even talk about such things without using the word "set" (or a synonym), with the assumption that if there is an infinite number of jumps, then we may talk about the set of points at which these jumps occur. (The 1911 *Encyclopedia Britannica* uses the word "aggregates" in a more than up-to-date article signed A. N. W.—Alfred North Whitehead.) Having made the conceptual leap to infinite sets or assemblages, Cantor became interested in basic questions about real numbers. This was in the air. Richard Dedekind (1831–1916) was working on a construction of the real numbers using the idea of "Dedekind cuts" (which are infinite sets themselves of a certain kind). Infinite sets were also entering algebraic number theory in the form of "ideals."

For illustration, consider the numbers between 0 and 1. A great deal about this set of numbers was not understood at the time Cantor commenced. What kinds of numbers are there greater than 0 and less than 1? All are real numbers, and any real number can be associated with a point on the "number line." There are all the whole number fractions. There are square roots that aren't rational, like $\sqrt{½}$. There are the algebraic numbers, a larger class that includes, but is not solely composed of, the rational numbers as well as all numbers that can be written as square roots, cube roots, etc. A number is algebraic if it is the solution to a polynomial equation with whole number coefficients. It was once hoped that all algebraic numbers could be written using square roots, cube roots, etc., but this is not so. Evariste Galois (1811–1832) and Niels Henrik Abel (1802–1829) each proved that they couldn't always be. This still left open the question of whether there exist numbers that are not algebraic. Are there points on the number line that are not algebraic? Leonhard Euler (1707–1783) conjectured such numbers existed and had an idea what some of them were, but could not prove any number was "transcendental." In 1844 Joseph Liouville (1809–1882) proved that such transcendental numbers not only existed but were infinitely abundant. He gave an example of a specific number he could prove was transcendental. When Cantor was first working, Lindemann had not yet proved that π was transcendental.

Cantor entertained the idea that there could be different sizes of infinite sets. He made the startling discovery that there were more real numbers, or points on a line, than there were fractions, though both sets of numbers were infinite. A key distinction he made was the difference between a countable (or denumerable) and an uncountable (or nondenumerable) infinite. A set is countable if there exists a one-to-one pairing of its members with the natural numbers—such a pairing would essentially be a list of the members of the set. (Giving a method of listing all the members of a set proves the set is countable.) The set of natural numbers is the quintessential countable set; we can list them all simply by counting. The set of fractions between 0 and 1 would appear to be a different situation. No matter how small a region or where it is, there is an infinite number of fractions within it. How could we possibly list them? But we can list the fractions easily if we impose an order. The whole numbers come ready-made for this. To list the fractions between 0 and 1 all we have to do is start with 1/2, then list the thirds (1/3, 2/3), then move on to the quarters, etc. If we come to a fraction that can be reduced, like 2/6, we don't list it again because we already listed it as 1/3—that is, we can make a list without duplications. If we proceed forever, we obviously will list all the fractions between 0 and 1.

More surprisingly, it turns out that the set of all algebraic numbers is a countable infinite as well, though the listing process is more complicated. At the heart is the same trick. Each algebraic equation has a finite number of solutions. First list the solutions to all equations where the largest number to appear is 1, then 2, then 3, etc.—including exponents as well as coefficients. We eventually will list every equation this way, again playing off the natural numbers. If we have forever to make the list, we have a lot of time to keep doubling back. Surprisingly many sets are "countable," though infinite. However, the set of real numbers isn't countable. Cantor's original proof, published in 1874, uses analysis, but the argument I will offer is in essence one that Cantor himself originated in 1891. Simple and powerful, it has become a staple in many situations since and in different guises is at the heart of the methods of solution of Hilbert's first, second, and tenth problems. It is called a "diagonalization" method, because it proceeds by following the diagonal down an infinite array of numbers.

With numbers ending in an infinite series of 9s excluded by convention, every real number can be written uniquely in decimal notation, with the decimal continuing forever if necessary. One-third is expressible as .3333333333333333 . . . going on forever. One-half is .5 (but one could write in 0s if one had the time). The number .49999 . . . would converge to 1/2 and represent it equally well. This is why it is excluded—having two names for a number would add confusion. Fractions fall into a repeating sequence: 5/6 = .833333 . . . , 1/7 = .142857142857142857. . . . Some algebraic numbers are expressed as a sequence that doesn't repeat. The square root of two, for instance, starts out 1.4142 . . . and goes on forever without falling into repetition. The real numbers that are not algebraic are by definition transcendental, but they too can be expressed as decimals. Pi starts out 3.14159 . . . and goes on forever without repeating. Every sequence of numbers in decimal notation corresponds to a distinct real number. To be more precise, the infinite series that corresponds to the decimal notation converges to (or tends toward) its own unique value. Every number is as unique as its string of digits. Every string of digits is a real number.

Now, suppose we can make a list of them (again talking about real numbers between 0 and 1, for our convenience):

1 **.2** 4 6 2 9 0 6 7 4 6 7 8 7 9 8 9 6 4 3 . . .
2 .8 **3** 8 3 8 7 5 9 0 2 0 1 8 6 5 0 3 5 7 . . .
3 .0 2 **8** 5 4 8 7 3 8 7 2 8 9 4 7 4 3 6 2 . . .
4 .9 3 8 **3** 6 4 5 6 3 7 6 2 8 2 9 1 9 2 8 . . .
5 .7 8 5 9 **3** 9 9 2 0 2 9 3 8 4 7 5 4 7 2 . . .
6 .2 5 0 0 0 **0** 0 0 0 0 0 0 0 0 0 0 0 0 0 . . .
7 etc.

I have made up numbers, typing almost randomly, to convey the feeling for such a list. The important thing about this supposed list of all the real numbers between 0 and 1 is that not only does it go on forever down the page, each entry individually goes on off to the right forever. This is where the wiggle room is that lets in a proof that at least one number has been left off the list. I can tell you a method to write such a number. Such a number is itself an infinite decimal with a first digit, a second digit, a third digit, . . . on forever.

Look at the first number on the list. The first digit is a 2. Well, we will pick the first digit of our number to be something different than 2. We have plenty of choices—we avoid choosing 9 to avoid choosing a number that ends in all 9s. Capriciously, I pick 5. So our number starts .5 —and we already are certain it is not the same as the first number on our list. Now we will select the second digit. To do that we look at the second number on our list. It starts out .83 . . . , so I pick something other than 3 for the second digit of our number. Again capriciously, I pick 2. (I could invent a definite rule for picking the numbers, but I think that distracts from what is really happening. The only important thing is that the number I pick be different.) Our number now starts out .52, so already we are certain it is neither the first nor second number on the list. Our number still has an infinite number of digits to choose, so we just continue picking the third digit, the fourth digit, etc., in the same way, so that we are sure our number is different from the third number, the fourth number, the fifth number on our list, on off forever. The number we have picked with this method is different in at least one digit from every number on our list. Therefore it isn't on our list.

Suppose we include it—we can still go back and find another number that isn't on the list, using the same technique. Since we can produce this objection to every list offered, we must conclude there is no such list, and hence the set of all real numbers is not countable. The mere fact that we cannot think of a method of listing a set does not mean it is uncountable or unlistable. In the above argument, we have shown that no list could ever be made.

The real numbers are an uncountable infinite, unlike the countable infinites of the whole numbers, the fractions, or for that matter, all algebraic numbers. The transcendentals, most of them dark and unknown, populate this uncountable infinite. In his classic, *Men of Mathematics*, E. T. Bell, who is in other ways unreliable on Cantor, came up with a good metaphor for the transcendental numbers: "The algebraic numbers are spotted over the plane like stars against a black sky; the dense blackness is the firmament of the transcendentals."[3]

The diagonalization argument above, that the real numbers are more abundant than the algebraic numbers, does not supply much intuition about the vastness of the difference in size between the fractions or the algebraic numbers and all the real numbers between 0 and 1. All it says is that if I think I have listed the real numbers I must have left at least one off the list. Measure theory, begun in 1901 by Henri Lebesgue (1875–1941), uses a topological argument that gives some sense of the vastness of the transcendentals. Lebesgue was able to show that any countable subset of the reals must have measure (a generalization of length) zero. Hence, any set of positive measure must consist mostly of transcendentals. He used a procedure of "covering" each point in a countable set with one of an infinite number of progressively smaller pieces of an original line segment. Since the segment supplying the pieces could be arbitrarily small, the amount of "measure" contained in any countable set of points is 0. Ideas of measure are of direct practical use in many problems in probability and statistics.

When we say an infinite series converges to a number, we are saying that the completed infinite of the series equals the number. That idea has had logical persuasiveness since the time of Zeno, though it still feels mysterious. The idea that different infinite sets of points can have different sizes (in the sense of countable or uncountable) also feels mysterious. A frequent response to these arguments is that they must be some kind of trick. But there is no trick. The development of a generalized measure theory as part of set theory showed that Cantor's set theory had the flexibility to allow mathematicians to apply well-accepted ideas about completed infinites in calculus to geometrically and topologically more complex sets of points. That expanded the number of problems that could be solved. We will find the great Russian mathematician A. N. Kolmogorov (1903–1987), who solved the thirteenth problem and part of the sixth, using measure theory framed in the language of set theory to find a rigorous mathematical treatment of how a floating grain of pollen is jostled by almost countless collisions with water molecules—Brownian motion.

If Cantor had stopped with the ideas and notation we have been talking about, he would have excited less controversy. As children, many of us have set out counting, often at bedtime, with the goal of counting very, very, very high. We might have counted by tens or even millions to speed up the process, but alas, none of us actually reached infinity, unless we partook of it in sleep. Cantor reached infinity and kept going, or at least created a system that kept going: smallest countable infinity and 1, smallest countable infinity and 2, smallest countable infinity and 3, and so on. His system would go on, collecting and continuing, until one eventually would collect all the countable infinities and then go: smallest uncountable infinity, smallest uncountable infinity and 1, etc. Then there would be the next-to-the-smallest uncountable infinity and 1, and so on. These are called the transfinite ordinal numbers, and Cantor defined an arithmetic that one could do with them. When he talked about size he defined another powerful concept, cardinality. All sets can be compared using the concept of cardinality. One can imagine what the passionate finitists thought of this. As mentioned, Kronecker was particularly livid.

The vast expanse of infinite sets, infinite ordinals, different cardinalities, the huge entirety of the "transfinitum," as Cantor referred to it, does have an inescapable connection to mystical thought and philosophical speculation. Religious philosophers have always dwelt on God's infinite nature, and one aspect of religious experience is an attempt to make contact with the infinite. Now, here was Cantor talking about a whole vast structure of the infinite. The potential sacrilege of these ideas was not lost on him. He believed that positivism and materialism had done great damage to religious culture, that Newton's *Principia* had played a central role in this (as I am sure it has), and that the philosophical understanding of a scientific discovery, which its discoverer has an opportunity to supply, greatly affects its role in the spiritual history of the world. In Cantor's own words:

> . . . that the greatest achievement of a genius (like Newton) despite the subjective religiosity of its author, when it is not united with a true philosophical and historical spirit, leads to effects (and, I assert, necessarily does so) by which it seems highly questionable whether the good in them simultaneously conveyed to mankind won't be significantly surpassed by the bad. And the most detrimental (of these effects) it seems to me are the errors of the "positivism" of Newton, Kant, Comte and others upon which modern skepticism depends.[4]

Cantor was laboring lest his discoveries be wrongly understood. Now he also began to show the mental illness that would send him increasingly for stays at the *Nervenklinik* in Halle. Cantor came to believe that God had revealed set theory to him and that the sets he talked about existed as fully realized entities in the mind of God. In fact, the whole transfinitum existed perfect and perfectly realized in God. He contacted the Catholic Church to supply the correct interpretation. The church was interested and considered possible pantheistic influences lurking in this theory—all those different infinites. It was decided set theory and the transfinitum were good philosophy, compatible with the church's position.

Cantor's first major breakdown came in 1884. He had just returned from Paris, where he had friendly and respectful visits with Poincaré, Hermite, Émile Picard (1856–1941), and Paul Appell (1855–1930). He went to the Opera and the *Comédie Française* and visited galleries and museums. He was deeply embroiled in his quarrel with Kronecker and the other finitists, however, and was also having trouble with "the continuum hypothesis," which deals with the size of all the real numbers on the number line (or "continuum") in relation to the size of the counting numbers. Proving the continuum hypothesis would, to use Hilbert's words in his formulation of the first problem, "form a new bridge between the countable assemblage and the continuum" and would have been a high point in Cantor's ordering of the transfinitum. He would work on it the rest of his life without succeeding. Cantor had six children, the last born in 1886, and family life was important to him. His breakdown was a blow to the children who were old enough to understand, particularly his oldest child, Else, who was nine at the time. Most people trust their sanity. They go to bed, wake up, day passes into day, and their thinking, behavior, and view of the world are the same, or vary within a pattern. Brilliant, original, passionate, nervous, sometimes insecure, now the ground of Cantor's being was changed.

He looked for reasons for his breakdown, as is natural, and managed to make up with Kronecker, at least partially. Kronecker expressed surprise at the thought that there might have been personal animosity between them. In this it would appear he was disingenuous or imperceptive. Cantor blamed the strain of his mathematical work and took up other interests, becoming more passionate in his study of the Scriptures and interested in different spiritual paths: freemasonry, theosophy, and Rosicrucianism. He also became intent on proving that Bacon was the real author of Shakespeare's works. On this last subject he became very knowledgeable and published a great deal. He requested that he be allowed to switch his professorship to philosophy. However, none of these changes helped in the long run.

In the years that followed, Cantor discovered a clearer proof of the nondenumerability of the real numbers, his diagonalization procedure, which shows very generally that the set of all subsets of a set is larger than the set itself. This is one of his most important results. He presented set theory in its most organized way from 1895 to 1897. Sometimes a person can have a mental breakdown but recover and live long enough to essentially forget the attack. This had not happened with Cantor, and life became a balancing act. He became aware that there was a contradiction or "antinomy," as it is sometimes called, in set theory. If the set of all subsets of a set is larger than the set, what about the set of all sets? Can that be larger than itself? It would appear not. The first published paradox came from Cesare Burali-Forti (1861–1931) in 1897 and is similar to Cantor's paradox, though more technically grounded in the details of the ordinal numbers.

At first Cantor was not disturbed. Set theory had been revealed to him. Paradox and mystery are perhaps inevitable when talking about God and the infinite. However, the presence of contradiction in a mathematical system is disastrous. It took awhile for this to sink in. Cantor set about attempting to remove the contradiction. Others were optimistic at first also. Cantor told Hilbert of his problems in 1895 in correspondence. Hilbert did not publicize the contradictions. He himself had fought battles against Kronecker and the finitists and was nothing if not shrewd in his maneuvering in the world of mathematics. This problem altered the course of Hilbert's career. He once said, "No one will drive us out of this paradise Cantor has created for us."[5] Hilbert's work on the foundations of geometry in 1898–1900 can be understood in this light.

However, in 1902 Bertrand Russell (1872–1970) discovered what is known as Russell's antinomy—the set of all sets that don't contain themselves. Is this set a member of itself? Russell was not afraid of controversy and broadcast his views. He also believed Cantor's results were probably true and important and set out to fix things, as did others. In 1904 at the Third International Congress of Mathematicians in Heidelberg, Julius König (1849–1913), a mathematician from Budapest, delivered a paper that "proved" that Cantor had some of his theory of cardinality wrong. Cantor was in the audience. He thought König must have made an error, which he had, and took König's talk as a great insult. The effort to get the foundations of set theory straightened out continued through the rest of Cantor's life. The contradictions resulted from injudicious use of the word "all," or self-reference in defining a set. The remedies restrict the way sets can be defined or specified.

In the years before the turn of the century, three deaths had hit Cantor very hard, beginning with the death of his mother in 1896. His younger brother, who had been an officer in the Hessian Dragoons and married an Italian baronessa, died in Capri in 1899. Then, later that year, Cantor's youngest son died suddenly at the age of twelve. Cantor had already been hospitalized again in 1899, and this tragedy appears to have sealed his fate. After the winter of 1902–03, Cantor would be in and out of the *Nervenklinik* for the next fifteen years, with the periods of hospitalization becoming more frequent and longer. His lucid periods were interspersed with madness. Cantor was hospitalized for the last time in May 1917. He wrote his wife again and again asking to come home. In one of his last letters Cantor enclosed a poem he had written. It starts:

> That was a winter cold and wild,
> Like none one can recall;

and ends:

> Cold would my winter've been;
> To suffer gladly, pen a poem,
> To escape the world I'm in.[6]

I Am Lying (Mathematics Is Consistent)

The Second Problem

Gödel

In English or any natural language argument can be riddled with contradictions, yet language itself is not damaged and issues are often provisionally resolved. Discussion continues. Logical reasoning in pure mathematics, however, must be reliable. Though our cultural faith in the method might be, in part, empirical, certitude is intrinsic to the method itself. If there is a contradiction in mathematics the entire structure loses its meaning, because any statement can be proved reasoning from a contradiction. The following displays this style of reasoning: Suppose I have proved a statement, "A," and I have also proved "not A." That is, there is a contradiction. Suppose I have another statement, "B." Let us examine the combined statement, "A is true or B is true or both are true." Since "A" has been proved, I can make and prove this statement. But I also know "A" is not true, having proved "not A." Therefore, in the statement under examination, it follows that "B" is true. But since "B" could be any statement in the language, I have proved every statement, an absurdity. The contradiction has created a shell game in which truth floats from "A" to "B" to "not A" to "not B." The question isn't whether this or that actual statement may be true; this is a demonstration that valid reasoning is not possible when there's a contradiction.

Hilbert's second problem, "The Compatibility of the Arithmetical Axioms," is an invitation to remove the possibility of contradiction from mathematics. It includes the statement: "To prove that they [the axioms for arithmetic] are not contradictory, that is, that a finite number of logical steps based upon them can never lead to contradictory results." With "axioms for arithmetic" Hilbert is referring to the axioms of the real numbers. He says, "The axioms of arithmetic are essentially nothing else than the known rules of calculation, with the addition of the axiom of continuity."

What would a proof of the consistency of arithmetic look like? Hilbert is vague, stating: "A direct method is needed for the proof of the compatibility of the arithmetical axioms." Any possible proof would be finite, even if it asserted something about an infinite search among all numbers or made statements about infinite sets. It might be possible to analyze the pattern of what one could deduce from a system of axioms and demonstrate that the process could not yield a contradiction. Hilbert envisioned a formal program that started with the arithmetic of the natural numbers but soon moved to the reals, then analysis, and eventually all of mathematics. For Hilbert, a proof of the compatibility of the arithmetical axioms would be a proof of "the mathematical existence of the complete system of real numbers or of the continuum." An ironclad formalism provably incapable of generating contradictions could inoculate Cantor's continuum and, from there, mathematics against assaults from narrowly constructivist and intuitionist quarters. Creating a reliable formalism would be profoundly constructive and would enable mathematicians to operate in realms that they could not literally inspect. Hilbert's overall strategy is clear in the following:

> If contradictory attributes be assigned to a concept, I say, that *mathematically the concept does not exist*. . . . But if it can be proved that the attributes assigned to the concept can never lead to a contradiction by the application of a finite number of logical processes, I say that the mathematical existence of the concept . . . is thereby proved.[1]

This took place in the context of specific personalities. Kronecker died in 1891. For the next twenty years no finitist mathematician stood with such force against Hilbert's desire that the expansion of mathematics go on unimpeded by philosophical reservations. Russell trumpeted forth the existence of contradictions, but he tried to vanquish them with a formal program not at odds with Hilbert's goals. Poincaré didn't accept Hilbert's formalism, but his objection to Hilbert's strategy didn't mean he was, in more general terms, a partisan of restricted mathematics. In his actual practice, Poincaré constantly extended mathematics beyond the limits of the mathematical rigor of his time. He died in 1912, leaving the debate. As Hilbert delivered his list of problems and turned his attention to analysis during the first decade of the century, however, L. E. J. Brouwer, a young Dutchman schooled in the town of Hoorn in north Holland, was studying mathematics at the University of Amsterdam.

If Cantor had seen his work in mathematics as part of a larger worldview, Brouwer saw mathematics as flowing from philosophy and mysticism. Cantor was an enthusiast. He discovered a beautiful mathematical edifice and decided only God could have led him to it. Brouwer saw Cantor's set theory as rife with error. In Brouwer's *Collected Works*, edited by his student Arend Heyting (1898–1980), the first item is a group of fragments from a 1905 pamphlet of lectures Brouwer gave for a students' society, *Life, Art and Mysticism*. It is a curious admixture of Calvinism and theosophy, mathematics and mysticism, with many digressions that Heyting omitted. Brouwer wrote:

> Science does not remain confined to serving industry, again the means becomes an end in itself and science is practiced for its own sake. A further aberration has been the concentration of all bodily awareness in the human head thereby excluding and ignoring the rest of the body. At the same time man became convinced of his own existence as an individual and that of a separate and independent world of perception. At that stage the full extent of the deviation of human scientific thinking became clear, for scientific thinking is nothing but a fixation of the will within the confines of the human head, a scientific truth no more than an infatuation of desire restricted to the human mind. Every branch of science, as it proceeds, will therefore always run into deeper trouble; when it climbs too high it becomes blindfolded in even more restricted isolation, the remembered results of that science take on an independent existence. The 'foundations' of this branch of science are then investigated and this soon becomes a new branch of science. . . . As they climb higher and higher trouble increases and in the end everyone is thoroughly confused. Some in the end quietly give up. . . . But there are others who do not know when to stop, who keep on and on until they go mad; they grow bald, shortsighted and fat, their stomachs stop working properly, and moaning with asthma and gastric trouble, they fancy that in this way equilibrium is within reach and almost reached. So much for science, the last flower and ossification of culture.[2]

If Brouwer did not arrive at mathematics unencumbered by philosophical baggage, he was indisputably an excellent mathematician. That's what made him dangerous from Hilbert's point of view. In the period 1907–1913 Brouwer achieved results of great importance, particularly in the emerging field of algebraic topology. He gave the first rigorous demonstration of the topological invariance of dimension. Mathematicians had long felt that the number of dimensions a space had was in some sense objective—that is, one couldn't set up complete local coordinate systems in the same space that had different numbers of axes. It seems obvious, but for a mathematician, if you can't prove it you don't really understand it. Cantor destroyed the arguments of his day in favor of the invariance of dimension and then, along with others, offered demonstrations of his own that were not valid. Brouwer invented and adapted machinery to prove the argument that was of great use in years to come on other problems as well, and he occupies a place of honor in the history of algebraic topology. His fixed-point theorem is another classic textbook result, though eventually his philosophy caused him to reject it.

Brouwer offered concrete alternatives to mathematical treatments he didn't agree with: his own version of real numbers, set theory, measure theory, and of the types of logical argument that were acceptable. In the 1910s and 1920s some of Hilbert's best students, including Weyl, defected to Brouwer's intuitionism. Weyl investigated how much nineteenth century analysis would be lost if one changed to an intuitionist program, akin to determining how much of the furniture one would have to give up if one sold a big house and moved into a smaller one. In Russia, Kolmogorov began his career in the 1920s as an intuitionist.

Here is an explanation of intuitionism from Heyting's student A. S. Troelstra:

> In Brouwer's intuitionism, mathematics consists in the mental construction of mathematical systems, an activity that is supposed to be carried out in the mind of an idealized mathematician, in principle without the use of language: language enters only in attempts to suggest similar constructions to other persons. Something is true in intuitionist mathematics only if it can be shown to hold by means of a construction.[3]

What's essential about this is the rejection of language as an instrument.

The intuitionists rejected existence proofs that were non-constructive, such as Hilbert's original proof that invariant theory had a finite basis. They didn't believe that a statement about infinite sets or infinite processes could be assumed to have meaning. In particular, they didn't believe that the law of the excluded middle (either a statement or its negation is always true) would apply in all cases. Further, what is known as Hilbert's "program" (to formalize and prove mathematics consistent) dealt with systems of axioms in a neutral way. If they were consistent, the axioms did not necessarily have to refer to whatever they were created to refer to. Implicit in this is that one consistent system is as good as another. The intuitionists were philosophically at odds with systems of formal axioms that might take on lives of their own. Mathematics was about things like numbers or lines that we know clearly through wordless mathematical "intuition." All more complex mathematics had to be built through precise construction from these elements.

Hilbert's formalist gambit put him in an odd position. The real content of his philosophy, as I see it, was that he wanted to protect the greatest possible range of mathematical discourse from criticism. He was willing to do some philosophical bending and stretching to achieve his goal. Hilbert thought Cantor's transfinitum convincing and beautiful mathematics; the intuitionists didn't. He believed it was important, even a "paradise." It was a judgment. So Hilbert expressed the warmth of his own intuitions by coldly asserting that formal consistency implied mathematical existence and that mathematics is understanding how formal systems work.

Within this frame, how can one system be more important than another? Without any appeal to intuition or aesthetic judgment, which are inherently informal, all mathematical problems are equal. Yet in listing twenty-three important problems, Hilbert followed his nose, highlighting some areas and eliminating others. This remains an issue today. Here is Kolmogorov's student V. I. Arnold, commenting on the 1994 International Congress of Mathematicians* in an article with the dire title "Will Mathematics Survive?":

> At the beginning of this century a self-destructive democratic principle was advanced in mathematics (especially by Hilbert), according to which all axiom systems have equal right to be analyzed, and the value of a mathematical achievement is determined, not by its significance and usefulness as in other sciences, but by its difficulty alone, as in mountaineering.[4]

Until the 1920s not a great deal had been done to move Hilbert's program forward or even clearly define what it was. Russell and Alfred North Whitehead (1861–1947) had created in *Principia Mathematica* (1910–1913) a formal system basing mathematics on logic that seemed consistent but was not proved so. Proceeding in great detail, *Principia* asserted that mathematics was consistent because logic was consistent. Hilbert himself had been working on analysis and mathematical physics. But as intuitionism became increasingly popular with younger mathematicians, Hilbert must have sensed the ghost of Kronecker rising from the grave. The following excerpt from Weyl's official and polemical entry into the debate, a paper titled "On the New Foundational Crisis in Mathematics," should make clear how this must have felt to Hilbert:

> The antinomies of set theory are usually regarded as border skirmishes that concern only the remotest provinces of the mathematical empire and that can in no way imperil the inner solidity and security of the empire itself or of its genuine central areas. Almost all the explanations given by highly placed sources for these disturbances (with the intention of denying them or smoothing them over), however, lack the character of a clear, self-evident conviction, born of totally transparent evidence, but belong to that sort of half to three-quarters honest attempts at self-deception that one so frequently encounters in political and philosophical thought. Indeed, every earnest and honest reflection must lead to the realization that the troubles in the borderland of mathematics must be judged as symptoms, in which what lies hidden at the center of the superfi-

* These conferences lost their numbers after the First World War. Was it an international congress if the Germans were not allowed to come?

> cially glittering and smooth activity comes to light—namely the inner instability of the foundations upon which the structure of the empire rests.[5]

Strong words. Writing in Germany in 1920, Weyl went on to compare existential statements to paper money.

Hilbert now devoted much of his time to his program. With able students, it moved forward nicely. By breaking "arithmetic" down into small steps, much was genuinely clarified about the foundations of mathematics. Hilbert's formalism was radically more simple than Russell and Whitehead's in *Principia*. At the International Congress in Bologna in 1928, Hilbert delivered an address to "stormy applause at the beginning as well as the end" and announced that Wilhelm Ackermann (1896–1962) and John von Neumann (1903–1957) had proved the consistency of whole number theory.[6] Further, he said Ackermann had proved the consistency of real number arithmetic, with only minor details to be filled in. Those minor details involved infinite searches. He listed four open questions in his formalist program—the first was to finish Ackermann's proof. By 1931 all four questions would be answered by a young man in Vienna, but in a way that destroyed Hilbert's program and solved Hilbert's second problem in the negative. There are those who would differ and still hope for a consistency proof for mathematics, either along Hilbert's lines or from new methods. However, as an enterprise Hilbert's program was effectively stopped. For just as geometry could be reduced to pairs of numbers and made to exist entirely in the world of arithmetic, something resembling the paradoxes that occur in a natural language could be encoded into the numbers and functions of arithmetic when it was treated as a formal system. This did not damage the consistency of arithmetic itself, but put fundamental limitations on what could be proved in formal systems.

Kurt Gödel was born on April 28, 1906, into a prosperous German-speaking family in the textile manufacturing town of Brunn, Moravia (now Brno in the Czech Republic). Gödel's father had not completed his academic schooling, but studied weaving instead and went to work for the Friedrich Redlich factory, where he rose to become managing director and part-owner. Gödel's grandfather on his mother's side had also risen from a poor background to a high position in a textile firm, after his own father, who had come from the Rhineland to work in Brunn as a hand-loom weaver, lost his livelihood with the advent of machine looms. Gödel's grandfather educated himself, hand-copying books when he couldn't afford to buy them, and provided a good education to his children. He sent Gödel's mother to a French school, which was not common among families of their social position. A good gymnast and ice skater, she had a happy childhood, with many friends, and loved the romantic German and French literature she was schooled on. Rudolf Gödel, Kurt's brother, wrote that their father and mother's marriage was not a "love match" but described it as "built on affection and sympathy." Their mother was the dominant figure for the boys.

In 1913, when Kurt was seven, the family built a well-situated villa with a beautiful garden. Rudolf fondly remembered the two dogs they had. At Christmas there was a round of parties, one stop memorable for the decorated tree that was hung from the ceiling so it could spin around. Presents were purchased by catalog from a Vienna toy shop, though Gödel's mother found this "unpoetic." The family read political and historical novels, and Gödel's mother had great sympathy after the First World War for the deposed Hapsburgs. She was fond of fairy tales. According to Rudolf, "Even in old age she could recite poems by Goethe, Heine, Lenau and others by heart. As a young woman she composed poems herself." She loved the garden and is reported to have said, "It would not be so terrible having to die, but not being able to experience Spring any more. . . ." Kurt spoke German at home and at gymnasium and did not study Czech. His mother had childhood memories of tension in Brunn between the German community and the "Czech element." The "clatter of horse's hooves" could often be heard as dragoons broke up clashes in the cobblestone streets.

Gödel was unusually attached to his mother as a child and was in a state of anxiety when she left the house, according to Rudolf. Kurt got along well with Rudolf, playing "quiet games: bricks, model trains, also board games and chess." Kurt was not a good loser. His intelligence and persistence were recognized early, and he acquired the nickname "Mr. Why." At about age eight he fell seriously ill with rheumatic fever and after he recovered, upon reading a medical book, became convinced beyond any reasoning that he had a bad heart. No doctor ever won a logical argument with Kurt Gödel. Sureness of opinion was a lifelong trait. In Rudolf's words, "My brother had a very individual and fixed opinion about everything and could hardly be convinced otherwise."[7]

In school Kurt was precise, thorough, and excellent. His favorite subjects were mathematics and languages, including Latin, French, and English. At his death he owned dictionaries in many languages. He also excelled in theology and showed some interest in history, but this faded. Gödel enjoyed "light" music—operetta and, later in life, American show tunes. He studied a now-obscure type of German shorthand, the Gabelsberger script, in order to write more efficiently (and possibly more privately). Gödel's collected personal papers and his many notebooks are page after page in this shorthand. John W. Dawson Jr., who catalogued Gödel's *Nachlass*, or literary estate, was assisted by his wife, who learned the shorthand, and by an aging German emigrant who knew it. In 1997 Dawson published *Logical Dilemmas: The Life and Work of Kurt Gödel*. In any conflict of biographical detail I have followed Dawson.

It was rumored that Gödel managed to make it through gymnasium without making a single mistake in Latin grammar. He always received the highest possible marks except once, in mathematics. Surviving samples of his homework show his meticulousness. According to Rudolf, he surprised everyone by learning most university mathematics on his own in gymnasium. Gödel himself dated his interest in mathematics from age fourteen and certainly studied calculus before university. He had an unusual number of excused absences, both for the full day and from exercise. At home he did not share his mother's and brother's love of nature and the garden, and when the family took Sunday walks he tended to stay behind, cooped up with a book. While still in school he formed a relationship with the daughter of family friends, described as an "eccentric beauty"[8] ten years his senior. It was stopped by opposition from his family.

Gödel went to the University of Vienna in 1924 intending to study physics but was wooed to mathematics by the three-year lecture cycle given by Philipp Furtwängler (1869–1940). The lectures were undoubtedly excellent. But logician Georg Kreisel, who had a lifelong friendship with Gödel, writes: "Another singular aspect of those lectures (which Gödel did not mention, possibly because of the medical history involved) may have had equal weight. Furtwängler was paralyzed from the neck down, and lectured from his wheelchair without notes, while a scribe wrote the proofs on the board."[9] If indeed Furtwängler's plight cemented Gödel's attraction to mathematics, this is reminiscent of Flannery O'Connor's "Good Country People," in which the real goal of the suitor is to make off with the female protagonist's wooden leg. Gödel attended Heinrich Gomperz's lectures on the history of European philosophy and Moritz Schlick's (1882–1936) seminar on the philosophy of mathematics. Philosophy united Gödel's interests in language, theology, and the foundations of science and mathematics. But whatever the reasons, Gödel studied mathematics and from all accounts was knowledgeable in wide areas.

Logical positivism was developing in what was called the Vienna Circle, or *Wiener Kreis*. Schlick was a leader of the circle, dubbed "a band of positivist philosophers"[10] by Kreisel, which met in a seminar room near the department of mathematics. Gödel formally switched his interest to mathematics in the fall of 1926, at the beginning of his third year. Through his primary teacher, Hans Hahn (1879–1934), who was a member of the circle, Gödel started attending meetings, sometimes irregularly, between 1926 and 1928. This undoubtedly brought questions of the foundations of mathematics to his attention. The circle was involved in a close reading of Ludwig Wittgenstein's *Tractatus Logico-Philosophicus*, which took up the question of what can be said, its short paragraphs numbered like the equations in a mathematics book. Those in attendance would read a sentence aloud and discuss it. Olga Taussky-Todd (1906–1995) wrote that Wittgenstein's stature was such that arguments could be settled by invoking his name. (At this point Wittgenstein had sworn off philosophy and was not present.) Later in life, Gödel took pains to separate himself from the views of the circle and said he had never really studied the *Tractatus*. He told Hao Wang (1921–1995) that he "did not like the *Tractatus* because it proposes to show that philosophy is not possible."[11]

Gödel shared an apartment with his brother, though they didn't see each other often, as Rudolf was much at the hospital in his medical studies. It was a happy, relatively untroubled time. The following vivid passage, from Taussky-Todd's contribution to *Gödel Remembered*, is about an incident that probably took place in 1925, when Gödel was in his second year:

> There is no doubt about the fact that Gödel had a liking for members of the opposite sex, and he made no secret about this fact. Let me tell a little anecdote. I was working in the small seminar room outside the library in the mathematical seminar. The door opened and a very small, very young girl entered. She was good looking, with a slightly gloomy face (maybe timidity) and wore a beautiful, quite unusual summer dress. Not much later Kurt entered and she got up and the two of them left together. It seemed a clear show-off on the part of Kurt.
>
> That same girl changed quite a bit later; maybe she became a student. She came to talk to me occasionally, and she complained about Kurt being so spoiled, having to sleep long in the morning and similar items. Apparently she was interested in him, and wanted him to give up prima donna habits.[12]

This girl later appeared at Taussky-Todd's door with a topology manuscript of her own.

In about 1928 Gödel met Adele Porkert, who would eventually become his wife. She worked in some capacity, probably as a dancer or singer, in a Viennese nightclub called *Der Nachtfalter* (The Nightmoth). She was six years older than Gödel and had a facial birthmark that was considered a disfigurement. She had been married before and had little formal education but possessed wit and playfulness. Again Gödel's family was not in favor of the relationship, particularly Gödel's father, but Kurt was now an adult. He felt comfortable with Adele. They continued to see each other and finally married in 1938. At the end of their lives, when she fell ill and could no longer take care of him, Gödel died.

Gödel's father died suddenly at fifty-four, in 1929, of an abscess of the prostate, leaving a fairly sizable estate. Gödel's mother moved to Vienna, where she rented a large apartment. Her sons lived with her. She delighted in the theater and music, and Kurt often went out with his mother and Rudolf. Rudolf remembers the time fondly: "We and my brother often enjoyed pleasant evenings together which concluded with long discussion about what we had seen." He adds: "At that time my brother was very interested in the theater; sadly, it was later obviously different."[13]

The stage was set for Gödel's great series of contributions. Hilbert had stated four specific problems in his Bologna address[14] (and in his 1928 book with Ackermann, *Grundzüge der theoretischen Logik*, which Gödel studied) and identified them as the next and nearly crowning steps in his program. These problems provided the perfect objects to which Gödel could direct his mind. He was also influenced by lectures given by Rudolf Carnap (1891–1970), a member of the Vienna Circle.

We have seen how early algebraicists phrased their problems with words, but eventually a code for writing the words as symbols greatly simplified the problems and allowed the expansion and clarification of our concept of number. In a similar way, many mathematicians set out to reduce the rules for logical inference to a formal language, that is, a collection of symbols with clear mechanical rules for the formation of strings of symbols (sentences) and for constructing (deducing) new strings starting from an initial collection of strings (axioms). One contemporary version of these symbols, which I take from Paul Cohen's *Set Theory and the Continuum Hypothesis*, is in its entirety:

$R\ R' \ldots$	$c\ c' \ldots$	$\neg$	$\wedge$	$\vee$	$\Rightarrow$	$\Leftrightarrow$
relations	constants	not	and	or	implies	if and only if

$\forall$	$\exists$	$=$	$x\ x'\ x'' \ldots$	( )	,
for all	there exists	equals	variable symbols	parentheses	commas.

An example of a string would be $\forall\, x,x'\ (x = x' \Leftrightarrow\ \forall x''(R(x'',x) \Leftrightarrow R(x'',x')))$, which is the first axiom of Z2, an axiom system for arithmetic. It says that if two things are equal, they both bear the same relationship to a third object.

The point of translating arithmetic or any other mathematical system, like set theory, into a formal language of this degree of purity is not, at least initially, to render natural problems easier to solve. However, making the system formal allows recognition of more general patterns. Emil Artin's (1898–1962) solution of the seventeenth Hilbert problem makes use of this, and the tenth problem was solved using ideas and machinery developed in this context. Once a system is made formal it might be possible to prove that its axiom system is consistent. Hilbert used the method of models to accomplish this in his analysis of the axioms of geometry. A model is a mathematical system or object, either already at hand or created with this purpose in mind, that embodies a set of axioms and all statements that can be derived from them. Any identifying reference to a specific thing in our formal language must correspond to a specific thing in the model. If the model is a good mathematical object that we trust, then we know the axioms it embodies are consistent.

The fourth of Hilbert's Bologna questions was answered by Gödel in 1929 in what is known as Gödel's completeness theorem for the predicate calculus. The predicate calculus is the set of rules for logical deduction in formal systems—mechanical rules that allow one to write new statements based on statements already established. There is a rich forest of historical interpretation of the word "completeness." We will take a modern mathematician's approach to Gödel's theorem, again following Cohen, who will return in our next chapter. The theorem as Cohen states it is:

> *Let S be any consistent set of statements. Then there exists a model for S* whose cardinality does not exceed the cardinality of the number of statements in S if S is infinite, and is countable if S is finite.[15] [My italics]

This states that any consistent set of statements in a formal language can be realized, or embodied concretely, in some kind of mathematical object or structure. The key here is that S is *consistent*, that is, can't generate a contradiction. The model is so constructed that it clearly does not have a greater cardinality or size (in the way we compare sizes of infinite sets in set theory) than the original set of statements, S. According to this result, any reasonable axiom system for the real numbers (i.e., finite or listable according to some definite rule) has a model that is countable. Yet the real numbers themselves, we have seen, are not countable. This is surprising, and in the next chapter we will explore it further.

Why is Gödel's result called the completeness theorem? Let me repeat two corollaries, as Cohen states them:

> 1) If A is not derivable from S, then there is a model for S in which A is false.
>
> 2) If a statement is true (i.e., true in every model), then it is provable.[16]

These two results say that all statements that hold universally in every interpretation or model of a set of axioms are, in fact, formally deducible from the axiom system. Further, if a statement is provable formally, it is true in any possible interpretation or model of the axioms. This says that formal logical deduction does capture truly and fully what we mean by logical deduction—in the sense of Leibniz, "true in all possible worlds." These systems are, in a word, complete. This represents a step forward on Hilbert's program.

In 1930 Gödel proved results that caused a good deal more surprise, his incompleteness theorems. Gödel's first incompleteness theorem says that if we take the axioms we have for the arithmetic of the non-negative whole numbers, including the operations of multiplication and addition, then there is a statement in the formal language that cannot be proved true and cannot be proved false by arguments made in the formal language; that is, there is a statement that is "undecidable" in terms of the formal language. Yet the statement itself is such that it is true as a statement about the non-negative whole numbers as we know them, outside the context of the axiom system. The existence of such a statement, unprovable in the formal system yet true as a statement about the non-negative whole numbers, is what kills Hilbert's program. The fish is larger than the net.

Gödel proves his theorem by actually constructing such a statement. He took the well-known liar's paradox, i.e., "I am lying," or "All Cretans are liars" uttered by a Cretan (in a logically softer but more colorful version), and made it into a statement in the formal system of arithmetic. The statement Gödel constructs has a clear interpretation, from the outside, of "this statement is not provable." It turns out that logical complexities arising from self-reference exist even in arithmetic. The idea is simple. We make a list of what we can prove and, using a diagonalization process like Cantor's, construct a statement that says, "This statement is not on the list of provable statements." In effect it says, "I am lying."

Here is a sketch of how Gödel did this:

The method starts by assigning a unique non-negative whole number, a code number, to every statement in the formal language (provable or unprovable—even incoherent nonsense), that is, every possible finite string of symbols in our formal language for arithmetic. Now we make a list of the code numbers of all the statements that can be proven from the axioms for arithmetic. (We make the list by selecting first the proofs that are one symbol long—if there are any—then two, etc.) A list of numbers can always be viewed as a function in arithmetic: $f(1)$ = first number on the list, $f(2)$ = second number, and so on.

Next we make an array, which we are going to use to generate a second list using Cantor's diagonalization method. In the first column of the array list all statements in the language in which x appears as a free variable (call them $A_1(x)$, $A_2(x)$, $A_3(x)$, . . .). These are statements that assert something about x, true or untrue, e.g., $x = x + 0$, $x = x + 2$. They could talk about the various functions of arithmetic, e.g., $f(x) = 2$. Then in parallel columns we substitute first 1, then 2, then 3, etc., for x. It amounts to an array of all possible examples of the general statements, e.g., $1 = 1 + 0$, $2 = 2 + 0$, etc., or in the false example, $1 = 1 + 2$, $2 = 2 + 2$. Most of these examples would be false; some would be true. A picture might help:

$$
\begin{array}{l@{\qquad}llllllll}
A_1(x) & \mathbf{A_1(1)} & A_1(2) & A_1(3) & A_1(4) & A_1(5) & A_1(6) & A_1(7) & \ldots \\
A_2(x) & A_2(1) & \mathbf{A_2(2)} & A_2(3) & A_2(4) & A_2(5) & A_2(6) & A_2(7) & \ldots \\
A_3(x) & \mathrm{A}_3(1) & A_3(2) & \mathbf{A_3(3)} & A_3(4) & A_3(5) & A_3(6) & A_3(7) & \ldots \\
A_4(x) & A_4(1) & A_4(2) & A_4(3) & \mathbf{A_4(4)} & A_4(5) & A_4(6) & A_4(7) & \ldots \\
A_5(x) & A_5(1) & A_5(2) & A_5(3) & A_5(4) & \mathbf{A_5(5)} & A_5(6) & A_5(7) & \ldots \\
A_6(x) & A_6(1) & A_6(2) & A_6(3) & A_6(4) & A_6(5) & \mathbf{A_6(6)} & A_6(7) & \ldots \\
A_7(x) & A_7(1) & A_7(2) & A_7(3) & A_7(4) & A_7(5) & A_7(6) & \mathbf{A_7(7)} & \ldots \\
\cdot & & & & \cdot & & & & \\
\cdot & & & & \cdot & & & & \\
\cdot & & & & \cdot & & & & \\
\uparrow & & & & \uparrow & & & &
\end{array}
$$

List of all statements with x as a free variable

Array of all statements about specific values for x, derived from list on left

We look at the statements on the diagonal, going down the array. Now we generate another list of statements—some true, some false—referring to the second list we made, which assigned code numbers to the provable statements in arithmetic: 1) "The code number of the first statement in the diagonal is not on the master list of provable statements." 2) "The code number of the second statement on the diagonal is not on the master list of provable statements." 3) "The code number of the third. . . ." 4) "The code number of the fourth . . . ," etc. Key is that this new list of statements is constructed in a regular fashion from the diagonal, so we can write a general formula using a free variable, x—that is, a sentence in the formal language that talks about x. That formula is: "The code number of the xth statement on the diagonal is not on the master list of provable statements." It captures all the statements on our new list as instances. But this general statement is itself a statement of one free variable, x, and so must appear on our list as $A_n(x)$ for some n! Further, Gödel can find this number, n, so the proof is entirely constructive.

This is a marvelous flip of Cantor's method. Cantor used the diagonal of his array to find a real number that wasn't on the supposed list of all real numbers. The list of all statements with x as a free variable that Gödel makes isn't bogus, nothing is left off. The statement we just generated with the help of the diagonal is on the list. Let us evaluate this nth statement, substituting n for x. It is, by our definition, the statement "The code number of the nth statement on the diagonal is not on the master list of provable statements." Since it itself is the nth statement, it says, "I am not provable." It is entirely a statement about well-defined functions in the arithmetic of the whole numbers. If we could prove it in our formal system, it would contradict itself by stating that it was not provable. If we could prove its negation, the negation would contradict itself. We conclude that if our system is consistent, we can prove neither our statement nor its negation.

Is our statement actually true about the arithmetic of the whole numbers, as opposed to provable? Our statement is that its own Gödel code number is not on the list of provable statements. If we went through the list forever and did not find the code number, our statement would be obviously true. On the other hand, if we went through the list and found this code number, we would be told, in effect, "You can't have found me. There is a proof that I am not here." This is impossible. Common sense is allowed when we think informally. Since all of our experience leads us to believe arithmetic really is consistent (as opposed to being proven consistent in a formal system), we do believe our statement is true and that we will never find it on the provable list. Though our informal, natural language is rife with contradictions, it is actually more mathematically powerful than a formal language in this instance.

Historically, the second incompleteness theorem killed Hilbert's program to prove mathematics consistent. It states that a coded version of the statement "This axiom system is consistent" cannot be proved in the axiom system if the system is large enough to include non-negative whole number arithmetic. If a formalism is complicated enough to include whole number arithmetic (which any plausible candidate would include), then it can't even prove itself consistent, much less all of mathematics. I think this obscures the issue. The first incompleteness theorem killed Hilbert's program when combined with the observation that the undecidable statement in the formalism was true as a statement about the non-negative whole numbers themselves. Hilbert's program was based on the assumption that the true statements of arithmetic could be captured in a formal system. If there are theorems about whole numbers that are lost through the act of making the system formal, then Hilbert's program is over before it starts. Mathematics is larger than what can be captured in a completely formal system.

Hilbert's second problem, the request for a proof of the consistency of arithmetic, and his Bologna questions were posed in the context of a philosophical argument, with the motivation of defending the scope of mathematics in general and set theory in particular. What did Gödel's negative solution do to the larger goal?

First, since it is entirely constructive, something like Gödel's incompleteness theorem applies to intuitionist mathematics as well (as formalized by Heyting), so that incompleteness does not supply a reason for preferring intuitionist mathematics. A full intuitionist would say that mathematics can't be formalized in this way. "Incompleteness is a truism for them,"[17] as Martin Davis commented to me. Second, the study of these formal systems led to the creation of interesting mathematics. Mathematicians create mathematics. That is what they were doing and continue to do. The insights of incompleteness led directly, though with many stops, to the solution of the tenth problem. As the philosophical energy of Hilbert's program and Brouwer's opposition to it became a memory, mathematics was left with a new and robust branch, the study of formal systems. Set theory was not harmed by the failed attempt to protect it. It is elegant, useful, and simply too good to pass up. It has therefore, to a large extent, become the language and foundation of modern mathematics. Proof theory, analyzing what can be formally deduced from systems of axioms, was mathematically interesting in its own right. Results in this area are useful in computer science, and Gödel's own short 1935 paper, "On the Lengths of Proofs," is now seen "despite some ambiguities in its statement," as Dawson says, "as an early example of what are known today as 'speed-up' theorems—a major topic in theoretical computer science."[18] There has been an ongoing interest in expanding the kinds of arguments accepted, and there have been proofs of the consistency of arithmetic that, at least, are not wildly infinite and can be seen as buttressing our belief in the consistency of mathematics.

As a practical matter, having Kolmogorov take the intuitionists' part was a Pyrrhic victory for them, and he is a good example of how a practicing mathematician dealt with these questions. Brouwer and Weyl had a deep belief in the importance of philosophy, as did Cantor and Gödel, and all of them brought motives to the debate that were not narrowly mathematical. Kolmogorov was twenty-one, already known for his work on Fourier series (in an interesting parallel with Cantor), and still being educated as a rigorous set-theoretic mathematician—he was fulfilling his examination requirements by publishing papers. He read Brouwer and concluded that his criticism was valid. Kolmogorov's 1925 paper (often referred to as legendary, partly because it is written in Russian and known to many only through hearsay), "On the Principle of Excluded Middle," begins with the sentence: "Brouwer's writings have revealed that it is illegitimate to use the principle of excluded middle in the domain of transfinite arguments." So far so good for the intuitionist position. However, the second sentence is: "Our task here will be to explain why this illegitimate use has not yet led to contradictions and also why the very illegitimacy has often gone unnoticed."[19] So Kolmogorov accepts the criticism, but feels that if the transfinite methods were really poisonous to mathematics, mathematicians should have encountered more problems. He has found out why: "We shall prove that all the finitary conclusions obtained by means of a transfinite use of the principle of excluded middle are correct and can be proved even without its help." End of debate, as a practical matter, for Kolmogorov. Kolmogorov's paper was in many ways more of an outline for a program of research than a completed result, but it satisfied him and he had other work to do. He spent the next few years transforming probability and statistical mechanics into a rigorous branch of mathematics, using wholesale the general language of set theory, specifically measure theory. In 1933 in the forward to the Russian translation of Heyting's book on intuitionism, Kolmogorov stated: "We cannot agree that mathematical objects are simply a product of constructive mental activity. For us, mathematical objects are abstractions of actually existing forms of reality which are independent of thought."[20]

As for the contradictions that had been found in set theory, this is what Gödel wrote in 1961—compare to what Weyl wrote forty years earlier:

> . . . it was the antinomies of set theory, contradictions that allegedly appeared within mathematics, whose significance was exaggerated by skeptics and empiricists. . . . I say "allegedly" and "exaggerated" because, in the first place, these contradictions did not appear within mathematics but near its outermost boundary toward philosophy, and secondly, they have been resolved in a manner that is completely satisfactory and, for everyone who understands the theory, nearly obvious.[21]

Gödel had not yet turned twenty-five when he solved Hilbert's 1928 Bologna problems (and the second Hilbert problem). During the next dozen years he found one important result after another. He did not publish a great deal, but what he published was of first importance. He used his incompleteness theorem as his habilitation to become a privatdozent at the university. According to Gregory H. Moore, he delivered the first of his lectures entirely facing the blackboard, due to shyness.[22] This method attracted few students, which was perhaps a relief. The estate Gödel's father had left was sufficient to support the family. Kurt neglected to tell his mother and brother that he had become a world-famous mathematician. Gradually they became aware of his success.

Rudolf reports that in late 1931 Kurt greatly worried the family with a crisis that involved thoughts of suicide. Dawson argues convincingly that Gödel's first breakdown occurred in 1934. In any event, the 1930s saw a deterioration of Gödel's mental health. When he worked he did so with great concentration. Gödel actively participated in the colloquium of Karl Menger (1902–1985). Almost a dozen of his papers appeared in the colloquium proceedings, and this support undoubtedly allowed him to publish more than he might have if left to his own devices. Gödel never wanted to publish anything that wasn't perfect and later in life had a tendency to revise commissioned papers well past their intended publication date.

Mathematicians are noted for not telling tales out of school. There is a curtain over the personal. Perhaps this reserve represents a respect for privacy, or perhaps it is an attempt to keep the institutions of mathematics above the vagaries of human character and weakness. Perhaps it is simply a distaste for disorder. Gödel was severely hypochondriacal. In his memoir, Kreisel admits to sharing this affliction and reports that the two of them spent considerable time in rapt contemplation of their illnesses and the inability of the doctors to quite capture what was wrong with them.[23] Furtwängler is reported to have asked, of Gödel: "Is his illness a consequence of proving the nonprovability or is his illness necessary for such an occupation?"[24] Gödel was much involved in examining his own condition. However, "depression" and "paranoia" are words that also come up.

From 1929 to 1937 Gödel lived with his mother and brother while seeing Adele. He made his first trip to the United States and the Institute for Advanced Study in Princeton in 1933–34. Coming home, he suffered a "nervous breakdown" when he got off the boat. Kreisel reports that "more than twenty years later he still spoke of the frustrations of—tacitly—his bachelor life in Princeton."[25] Kreisel, however reliably I do not know, seems to be attributing the breakdown, in part, to sexual frustration. He talks of a "second stress" but can't seem to bring himself to state it in a closed form—likely Gödel's family's lack of acceptance of Adele.

The Nazis had come to power in Germany in 1933 and Vienna was destabilizing. Colleagues moved away. Chancellor Engelbert Dollfuss was assassinated in 1934. Schlick, the leader of the Vienna Circle, was assassinated in 1936 on the way to teach a class, by a former student who was a National Socialist. Rudolf says that at first the family was not much affected by or aware of the significance of the Nazi rise to power in Germany. I don't know if this applies to Gödel. After the war he never went back to Austria to visit, though Adele did, nor did he accept any of the many academic honors that were offered to him in his former country.

The Institute for Advanced Study was becoming a haven for German-speaking scientists and mathematicians fleeing Europe. Von Neumann was at the Institute and was one of the people behind Gödel's invitations. A polymath of theoretical and applied mathematics who contributed to the atomic bomb project and the development of the first computers, von Neumann had started his career in the 1920s with set theory. When Gödel proved his first incompleteness theorem in 1930, von Neumann very nearly saw the implications that led to the second incompleteness theorem before Gödel did. Von Neumann passed through Vienna on his honeymoon with his first wife; she was astounded at the amount of time her new husband spent talking to Gödel.

Gödel visited the Institute again in the early fall of 1935. Because of another mental breakdown, his stay lasted only a little more than a month. He spent much of 1936 in mental hospitals. He had introduced the idea of "the constructible sets," with which he was able to prove the consistency of the axiom of choice in set

Wedding photo of Kurt and Adele Gödel (courtesy of the Archives of the Institute for Advanced Study).

theory, before his 1935 breakdown. In 1937 his health returned again somewhat, and he taught a course at the university in the spring. In the summer he saw how to use his concept of the constructible sets to prove the consistency of the continuum hypothesis, part of Hilbert's first problem.

The Gödel family wealth was in Czechoslovakia, and Vienna was becoming increasingly costly. Gödel's mother returned to the family villa. (She remained there in a dangerously outspoken state—she was not among the German-speaking element that approved of Nazi confiscation of Czech property—until 1944, when Rudolf managed to bring her back to Vienna. At the war's end the Germans were expelled from Czechoslovakia and many of her friends died on the march out. She was shocked that her villa should then be confiscated by the Czechs.) The *Anschluss* occurred in March 1938. In September Gödel married Adele, something made easier by his mother's absence. He then left for another visit to the Institute and then Notre Dame—Menger had fled to Notre Dame. Money earned on these visits added to Gödel's independence from his family.

When Gödel returned to Vienna in the summer of 1939 he was promptly called in for a military physical. He said, "I was mustered and found fit for garrison duty."[26] This is somewhat astounding, given the amount of time Gödel had been spending in sanatoria, but Gödel believed it to be true. At about the same time, Gödel was set upon in the street and beaten by a group of Nazi thugs who, based on his appearance, believed he was Jewish. Gödel was thin, had an oddly shaped head, and wore thick glasses. Whatever a Jew was supposed to look like, Gödel was clearly an "other." As Kreisel tells this, the thugs were driven off by Adele swinging her handbag. Most privatdozents were automatically changed to *Dozent neuer Ordnung*, "(paid) lecturer of the New Order," under the Germans. Gödel was not, because he had many close Jewish colleagues. Now he had to apply to the Nazi government for permission to keep teaching at the university and fill out forms saying he wasn't Jewish.

Gödel was culturally and "ethnically" German and not Jewish, but he didn't volunteer this information, in Austria or later, as he didn't think it relevant. As a gymnasium student, he had written a paper "on the superiority of the austere life led by Teutonic warriors over the decadent habits of civilized Rome."[27] He saw the way the Nazis associated themselves with this romanticized view of the past as a particular affront. Taussky-Todd wrote, "Kurt had a friendly attitude towards people of the Jewish faith. And once he said out of the blue that it was a miracle (*Wunder* in German) how, without a country, they were able to survive for thousands of years, almost like a nation, merely by their faith."[28] Things kept getting worse in Austria. Gödel wasn't at all certain he wouldn't have to go into the army or that he would be allowed to teach. He did not find the Nazis to be logical. Kreisel wrote, "The fact is that he was bitterly frustrated. Once again, despite great care he had not escaped trouble."[29] In January of 1940, Gödel and Adele left for America. It was no longer possible to travel by boat, due to the British blockade. They went north though Lithuania and Latvia, then boarded the Trans-Siberian railway at Bigosovo. They crossed Russia and Manchuria on the railway, eventually making their way to Yokohama. From there they took a boat to San Francisco and arrived in Princeton sometime in March. Gödel seldom left Princeton again.

At first Gödel continued to work in the foundations of mathematics. He mounted an effort to prove the independence of the continuum hypothesis, and his attempt was significant. Some believe it would have eventually proved successful. However, probably in late 1942, he gave it up. After 1943 he no longer worked in mathematical logic and set theory in a concentrated way, so at age thirty-seven this great logician essentially departed the field. In the late 1940s he achieved significant results in the theory of general relativity that Einstein thought highly of. They involved general solutions to the field equations which, among other things, seemed to permit time travel. For a long time afterwards Gödel maintained an interest in observational data on the general theory, particularly data that might support his solutions. Gödel wrote commissioned essays about philosophers for the *Library of Living Philosophers* series, on Russell, Einstein, and Carnap. All were late, due to his care in revision, but the essays on Einstein and Russell made it into the series. After 1940 he delivered only a handful of public lectures. He spent more time in the study of philosophers, including Kant, Leibniz, and Edmund Husserl. He wanted to make some contribution to the clarity of philosophical thought that would parallel his achievements in logic and set theory but never published substantially in this area.

After the war Gödel corresponded with and was visited by his mother and brother, who had grown more resigned to his marriage. His mother was much taken with the beauty of her son's garden, though he continued to spend more time in his study than out of doors. Maybe he was cold. Even in summer he could be seen walking to and from the Institute in a heavy wool coat. Gödel was solicitous of his family's comfort during their visits and went so far as to guide Rudolf to the New York Public Library. He continued to have health problems, real and imagined. In 1951 he had a bleeding ulcer that nearly killed him and required blood transfusions. There were the visits to cardiologists who could never properly find what the matter was.

Gödel didn't become a permanent member of the Institute until 1946 and wasn't made a professor until 1953, despite the importance of the work he had done. If a curtain is often drawn before the personal in mathematics, a podium stands in front of it: Achievement is objective and recognition should be objective. Stanislaw Ulam and Freeman Dyson have mentioned these delays, and Dawson states that "others have called for an explanation."[30] Gödel never complained and was given a good home at the Institute until his death. He was eventually given a professorship. Dawson probably captures the actual situation in the following: "Some felt that Gödel would not welcome the administrative responsibilities entailed by faculty status, while others feared that if he were promoted, his sense of duty and legalistic habit of mind might impel him to undertake such responsibilities all too seriously, perhaps hindering efficient decision-making by the faculty. In the event, such fears seem to have been justified. . . ." Dawson also writes, "Princeton society proved unreceptive (not to say hostile) toward Adele, and she appears to have led a very lonely life there."[31] One is saddened to hear of her, according to Wang, begging Gödel to accept an offer to move to Harvard, because they are "so much more friendly."[32]

Gödel himself had friendships at the Institute. Sometimes he would put off fools by making ambiguous pronouncements in a manner that reminded Kreisel of De Gaulle. But the Institute harbored another philosophically minded German-speaking refugee who was past his scientific prime, Albert Einstein, and these two most unusual men found a friendship that gave both of them solace. Kreisel writes, "Einstein was enchanted by Gödel's combination of elegance and precision, and they saw each other constantly till the death of Einstein."[33] Ernst Straus called Gödel "by far Einstein's best friend."[34] Economist Oskar Morgenstern (co-inventor of game theory with von Neumann) wrote:

> Einstein has often told me that in the late years of his life he has continually sought Gödel's company, in order to have discussions with him. Once he said to me that his own work no longer meant much, that he came to the Institute [building] merely "*um das Privilege zu haben, mit Gödel zu Fuss nach Hause gehen zu dürfen*" ("to have the privilege to be able to walk home with Gödel").[35]

Kurt Gödel and Albert Einstein (photograph by Richard Arens, courtesy of the Archives of the Institute for Advanced Study).

This happiness in friendship was clearly shared by Gödel. His correspondence with his mother includes Einstein often. Like many people, Gödel's mother wanted to understand what it was exactly that Einstein had done. Einstein was ill during much of this period, and Gödel was always very concerned with his health. There was a feeling of equality between them. They talked on their walks. Gödel was interested in ghosts, demonism, and theology. He was particularly hostile to the Catholic Church and according to Kreisel had "a soft spot for new sects in the New World, of which he spoke often in conversation, and also wrote at some length to his mother."[36] Einstein was such a legend that most people were unwilling to challenge him. Gödel wasn't. Their debates ranged from the trivial to the profound. Gödel was skeptical about the idea of a unified field theory and said so. Einstein loved classical music, whereas Gödel was most fond of light opera. Gödel's favorite film was "Snow White," and he was fond in general of Walt Disney's work.

Straus writes:

> Gödel had an interesting axiom by which he looked at the world; namely that nothing that happens in it is due to accident or stupidity.

> If you really take that axiom seriously all the strange theories that Gödel believed in become absolutely necessary. I tried several times to challenge him, but there was no out. I mean from Gödel's axioms they all followed. Einstein did not really mind it, in fact thought it quite amusing. Except the last time we saw him in 1953 he said, "You know, Gödel has really gone completely crazy." And so I said, "Well, what worse could he have done?" And Einstein said, "He voted for Eisenhower."[37]

The 1952 presidential election was much discussed by these two. Einstein was, of course, for Adlai Stevenson. Gödel liked Ike. (Adele was also for Ike but was more genuinely a partisan of Douglas MacArthur.) Einstein also played a role in Gödel's naturalization as an American citizen. Solomon Feferman repeats a story, which I will quote verbatim.

> Morgenstern had many stories to tell about Gödel. One concerned the occasion when, in April 1948, Gödel became a U. S. citizen, with Einstein and Morgenstern as witnesses. Gödel was to take the routine citizenship examination, and he prepared for it very seriously, studying the United States Constitution assiduously. On the day before he was to appear, Gödel came to Morgenstern in a very excited state, saying: "I have discovered a logical–legal possibility by which the U.S.A. could be transformed into a dictatorship." Morgenstern realized that, whatever the logical merits of Gödel's argument, the possibility was extremely hypothetical in character, and he urged Gödel to keep quiet about his discovery at the examination. The next morning, Morgenstern drove Gödel and Einstein from Princeton to Trenton, where the citizenship proceedings were to take place. Along the way Einstein kept telling one amusing anecdote after another in order to distract Gödel, apparently with great success. At the office in Trenton, the official was properly impressed by Einstein and Morgenstern, and invited them to attend the examination, normally held in private. He began by addressing Gödel: "Up to now you have held German citizenship." Gödel corrected him, explaining that he was Austrian. "Anyhow," continued the official, "it was under an evil dictatorship. . . but fortunately, that's not possible in America." "On the contrary," Gödel cried out, "I know how that can happen!" All three had great trouble restraining Gödel from elaborating his discovery, so that the proceedings could be brought to their expected conclusion.[38]

Einstein died in 1955. Gödel became older; his health did not improve, he did not become more sane, and he did not become happier. His last published paper was in 1958. When Cohen proved the independence of the continuum hypothesis in 1963, he journeyed to Princeton and submitted his proof to Gödel. Gödel gave his approval, made comments on the manuscript, and sent it under his signature to the National Academy of Sciences for publication. In the late 1960s, Gödel's health began to decline more dramatically. His depressions grew. In the 1970s he refused surgery for a prostate problem. In 1970–71 he circulated "Scales of Functions"—an attempt to get a handle on the continuum hypothesis—but errors were found and he withdrew it. Kreisel wrote, "[It] soon became too much for me to watch."[39] Adele's health began to fail also. She was in the hospital and then a nursing home for most of the six months leading up to Gödel's death. Gödel became more afraid of food poisoning than ever and didn't eat. Kurt Gödel died on January 14, 1978, at the age of seventy-one. The death certificate listed the cause of death as "malnutrition and inanition" caused by "personality disturbance."[40] Wang heard that Gödel weighed sixty-five pounds at the end.[41]

But consider the following from Feferman:

> Gödel's main published papers from 1930 to 1940 were among the most outstanding contributions to logic in this century, decisively settling fundamental problems and introducing novel and powerful methods that were exploited extensively in much subsequent work. Each of these papers is marked by a sense of clear and strong purpose, careful organization, great precision—both formal and informal—and by steady and efficient progress from start to finish, with no wasted energy. Each solves a clear problem, simply formulated in terms well-understood at the time (though not always previously formulated as such). Their significance then was in one sense prima facie evident, though their significance more generally for the foundations of mathematics would prove to be the subject of unending discussion.[42]

Gödel is often called the greatest logician since Aristotle. When I tried to find who had said this, it appeared to have sprung up spontaneously in more than one location—though perhaps von Neumann said it first. That the deep ideas Gödel left have proved worthy of "unending discussion" goes to the heart of his achievement.

The Perfect Spy: How Many Real Numbers Are There?

The First Problem

∝ *Cohen* ∞

The hierarchy of the cardinal numbers (measures of size in set theory) is key to Cantor's ordering of the transfinitum. The smallest infinite cardinal, $\aleph_0$, is the size of the counting numbers and is the only countable cardinal. Then there is the first uncountable cardinal, $\aleph_1$, the second, and so on. The infinite cardinals are written using the first letter of the Hebrew alphabet, aleph,

$$\aleph_0, \aleph_1, \aleph_2, \aleph_3, \aleph_4, \ldots \text{ and go on forever.}$$

What is an example of a set of cardinality $\aleph_1$, the next size up from the counting numbers? The real numbers are larger than the counting numbers. Are they of size $\aleph_1$? The continuum hypothesis is the statement that they are. Hilbert's first problem is titled "Cantor's Problem of the Cardinal Number of the Continuum."

The question can be phrased without mentioning the cardinal numbers. We have seen, using a simple method, that there are more real numbers than there are counting numbers. Does there exist a subset of the real numbers that is larger than the counting numbers and yet smaller than the reals? Two sets are the same size if their members can be matched up one-to-one. Making a list (or stating a clear method for making a list) of an infinite set establishes that it is the same size as the counting numbers. Finding a rule for a one-to-one mapping between a set and the real numbers would establish they are the same size. (One need not actually produce a rule or a list, only prove a one-to-one mapping exists.) Is there any size bigger than the counting numbers but too small to be matched up one-to-one with the reals?

The continuum hypothesis feels like a concrete statement that ought to be true or false. Cantor believed that the continuum hypothesis is true, that there are not any sizes between the counting numbers and the real numbers. He spent the second half of his life trying to prove it. Hilbert attacked the problem unsuccessfully in the mid-to-late 1920s. Many other mathematicians worked on it. Cantor did not create set theory as a formal system, but the drive to formalize logic and then set theory began long before Cantor and continued after him. Only by working in the context of the formal systems for set theory that slowly developed did Gödel and then Cohen make progress in understanding the continuum hypothesis.

Leibniz first asked for a *lingua characterica* that would escape the confusion of natural languages, with basic concepts represented by ideograms. Argument would be accomplished using a *calculus ratiocinator*, a clearly defined set of rules by which the ideograms could be manipulated and logical inference accomplished with mechanical precision. Once in possession of this new language, all of science could be built on one unified foundation.

Starting in the 1840s, Augustus De Morgan (1806–1871) made genuine progress in turning Aristotelian logic into a symbolic logic and expanding its range, particularly in dealing with statements involving quantity. (Aristotelian logic did not deal well with the word "most" or its like.) De Morgan legitimized symbolic logic in England, where it had been neglected (perhaps simply because Leibniz had suggested it and there was still rancor over the paternity of calculus). A good example of De Morgan's prose style, as well as a piquant example of the English attitude toward Continental theory, can be seen in his comment that he "had no objection to Metaphysics, far from it, but if a man takes a candle to look down his own throat, he must take care not to set his head on fire."[1] Around the time he began his logical researches, De Morgan was involved in a bitter controversy including charges of plagiarism, and into this controversy stepped his friend and correspondent, George Boole (1815–1864).

Boole was a self-made man in an era open to self-made men like Michael Faraday and Charles Dickens and of close to their order of talent. At age fifteen, far from London or the universities, he found himself teaching mathematics, among other subjects, in Lincoln—near where Newton emerged from the English countryside. Not content to teach a subject he didn't know, he read Newton's *Principia* and Joseph Louis Lagrange's (1736–1813) *Mécanique Analytique*, the latter noteworthy for its utter lack of diagrams. By 1849 Boole had become professor of mathematics at Queen's College in Cork, despite his lack of any degree.

Boole's defense of De Morgan was his *Mathematical Analysis of Logic*, published in 1847. This work was extended in 1854 in *An Investigation of the Laws of Thought, on Which Are Founded the Mathematical Theories of Logic and Probability*. Boole reduced a large part of logic to a formal algebra, a giant step toward Leibniz's goal. His work opened up the construction of an entirely new class of algebras that did not follow all of the general rules of real number algebra. This had been a stumbling block. Mathematicians and philosophers had wanted any new algebra or calculus, in this case of logic, to mimic the general form of conventional real number algebra—for example, to be commutative: $a + b = b + a$. William

Rowan Hamilton (1805–1865), Hermann Grassmann (1809–1877), and other mathematicians were beginning to break free of these restrictions, but Boole made his own break. His algebra of logic is similar to real number algebra, but in it $x^2 = x$, and this initial difference leads to vast consequences. It is analogous to the creation non-euclidean geometries, which had already occurred on the Continent. Boole, isolated even in England, does not appear to have been aware of the geometrical results. (Boole has another connection to our story. In two papers, in 1841 and 1843, he initiated the algebraic invariant theory quickly picked up by Cayley and Sylvester that Hilbert made his name on.)

The early history of symbolic logic receives an excellent treatment in Clarence I. Lewis's *A Survey of Symbolic Logic*, published in 1918. Lewis reconstructs the early attempts at formalizing logic along narrowly Aristotelian lines, and then suddenly there is Boole and everything is much clearer. Lewis writes:

> It is probably the great advantage of Boole's work that he either neglected or was ignorant of those refinements of logical theory which hampered his predecessors. The precise mathematical development of logic needed to make its own conventions and interpretations; and this could not be done without sweeping aside the accumulated traditions of the non-symbolic Aristotelian logic. As we shall see, all the nice problems of intension and extension, of the existential import of universals and particulars, of empty classes, and so on, return later and demand consideration. It is well that, with Boole, they are given a vacation long enough to get the subject started in terms of a simple and general procedure.[2]

William Stanley Jevons (1835–1882) presented Boole's logic in a less stridently mathematical form and extended it during the 1860s. Today we might view a computer as a natural home for formal language—mathematical logic classes are often also registered for credit in computer science. Jevons was quick to see this. In 1870 he exhibited a machine at the Royal Society that resembled an upright piano and was capable of calculating logical conclusions from sets of premises. In the same period, at Oxford, Charles Dodgson (1832–1898), a.k.a. Lewis Carroll, also drank from the bottle of logic (and paradox). He regarded much of his logical work as recreational and published logical games and problems difficult enough, perhaps, to send his countrymen running for Jevons's calculator.

An American, Charles Sanders Peirce (1839–1914), added a relation called "inclusion" to Boolean algebra in 1870 and extended it. He and his students introduced logical quantifiers (independently of Frege). Peirce was aware of Cantor's set theory as it came out and was interested in understanding the continuum. He used the word "multitude" for an infinite set and recognized that there were collections of things that were too numerous to be considered multitudes, like the set of all ordinals (this is now called the Burali–Forti contradiction). Peirce believed that there weren't enough real numbers to fill a geometric line or continuum and that a line was a supermultitudinous collection of infinitesimals, which were the glue that held it together. Abraham Robinson (1918–1974) later constructed a system of nonstandard analysis in which there are infinitesimals. Peirce is best known as a philosopher and the founder of pragmatism, which was extended and popularized by William James and John Dewey.

In Germany, Gottlob Frege (1848–1925) was in Göttingen just as Dedekind's cuts and Cantor's set theory were beginning to appear. He did not think the attempts at a rigorous treatment and construction of the real numbers were successful and set out to properly accomplish this. The first result was the *Begriffsschrift*, published in 1879. Frege was steeped in Leibniz. He made an "ideography"—*begriff* here means ideograph—intended to be a direct, formal embodiment of logic and language—a *lingua characterica* and *calculus ratiocinator* for the realm of the real numbers. Frege followed his program through with such care and detail that it was, in essence, an entirely formal system. However, his notation went up and down as well as right and left and had no life outside of his own work. Lewis comments: "Frege's notation, it must be admitted, is against him: it is almost diagrammatic, occupying unnecessary space and carrying the eye here and there in a way which militates against easy understanding."[3]

Frege's two-volume *Grundgesetze der Arithmetik* (1893–1903) was to cap his work. However, in 1902 Russell sent him a short letter that after some praise said, "There is just one point where I have encountered a difficulty."[4] In a few sentences Russell displayed his paradox "the set of all sets that do not contain themselves" in terms of Frege's system. Russell was possibly the first person ever to have fully understood and appreciated what Frege had done. Frege published his second volume but included an appendix in which he displayed the paradox. His system displays the contradictions very simply; Russell and Whitehead used it as a starting point for *Principia*.

In Germany, Ernst Schröder (1841–1902) added to the tradition of Boole and Peirce. One of the first mathematicians to endorse the work of Cantor, Schröder began to connect Cantor's theory to formal systems and symbolic logic. There is a natural connection between logic and set theory in that logic is about classes—the class of all bald men, for example—and "class" is a close synonym of "set."

At the University of Turin, Peano picked up the developing subject of symbolic logic. He had a gift for notation and a knack for writing axiom systems. The symbol $\in$, for membership in a set, came from Peano. His work on axioms for geometry, following Pasch, was a precursor to Hilbert's work. His axioms for the arithmetic of the non-negative whole numbers still survive. He did important work in analysis, the most famous example being his construction of a continuous space-filling curve. Peano was pragmatic, not a rigorous formalist, and appreciated but didn't fully understand Frege's deep analysis of logic. He increasingly saw his own work as being in the tradition of Leibniz's dream of a universal language. In 1892 he announced a project to create a *formulario* to contain all known mathematics, written in part in his version of symbolic logic. He viewed his language as a tool for the expression and analysis of mathematical ideas that don't themselves derive from the language and cannot be entirely captured by a set of axioms, in marked contrast to Russell's logicist position (mathematics is nothing more than logic) or Hilbert's formalist position (if a formal set of axioms is consistent, what they refer to exists, and what we know about these things is what we can prove from the axioms). The final edition of the *Formulario* appeared in 1908 and is encyclopedic, containing 4,200 theorems in only 516 pages. The commentary is written in an artificial language called Interlingua, which did not help it in obtaining readers. After 1908 Peano turned almost all of his attention to assisting the emergence of Interlingua as an international language. There was an existing society for this purpose, and he was elected its president. He believed Interlingua should be a version of Latin with little grammar, which he saw as an unnecessary complication. But Interlingua fractured into dialects. This helped the competing Esperanto. Interlingua was salvaged in the 1940s and 1950s and is in existence today.

Russell met Peano, then in his mathematical prime, at the Second International Congress of Mathematicians in 1900, where Hilbert delivered his problems. Russell emerged from this meeting with a clear belief that Peano's language and symbolism held the key to attacking problems in the foundation of mathematics. When Russell realized that his elementary paradox could be expressed in Frege's logical system, everything was in place for the great crisis in the foundations that Hilbert had anticipated. Russell was the first to resolve the paradoxes, with his type theory. He restricted the use of the word "all" to phrases like "all sets of an already specified type satisfying a given condition." The universe of discourse, or type, is already specified before it is used in a definition, thereby avoiding vicious circles like "the set of all sets that don't contain themselves."

In Germany, Ernst Zermelo (1871–1953) published his axiom system for set theory in 1908. With additions and refinements by Adolf Abraham Fraenkel (1891–1965) in 1919 and Albert Thoralf Skolem (1887–1963) in 1922, it is the standard axiom system in use today, called Zermelo–Fraenkel set theory. They likewise limited the use of "all" in the formation of new sets. At this point logic had been made formal and the contradictions of set theory eliminated. Arriving at an efficient notation, stumbling into logical error, and resolving the errors had taken eighty years, from De Morgan to the completion of the Zermelo–Fraenkel axioms. While this was happening, though, the foundation of an increasingly large part of mathematics was being constructed using set theory.

The first really concrete insight that there was something peculiar about the formal axiomatic method came from Leopold Löwenheim (1878–1957), a gymnasium teacher in Berlin. In 1915 he published the first version of what is now known as the Löwenheim–Skolem theorem. Löwenheim's paper was written using Schröder's notation, which was already falling into disuse, there were gaps in the proof, and the theorem was not stated in general terms. It is about model theory and different models for axiom systems and was not widely appreciated or understood. Skolem, a Norwegian who had gone on a mission to the Sudan to "observe zodiacal light"[5] in 1913–14, had the good fortune to be studying in Göttingen in 1915–16. He closed the gaps in Löwenheim's proof and generalized the theorem in a paper published in 1920. He returned to the topic in 1922, 1928, and 1929.

An axiom system in a formal language is a set of statements in the language. A model is a concrete mathematical realization of what these statements are talking about. Many people have had the experience of participating in a nominally coherent conversation, only to realize that each speaker was talking about something different. A set of statements can often be true in different contexts. This is true of mathematical languages and axioms also. Mathematicians like Hilbert were clearly aware of this but were slow to realize all the implications. When mathematicians wrote an axiom system to capture the real numbers, they assumed that is what they were talking about and that is what they would capture if they wrote a good system. Though axiom systems had had striking success, as far back as Euclid, mathematicians needed to think about other, different interpretations or embodiments of their language and axioms, particularly as the axioms increasingly referred to completed infinites, because alternative embodiments were not *a priori* ruled out.

The Löwenheim–Skolem theorem does this. It says that if we have a model, even an uncountably infinite model, for a finite or countably infinite set of axioms (which could be defined by a recursive rule), then there exists a submodel (i.e., part of the original model that is also itself a model) that is at most countably infinite. The real numbers are a model for the various axiom systems for the real numbers. They are what we are *trying* to refer to. The real numbers are an uncountable infinite. The Löwenheim–Skolem theorem says there is a subset of the real numbers that is countably infinite and yet is also a model for the axioms of the real numbers. Therefore any axiom system for the real numbers, in addition to describing the vast continuum of the real numbers, also describes a countable subset of them. What this meant in larger terms was not immediately obvious to anyone, including Skolem. To Skolem it was "paradoxical" that the axioms for the real numbers would also describe anything smaller than or other than the real numbers. It was even more paradoxical that the axioms of set theory, which describe all orders of the transfinitum, would also have a small, that is, countable, model. The Löwenheim–Skolem theorem, though not a paradox per se, is a surprising fact about formal languages and axiom systems.

Skolem's 1922 paper, "Some Remarks on Axiomatized Set Theory," is brilliant, prophetic, and on a certain point wrong—at least as a piece of prophecy. Not long, eleven pages, it shows how well he understood the mathematics of models and set theory. The paper contains a version of the Löwenheim–Skolem theorem that does not require the axiom of choice, a controversial part of set theory, to prove. It supplies the missing link to complete the Zermelo–Fraenkel axiom system—a clearly stated axiom of replacement. There is a remarkable statement: "It would in any case be of much greater interest if we could prove that a new subset of Z_0 [here, the non-negative whole numbers] could be adjoined without giving rise to contradictions; but this could probably be very difficult."[6] He adds in a footnote, ". . . it is quite probable that what is called the continuum problem . . . is not solvable at all . . . ." As we will see, this is at the heart of Cohen's solution to Hilbert's continuum problem, his "forcing" machinery being the "very difficult" realization of Skolem's idea. Skolem's paper ends with the following remarks:

> What does it mean for a set to exist if it can perhaps never be defined? It seems clear that this existence can be only a manner of speaking, which can lead only to purely formal propositions—perhaps made up of very beautiful *words*—about objects *called* sets. But most mathematicians want mathematics to deal, ultimately, with performable computing operations and not to consist of formal propositions about objects called this or that.[7]

This is a remarkably clear (and early) statement of the point of view that set theory is the exploration of a formal game played with language. Skolem concludes:

> The most important result above is that set-theoretic notions are relative [the real size of what is being talked about can't be specified in the axioms]. I had already communicated it orally to F. Bernstein in Göttingen in the winter of 1915–16. There are two reasons why I have not published anything about it until now: first, I have in the meantime been occupied with other problems; second, I believed that it was so clear that axiomatization in terms of sets was not a satisfactory ultimate foundation of mathematics that mathematicians would, for the most part, not be very much concerned with it. But in recent times I have seen to my surprise that so many mathematicians think that these axioms of set theory provide the ideal foundation for mathematics; therefore it seemed to me that the time had come to publish a critique.[8]

Most mathematicians continued to think set theory a wonderful thing to base mathematics on. They didn't continually work in a formal way with the axiom system, but having the axioms pinned down made the enterprise seem more secure, a different sort of "origin of the coordinates." Skolem's work was known in the community of mathematical logicians but was not well understood. The 1922 paper was not easy to get. In saying there is a countable submodel for any countable system of axioms, Löwenheim–Skolem says no formal system that attempts to describe a large mathematical object like the real numbers accurately describes only the mathematical system it was written to describe. Gödel was familiar with the

1920 paper but not the 1922 paper when he wrote his completeness paper—he had requested it, but it hadn't yet arrived.[9] Löwenheim–Skolem is explicitly about the size of models, while Gödel's result is about the completeness of Frege–Russell's formal version of logic and obtains the results about size of models as a side benefit of its method of proof. It is the door through which one passes to solve Hilbert's first problem.

Until 1967 the pieces of the later history of symbolic logic were hidden in papers written in many languages, sometimes published in obscure places. Skolem's 1922 paper was originally delivered to the Fifth Congress of Scandinavian Mathematicians in Helsinki. Kolmogorov's "legendary" paper, as we have said, was written in Russian and published in Russia, so for most non-Russians legend was the only way to find out about it. From 1936 on, Alonzo Church (1903–1995) documented what was happening through the review section of *The Journal of Symbolic Logic*, which he edited, but his attempt to extend the record back in time was limited. But in 1967 Jean van Heijenoort (1912–1986) published *From Frege to Gödel: A Source Book in Mathematical Logic, 1879–1931*, making forty-six well-chosen papers available, with commentaries.

I assumed van Heijenoort was merely a logician as I gratefully pored over his book and was astounded when I learned that as a young man, in the 1930s, he had followed Leon Trotsky from Turkey to France to Norway to Mexico, as his bodyguard and secretary. Anita Feferman's *From Trotsky to Gödel: The Life of Jean van Heijenoort* tells van Heijenoort's remarkable story. Trotsky and his camp were pursued by Stalinist agents; van Heijenoort always packed a gun. Van Heijenoort had an affair with the artist Frida Kahlo. He epitomized cool and was attractive to women, and this continued even after he became a logician. He had grown tired of the rigors of life with Trotsky and traveled to New York in 1939 on a somewhat vague mission. (In a later period he hung out with the painters Jackson Pollock, Willem de Kooning, and Franz Kline.) He blamed himself for Trotsky's assassination, because he felt he would have recognized that the assassin, a Spaniard named Ramon Mercader posing as a French-speaking Belgian, was not a native speaker of French, and hence he would have denied him access. (Van Heijenoort was French, not Dutch.) Van Heijenoort himself was murdered by his fourth or fifth wife, "depending in how one counts," in Mexico City in 1986—three bullets in the head as he slept.

In his book van Heijenoort placed Frege as the great originator of the vital period that climaxed with Gödel. Some critics thought this undervalued the Boolean tradition, and one could make this argument based on reading van Heijenoort's own book, which in itself is a tribute. All of Löwenheim's and Skolem's papers, arguably the deepest and most influential short of the closing papers by Gödel, are written in the tradition of Boole, Peirce, and Schröder and in their language. In politics, van Heijenoort started with an all-inclusive system that answered every question. He lost faith and moved to a critical position. His book starts with Leibniz's dream as embodied in Frege and ends with criticism that is at a higher level and more interesting than the original dream.

The strong feeling that the continuum hypothesis was a concrete problem with an answer—was it true, yea or nay—is one reason why no one immediately appreciated or was motivated to exploit Löwenheim–Skolem's odd results on the size of models. Gödel understood their usefulness more quickly than anyone, and we have already seen that he achieved as a side benefit similar results to Löwenheim–Skolem with his proof of the completeness of the predicate calculus. As early as 1930 Gödel was thinking about the continuum hypothesis.[10] These results supplied the opening for his half of the solution to Hilbert's first problem in 1937. Most of Gödel's results are not technically difficult when compared to other mathematics being done at the time. Gödel always had enough technique to accomplish what he wanted to with great elegance but never did significant mathematical work without a philosophical motivation. His proofs of the consistency of the axiom of choice and the consistency of continuum hypothesis are his most technically difficult work.

How do results on the size of models relate to the continuum hypothesis? We see that the axioms of set theory, the most infinite of mathematical structures, have a model that is itself countable. If we have a countable model for set theory, there has to be a lot of empty space where the rejected pieces of the larger transfinitum used to be. Both Cantor and Gödel believed that all the vast multitudes referred to in set theory actually existed. No finite statement in language can compel them to be in the model, though. Gödel and then Cohen used this wiggle-room to construct models for set theory that suited their purposes. The question in this game of models becomes: Is the continuum hypothesis like the parallel postulate in geometry? First, is it consistent with the other axioms of set theory? That is, will we introduce no new trouble if we accept it as true? But is it independent? Is it decided one way or the other by the other axioms, or are we free to assume it true or false? Just as Hilbert analyzed the axioms of geometry by making a model using pairs of numbers, Gödel and Cohen analyze set theory with models they construct. (My use of the word "model" glides over important mathematical distinctions. However, I believe it conveys the spirit as opposed to the letter of the technique.)

The Zermelo–Fraenkel axioms for set theory, and the axiom of choice in particular, are relevant here. Cohen lists nine axioms for Zermelo–Fraenkel, with one, the axiom of replacement, really being an infinite scheme of axioms based on a list of all possible formulas so that the axiom can be applied to any formula in the language. It is Zermelo–Fraenkel–Skolem's method of restricting but not eliminating the word "all" in the formation of new sets. All the axioms (except the axiom of infinity) are statements that are obviously true when applied to finite sets. One example would be the axiom of the sum set or union. It states that if we have a set, we can obtain a new set by combining the contents of each of its members. The axiom of extensionality says that if two sets have exactly the same members, they

(1) Sum set/union
(2) Extensionality
(3) Unorder pairs
(4) Power set
(5) Null set
(6) Regularity
(7) Continuum?
(8) Choice
(9) Infinity

are the same set. The axiom of unordered pairs says that if we have two sets, there exists another set that contains them and only them as members. The power set axiom asserts the existence of the set of all subsets of any given set. The axiom of the null set to some seems odd. Zermelo himself fudged it by calling it a "fictitious set."[11] It says there is a set with no member, also called the empty set. It is like a container with nothing in it, like 0 in arithmetic. The axiom of regularity ensures that all sets are built up starting with the empty set. It forbids infinite nests of brackets with nothing in them, like parallel mirrors in a deserted barber shop. The axiom of infinity asserts the existence of an infinite set, specifically the set of all non-negative whole numbers, and is what makes set theory volatile when it combines with the others. It injects the idea of a completed infinite into a logical system and as a part of that system applies all the other axioms to infinite sets. The axioms are "obviously" true when applied to infinite sets, but suddenly we are confronted with statements that cannot be constructively verified. At this point an intuitionist bails out, and thoughtful people realize the ground has shifted.

The axiom of choice is the most controversial and one of the most useful. Cantor's continuum hypothesis is not used often, if at all, in conventional mathematics. The axiom of choice is used often. Cohen uses it to prove the general version of Gödel's completeness theorem he states, and it is used in Löwenheim–Skolem in its more general formulations. It is used particularly often in analysis, to pick a more practical field. Gregory H. Moore wrote about its long and complex history in *Zermelo's Axiom of Choice: Its Origins, Development and Influence*.

The axiom of choice can be phrased in many ways. One formulation is: If we have a set that is not empty itself and composed of disjoint sets, each of which has at least one member, then there exists a second set that has exactly one member of each of the sets in the original set. "Disjoint" means the sets don't share members. In his well-known text on real analysis, Halsey Royden puts it well: "We may think of the axiom of choice as asserting the possibility of selecting a 'parliament' consisting of one member from each of the sets in C [the original set]."[12] This axiom is "obviously" true about finite sets, because it says that we can compile a list of exemplars. When combined with the axiom of infinity it allows an infinite amount of choosing to be done, even an uncountably infinite amount of "choosing," even if we can't specify a rule for how it is to be done.

This axiom tends to creep in, and logicians are alert to when it makes its appearance. It is powerful and convenient. In many cases the axiom of choice is necessary to a mathematical proof, and its use cannot be eliminated. Sometimes it will appear in a first version of a proof but with effort, often great effort, can be made to go away. To someone with intuitionist qualms, the axiom of choice does not seem obvious. The substance of the axiom had been used by nineteenth century mathematicians. Peano was one of the first to explicitly comment on it, in an 1890 article about analysis.[13] He rejected it. Zermelo recognized that it or something very much like it was necessary to the formation of Cantor's full transfinitum. He used it in a 1904 paper to prove that every set could be "well-ordered," and this usage legitimized the axiom and put it at center stage. Zermelo created his axiom system, in part, to defend this choice principle from criticism.

Gödel proved the relative consistency of the axiom of choice in 1935—that if the axioms of Zermelo–Fraenkel without the axiom of choice are consistent, then Zermelo–Fraenkel plus the axiom of choice is also consistent. To do this he constructs an "inner model" for Zermelo-Fraenkel (not including the axiom of choice and without using it) in which the axiom of choice is true. The inner model is made of what Gödel called "the constructible sets." The "set of all constructible sets" is too large to be a set, which is what a model actually is, straying into a realm tainted by the word "all." But one can define constructibility in the language of set theory as a property, so that there is a statement in the language that says "*x* is constructible." We get the inner model by looking at only sets that satisfy the statement "*x* is constructible" and verifying that the axioms are all satisfied when restricted to these sets.

The constructible sets are created incrementally by induction on the ordinal numbers, including the transfinite ordinals (i.e., first countable infinity and 1, first countable infinity and 2, etc.). The first level is a set with nothing in it. After that, at each level in the process, there are already-constructed sets. All possible formulas that refer only to the sets already constructed are listed and used to define the next group of sets. These newly constructed sets are then added back to the pile, and all are used in the next step. Each constructible set is by definition associated with a specific formula and a specific ordinal, which marks the stage in the process at which it was first constructed. This order can be used to accomplish the infinite amount of choosing that is required by the axiom of choice. The axiom of choice is true in the "model" of the constructible sets.

To prove that the continuum hypothesis is true in the "model" of the constructible sets took Gödel an additional two years to accomplish. In his introduction to Gödel's *Collected Works*, Solomon Feferman speculates that the stress of the difficulty of this problem is what caused Gödel to spend such a large part of this period in mental hospitals. The key twist is really marvelous. Any given set is first constructed at a given stage. Hence any real number is first constructed at a given stage, because real numbers are sets themselves in set theory. Prima facie, this could be many layers deep into the cardinal numbers, overshooting the first uncountable cardinal, and hence could seem to imply the continuum hypothesis false. However, we can now prove a version of Löwenheim–Skolem for the constructible sets. Gödel uses this modified Löwenheim–Skolem theorem to show that even if a real number was constructed by a large ordinal, many layers into the unℵuntable cardinals, it had also been constructed in the countable submodel that Löwenheim–Skolem says must exist. He then uses another more technical, but straightforward, result to shrink the ordinals in the countable model to show that the real number in question had already been constructed with a countable ordinal. This is the result. All the real numbers have been constructed before the accumulation that yields the first uncountable cardinal. Therefore the size or cardinality of the real numbers in this model is the first size larger than

$_{0,}$ or the countable cardinal. That is the continuum hypothesis, and here Gödel proved its consistency, i.e., that it would not cause trouble to assume it was true.

What about the question of the independence of the continuum hypothesis, i.e., is it provable from the other axioms? Even before he published his results, Gödel turned to this question, again using the method of models. He worked through 1942, when he left mathematical logic. He did make some progress in his efforts to prove independence, but since he didn't publish them it is difficult to know precisely how much he accomplished. Moore says:

> During the summer of 1942, while on vacation in Maine, he obtained a proof of a related result, the independence of the axiom of choice relative to the theory of types as well as the independence of the axiom of constructibility. But he did not succeed in showing the independence of the continuum hypothesis.
>
> Gödel never published his result on the independence of the axiom of choice. According to comments he made later to John Addison and others, he feared that such independence would lead research in set theory "in the wrong direction."[14]

The question of the continuum hypothesis lay like Snow White where Gödel left it, in the middle years of the Second World War.

We now meet our first American member of the honors class, Paul Cohen, currently a professor at Stanford University in California. He was born to immigrant parents in Long Branch, New Jersey, on April 2, 1934, and grew up in Brooklyn "feeling . . . a strange kind of combativeness that we were going to take on everybody."[15] Cohen reports he was a Giants rather than Dodgers fan—he was going to take on Brooklyn itself.

His parents had emigrated as teenagers from what is now Poland and was then Russia. His father had some religious education, while his mother's education consisted of one summer with a tutor hired by her parents to teach her to read the languages relevant to their situation, Polish, Russian, and Yiddish. Perhaps in Cohen's family one summer was enough for three languages in three different alphabets. She wished she had had more educational opportunities and attended night school as an adult, though she was never completely fluent in English. Cohen's father was a "fairly successful grocery jobber"[16] in the 1920s, but in the 1930s, the situation worsened considerably. Cohen said to me, "We were in terrible shape."[17] There was a period in a leather goods factory, work for the WPA, street vending. They were "poorer than most." His father moved to New Jersey and Cohen only saw him occasionally, though there was no official divorce. His mother had to work and worked hard. His father worked as a taxi driver at the end of his life. Money was very tight all through the 1940s.

Cohen was the youngest of four, after the eldest daughter Tobel, then Sylvia and Rubyn. There is a striking, posed photograph of the children with Cohen's interview in *More Mathematical People*. Standing with their arms at their sides, the older children have a look of happy confidence. Paul, perhaps five, sits on a cushion, an ankle tucked under a knee, in the scoop of a Renaissance-style chair. He addresses the camera, but with his father's eyes—inward, reserved. Sylvia was the first in the family to go to college. At nine or so Paul was fascinated by her high school algebra homework, by the patterns the letters and numbers made. From seeing how the problems were solved, he arrived at an understanding of algebraic manipulation. He managed to find some popular books on mathematics and was able to discover more. By the age of ten he was helping all his older siblings with their mathematics homework without having any particularly coherent grounding. He reports that it all had a "formal" quality. When Sylvia went to college he helped her with trigonometry.

Sylvia had no particular interest in mathematics, but she did have an interest in her baby brother. She bought him a book on geometry at a five-and-dime that contained a reasonable treatment of what a mathematical proof is. Cohen, however, reports that at that time he was more interested in calculation. He was exploring on his own at the Brooklyn Public Library. He knew the solutions of quadratic and cubic equations and had also heard that Galois theory proved that there was no solution to the quintic. Mathematics of the type he was interested in was considered an adult subject and thus closed to browsing children. Cohen sneaked into the mathematics section. Richard Feynman recounted the same problem in *What Do You Care What Other People Think?*; he solved it by saying the book he wanted was for his father. When Cohen's sister brought him Felix Klein's book on the quintic equation and the icosohedron, he felt "that there was a world that was permanently beyond me."[18]

Throughout his childhood, Cohen was interested in all kinds of science, from biology to chemistry to physics. He comments: "However, math especially appealed to me. If you read something about electricity, for instance, you find out that you need a lab to do anything yourself, but with math you can do problems right away."[19] This is a mathematician speaking. Setting up a lab without the right apparatus seems unappealing. The results would not match the equations. Feynman's account of his early experimental efforts displays the attitude, give me some wire, a light bulb and some aluminum foil and I can get this thing to work. Enrico Fermi, when he witnessed the first atomic bomb blast, didn't want to wait for the official results. He tore paper into little pieces and let them flutter to the ground as the shock wave passed him. By seeing how far they moved horizontally, he was able to calculate a good estimate of the force of the blast. On the other hand, the impatience with imprecision and the intuition about mathematical form that Cohen possessed allow a mathematician to solve something like Hilbert's first problem. Aluminum foil not required.

Sylvia continued to bring home books from the college library. One that Cohen remembers in particular is *Fundamentals of Mathematics* by Moses Richardson, a professor at Brooklyn College. The book was rich in scope if not detail, and it provided Cohen his first contact with set theory. His first reaction was that set theory was too "verbal."[20]

By sixth grade Cohen knew algebra and geometry "well" and knew something about calculus and number theory. He particularly liked number theory. Yet he felt isolated. He felt his teachers were threatened by his precocity and that he couldn't talk to his fellow students about mathematics, either. He was encountering things he couldn't understand, and there was no one he could go to for help or the name of an appropriate book. Cohen said his early isolation was a mixed blessing. His contact with mathematics was all his own. He discovered things that a child on a more organized program might have moved right by. For example, if one uses Cardano's method of solving cubic equations, the answer, even if it is 2, usually comes out in a complicated form that involves cubic roots of imaginary numbers. The method gives the answer, but the simplification is often not trivial. Cohen spent time thinking about this. Mathematics was his, and from the first, he wasn't a passive receiver.

Cohen's isolation lessened substantially when he went to Stuyvesant High School in lower Manhattan. Stuyvesant, along with the Bronx High School of Science and Brooklyn Tech, was a godsend for the kind of student Cohen was. These were schools with admission requirements that specialized in science and mathematics. The level of the students, of instruction, and the general level of intellectual excitement in them rivaled the best gymnasiums of Germany or the lycées of France. Many prominent scientists have come out of these three schools. When Cohen was at Stuyvesant, Elias Stein, Harold Widom, and Don Newman were in attendance. A defining moment occurred at the first mathematics team meeting, when Cohen met Stein, who was, he says, "about two years older, and he was reading measure theory and complex function theory."[21] (Stein won the Wolf Prize in 1999.) Interactions with other students at Stuyvesant let Cohen know what he didn't know. They gave him advice. There was a strong school spirit, conveyed well by Cohen's statement that he and his fellows felt they were "in the best high school in the greatest city in the world."[22]

Cohen graduated sixth in his class, based on overall grade-point average, at the age of sixteen. He had read and understood, at least substantially, advanced books by Birkhoff and Mac Lane on modern algebra, Titchmarsh on the theory of functions, and Landau's classic book on number theory. His only regret from high school was that he had never walked over to New York University and made connections there, where real mathematics was being done. He was one of forty national Westinghouse scholars, based on an essay he wrote about relativity in which he rediscovered a proof of the addition formula for velocity. All Westinghouse scholars receive scholarship offers, and Cohen did, but none offered a completely free ride, so he didn't accept any. He believed it would be too difficult for his family to make up the difference. His parents had now officially divorced, and Cohen said to me that he felt guilty that he couldn't help his mother more. Therefore he enrolled at Brooklyn College, following his sister.

Stuyvesant High School Math Team, Fall 1948. Top row, left to right: 3. Elias Stein (Wolf Prize) 4. Harold Widom 5. Paul Cohen (Fields Medal) Bottom row center: Martin Brilliant (courtesy of Martin Brilliant).

Though there were good mathematicians at Brooklyn College, this could have been a dangerous decision. Cohen was bored at school and, he said, "Strangely enough—I don't know why—again I didn't travel up to NYU or Columbia. I did go once to City College and heard Martin Davis lecture on logic."[23] About this reluctance to travel to where the mathematics was, shyness comes to mind. I have seen Cohen teaching and at seminars and one would not think of him as shy, but as a young outsider lacking some specific invitation, it is a convincing explanation. Two things helped him to move on. One is that while he was at Brooklyn College he took the Putnam exam, a very difficult national mathematics competition. Both years, 1951 and 1952, he placed in the top ten in the country, this at ages sixteen and seventeen. It hung a sign around his neck saying that he belonged someplace like Harvard, Princeton, or Chicago. Cohen didn't mention the Putnam in his interview. When I mentioned it to him, his reply was, "How did you find out about that?" and then, "I always thought I could have done better."[24] The second fortuitous occurrence was that friends of Cohen's had gone on to the University of Chicago and talked about him. Cohen received a letter from Adrian Albert, a professor at Chicago, suggesting he apply for graduate school and mentioning that finishing college would not be necessary. Cohen applied and off he went mid-year for the winter quarter in 1953.

The mathematics department at the University of Chicago was experiencing a renaissance period, challenging Harvard, Princeton, and the University of California at Berkeley. After a golden age in mathematics, 1892 until roughly 1910, with money to hire faculty including Eliakim Moore (1862–1932), Oskar Bolza (1857–1942), and Heinrich Maschke (1853–1908), Chicago had experienced some decline. George David Birkhoff (1884–1944) and Oswald Veblen (1880–1960) studied at Chicago but had then led the charge of American mathematics from 1910 to 1940 while teaching at Harvard and Princeton, respectively. Starting with Robert Maynard Hutchins's decision to host the Manhattan Project, which brought physicists of the caliber of Fermi and Harold Urey, among others, Chicago was ready for another efflorescence. In 1946 Hutchins hired Marshall Stone (1903–1989) to improve the mathematics department. This Stone did, through a series of brilliant appointments—Antoni Zygmund (1900–1992), Shiing-Shen Chern (1911–), Saunders Mac Lane (1909–), and André Weil (1906–1998), among others. Chern came from Kashing, China, and had studied in Germany. Mac Lane was a Chicago product at the cutting edge of modern mathematics. A talented string of graduate students appeared.

Unquestionably one of the most brilliant mathematicians in the world, Weil was perhaps the boldest appointment, because he was reputed to be unusually difficult to get along with. Weil was a student of the *Bhagavad Gita*, in the original. In 1938 he had decided his *dharma* was "to devote myself to mathematics as much as I was able. The sin would have been to let myself be diverted from it."[25] Since the First World War had killed most of an entire generation of French mathematicians, Weil decided he would not sit at the Maginot Line waiting for the Germans. He tried to flee through Scandinavia, but was caught, jailed, and shipped back to France, where he was in considerable danger. (This also explains why he was not immediately summoned back to France after the war.) Once in the U.S., he found a series of poorly paid, obscure appointments and had not remained silent about the injustice of it all. Stone found him teaching in Sao Paulo, Brazil. Weil was not patient with mathematicians less bright than he, that is, virtually all of his potential colleagues. However, the following quote from Irving Kaplansky (1917–), himself a notable contributor to Chicago's Stone Age, as it is sometimes called, lets us know what Chicago got.

> There were several times in my life that I've, one way or another, got that feeling, my gosh, here is a tremendous mathematician; for instance, Weil, von Neumann, Serre, Milnor, Atiyah. Well, those are obvious names. But, above all, André Weil. We were colleagues for about ten years. I can even pinpoint mathematics I did that I wouldn't have done if he had not been there. It's not that he showed me how to do it or anything like that, but by just a casual remark he would start me off on something. He was very impatient with what he re-

> garded as incompetence. I don't think he'll mind my saying that. Then there is his extraordinary quickness. You may know about that. You can take an area of mathematics that he presumably never heard of before and just like that he'll have something to say about it.[26]

So it was into a hotbed of mathematical activity that Cohen moved. The style represented by Weil, Mac Lane, and Chern was modern and abstract. It frequently involved investigating mathematical objects so large they would have resulted in set-theoretical contradictions if you called them sets, e.g., "categories," etc. Sophisticated algebraic methods were everywhere. Nothing was what it started out as. Understanding these things was like being a member of a gnostic priesthood. Cohen, with his problem-oriented, calculational style of mathematics and his habit of figuring things out for himself, was not drawn to this group and style. He gravitated to Zygmund, a Polish refugee who could best be described as a classical analyst persevering amid modernism. Zygmund was good enough so that no one was much bothered by the fact that he wasn't in the avant garde. He produced an impressive list of students. Cohen worked on things like Fourier series. He absorbed some of the modern style but overall was in Zygmund's line.

Cohen was also interested in number theory. He thought he might be able to find a way of deciding if an arbitrary equation of a certain type was satisfied by all possible substitutions for the variables. A fellow graduate student told him his quest was doomed to failure because of Gödel's incompleteness theorem. Cohen was directed to Stephen Kleene's formidable book, *Metamathematics*. He read the section on Gödel's theorem and distrusted it. It seemed more like philosophy than mathematics to him. When Kleene came to Chicago to give a talk, Cohen asked him about his problem. Kleene, after a bit of thought, said Gödel's theorem did apply. Cohen returned to Gödel's result and became convinced. He told me, "I was rather depressed when I realized Gödel was right."[27]

He completed his Ph.D. in 1958 at twenty-four and became an itinerant mathematician, teaching at the University of Rochester and the Massachusetts Institute of Technology and spending time at the Institute for Advanced Study before settling at Stanford in 1961, as an assistant professor. Cohen considered himself an analyst. His first notable success was a partial solution, published in 1960, of what is called the Littlewood problem. This is a problem in classical harmonic analysis involving Fourier series, though only stated in 1948. This work established him as a mathematician to watch. Later, in 1964, it won him the Bôcher Prize, presented by the American Mathematical Society. Cohen's early publication history proceeded at an easy pace. If he achieved a significant result, he published it.

During his first year at Stanford, Cohen listened to Feferman (a logician) and Royden (an analyst) at a departmental lunch talking about a "consistency proof" for mathematics, whether there might be a method that got around Gödel's results. This conversation set him to pondering various schemes. He briefly thought he had a consistency proof for analysis and gave an informal seminar on his ideas, but stopped when it became clear his proof wouldn't work.[28] Gödel's result always blocked the path. Cohen had no training as a logician but had been talking to logicians since graduate school—with fellow students Bill Howard, Raymond Smullyan, and Stanley Tennenbaum at Chicago; with Feferman at the Institute (in 1959–60); with Azriel Lévy at MIT. As early as 1959 he had asked Feferman what the outstanding open problems in the foundations were.[29] Cohen was looking for a new problem to work on. He says that because he had been thinking about set theory on his own, he felt he had certain intuitions about it. The lunchtime conversations continued. As Cohen says:

> I don't know who said it to me, perhaps it was Sol Feferman, but somebody eventually said something like, "Well, if you think you understand set theory and you think it's consistent, why not investigate the problem of the continuum hypothesis?[30]

The continuum hypothesis deals in a clearly defined way with the real numbers and sets of real numbers, whereas many of the problems logicians were interested in were "philosophical," and Cohen distrusted them. This attitude was present in another young Chicago-educated mathematician, Robert Solovay, a differential topologist. Solovay was attracted to the problem at about the same time. Moore writes, "Solovay knew more logic than he considered to be respectable for a mathematician."[31] Solovay told me that he was always interested in a lot of things and that he had already read Kleene's book on Gödel's incompleteness theorem in high school.

The contrast between the approaches and personalities of Cohen, Cantor, and Gödel couldn't be more striking. Things American often have a pragmatic, ahistoric quality. In Peirce, James, and Dewey's pragmatism, a belief is true if we are prepared to act as if it is true. We unburden ourselves of old preconceptions, drop everything and head west to look for gold, even if we are comfortable where we are, much less if we aren't. Cohen says, "As a matter of fact, I was told by many people that I should stick to one thing, but I have always been too restless."[32] Unlike Cantor or Gödel, he did not receive a classical education—he didn't even receive an undergraduate degree. He moved on. As an analyst he possessed a great deal of technical dexterity. No analyst succeeds without it. Cohen's lifetime pattern has been to roam and look for very hard, basic problems and attack them straight-on, in the hope of finding a "simple," direct solution. Immediately prior to picking up the continuum hypothesis, he had been working on a problem in differential equations: "Again it was one of those problems that involve lots of machinery [the modern machinery that, except for Zygmund, was the style at Chicago], and I am always trying to do things a little bit more primitively. I also wanted to do things completely independently, and that was bad. I later realized that the problem couldn't be done that way. I would like to have solved it, but I didn't."[33] When the problem was solved it was with the modern machinery, and no more elementary solution has emerged. If distrustful of philosophy, Cohen is adept with philosophical ideas and seems drawn to informal philosophical discussion when it is associated with mathematics. These traits, weaknesses when attacking his differential equation problem, put him in a good position to attack the continuum hypothesis, though at the time this wouldn't have seemed obvious to anyone. All he would literally need for his results were the axioms of set theory, Löwenheim–Skolem, good intuition, and dexterity.

Cohen, at first, was more interested in the axiom of choice than in the continuum hypothesis. He worked intensely for a number of months trying to prove the independence of the axiom of choice, until he decided some of his techniques were beginning to resemble what he had heard about Gödel's methods in his work on the consistency of the axiom of choice and the continuum hypothesis. This is amazing. By Cohen's own account he hadn't read the single major paper in the field. But people in the field had decided that the problem wasn't solvable with anything available. Something radically new was needed, and no one had a candidate idea. Therefore Cohen didn't think it necessary to look too closely at Gödel's work. As he rediscovered some of Gödel's ideas, he had no reason to believe that what he was doing was not likely to be germane. As in his childhood, Cohen's mode of study gave him ownership.

Though Cohen felt he had an idea, he couldn't get it to work in any concrete form, and in 1962 he became depressed about his lack of progress and stopped working on the problem for four or five months. Davis has a handwritten note from Cohen in which Cohen says he had given up on set theory because of the negative attitude he found among logicians and that it was Davis's encouragement (at the Stockholm Congress in 1962) that led to his going back to it.[34] In Sweden Cohen met his future wife, Christina Karls, and they began courting. Back in the U.S. the two made a driving tour of the Southwest at winter break in 1962. In the hours on the road, he began to think about the continuum hypothesis again and says he had a growing sense of confidence. There is a photograph of the young couple in *More Mathematical People*. It is hard to imagine a scenario more inclined to give a young man a sense of optimism than to be driving through one of the most beautiful (and "philosophical") landscapes in the world with a beautiful woman.

Cohen was back on the problem and making striking progress in the new year. His first paper, "A Minimal Model for Set Theory," was communicated to the *Bulletin of the American Mathematical Society* on March 22, for the July issue. Cohen wasn't aware that the existence of a minimal model for set theory had already been established by J. C. Shepherdson of Bristol in 1953. Shepherdson's papers don't focus on the minimal model but it is there stated clearly. In a remarkable turn, according to Moore, "Cohen recalls that both Kreisel and [Dana] Scott urged that the result on the minimal model be published. . . . Cohen was astounded when neither Kreisel nor Scott knew such an elementary result. . . . Henceforth, Cohen's relations with Kreisel were strained."[35] Cohen's article made it into print rapidly; he credited Shepherdson's result correctly in his 1966 book.

The existence of a minimal model gave Cohen a clue as to where to start when constructing a model in a different fashion. Start with the minimal model and add to it. Cohen has talked about his side thoughts: "I suddenly felt that I had the general idea of what this new notion of 'truth' was that I was looking for."[36] In inspecting his work I don't see a new notion of truth exactly. I see an ingenious method of remarkable flexibility and power that builds models that behave exactly as one wants them to. This is a child of Hilbert's method of models for proving independence results.

By the middle of April, Cohen had his solution. Forcing was so powerful that it seemed as if it could also solve most of the open questions in set theory. His local colleagues began to get excited, but they wondered if it was too good to be true. There were mistakes in Cohen early manuscript. In a state of some agitation he hand-carried a draft to Gödel in Princeton. Legend holds that Gödel awkwardly let him in, looked at the paper briefly and said the argument was correct. In fact, Gödel accepted the paper to study it, and Cohen received no such immediate response.

On the contrary, on May 3, 1963, thirty minutes before Cohen was scheduled to give a talk at Princeton, he received a telephone call from Dana Scott saying he had found an error in the proof.[37] Cohen quickly convinced himself that Scott was wrong and gave the talk, giving only a brief account of forcing, however. At least one person in the audience became convinced, Solovay, who went home, reconstructed the details, and "concluded that all the interesting questions had been settled."[38] News of Cohen's results spread. Some people accepted the results quickly; some didn't. Hilary Putnam told me that Cohen sent him a first draft of the proof. Putnam found a "glitch" about whether the power set axiom held in Cohen's model and telephoned Gödel. Gödel had noticed the problem himself and said it wasn't serious. His difficulties overcome, Putnam taught the proof in a course at MIT before Cohen's paper was published.

Cohen got Gödel's answer in a letter dated June 20, 1963. Gödel wrote, "I hope you are not under some nervous strain which hampers you in your work. You have just achieved the most important progress in set theory since its axiomatization. So you have every reason to be in high spirits."[39] Gödel passed Cohen's paper on with his approval to the *Proceedings of the National Academy of Sciences*, where Gödel himself had published his consistency proof twenty-five years earlier. Cohen still had to edit his paper with Gödel, so this period of frustration dragged on through September, at which time, "Cohen was telling Gödel to change anything that he wished."[40] For a more detailed history of forcing, read Moore's article, "The Origins of Forcing."

This is what Cohen put together (I follow his book rather than his original papers): He starts with the minimal model just referred to, which is made of actual sets.* Cohen states a remarkable theorem about the minimal model in a way that goes to the heart of the issue:

* This requires the assumption of what is known as the standard model axiom, which Cohen thought was uncontroversial. He has written, "All our intuition comes from our belief in the natural, almost physical, model of the mathematical universe."[42] He was wrong that this approach wasn't controversial, and early on there was controversy around his use of the standard model axiom. Cohen used this approach in his book because he thought it conveys the key ideas more clearly (and I agree), but explains how it could be avoided. He does describe avoiding it as "rather tedious."[43]

> For every element x in M [the minimal model] there is a formula *A*(y) in Zermelo–Fraenkel set theory such that x is the unique element in M satisfying *A*(x) restricted to M, *thus in M every element can be "named."*[41] [Paraphrased with my italics.]

We talk about infinite sets using finite language and a listable set of axioms. We can therefore, in principle, list everything that could ever be said about infinite sets, reasoning from the axioms we choose. If we can't talk about something, why would it have to be in a model of the formal system? If we can't refer to a set, that is, call out its name and ask that it step forward out of the ranks, why does it have to be in the model? The formal language talks about all the real numbers, and so in the minimal model there is a set labeled "all the real numbers." However, most of what we imagine are the members of this set are not included in the model. Our entire model is countable. The set of all real numbers, in effect, is never dumped out on the floor of the model. It stays in its container. If one asked about the size of our model, or the universe of sets, the answer would be that it is vast. But what that means is that there is a lot of inaccessible space in our model. If we can't actually go there, what does it matter to our sense that we have a good system if in fact nothing is there? The idea is then to use this fact (or possibility) and introduce sets into this small (but still infinite) model that blend in with the other members of the model, except when one wants to talk about the continuum hypothesis. These generic sets are like the perfect spy. Nothing unusual stands out about them. Then they suddenly stand up and contradict the continuum hypothesis, speaking rather than holding their peace.

To accomplish this, Cohen invented "forcing," a technique to show the existence of a set or sets that are, in his word, "generic" relative to the sets in the minimal model. A forcing condition is merely a finite set of statements that a given natural number is or is not in this generic set we are trying to select. A peculiar aspect of how Cohen defines forcing in his book can be seen in the fact that a statement "not B" is forced by a finite collection of forcing conditions if there is no larger collection of forcing conditions that includes the original collection for the set and forces "B" to be true. (Scott is actually responsible for this simple form.)[44] Statements are forced if they can't be contradicted by a larger finite set of statements about whether a given natural number is in the generic set or not. One way to get a feel for why forcing might work is to think of all the sets of natural numbers that can be both named and defined by a recursive rule. These are the sets that have to be in the minimal model because a definite rule says what numbers are in them. One doesn't want the generic set to be one of these. The generic set can be made different from each of these by saying a single natural number is or is not in the set, just so it differs from each of these recursive sets in one particular. This is Cantor's diagonal process again.

On the other hand, a subset of the natural numbers that is defined purely existentially, or "named," by a formula that does not supply a recursive rule or other more-partial information for what is actually in it is almost like a generic set itself. It is in the minimal model to satisfy its naming formula—but we really don't know much about it. Since we can't say precisely what is in either set, the statement "The generic set is not this named set" is forced by just about any forcing condition. No finite set of statements that a given natural number is or is not in the set can contradict this statement. The two sets are fundamentally incomparable and therefore can be taken to be different.

This peculiar aspect of forcing implies that any statement in set theory can either be forced or its negation can be forced by a *finite* forcing condition. Again, the statement that is forced is the one that can't be contradicted. This is the most surprising point in the whole argument and probably is close to what Cohen refers to as a new notion of truth—truth in the model is controlled by statements that one would think are so limited as to be irrelevant. Normally the character of an infinite set is not determined by a finite number of statements whether a number is or is not in the set.

Using the odd fact that every statement in set theory or its negation can be forced by a finite forcing condition, one proves the existence of an infinite chain of forcing conditions which, taken as a whole, forces every possible statement (or its negation) that involves in any way the new set one is proposing to add. The generic set itself can be thought of as the limit of the chain of forcing conditions.

One of the trickier aspects of forcing is that the generic set is referred to as it is being constructed. One takes the customized set or sets one is proposing to add and goes through a construction process that parallels Gödel's process for the constructible sets. The forcing process is braided into the construction. One proves that one does get a valid model out of this process by checking that the axioms are all true in the model in a way that closely follows Gödel. One also proves fairly easily that any statement that is forced is true in the model. One gets a model that has the properties that have been designed in. A single generic set added at the beginning doesn't change anything, except that one has sets in the model that are ringers, that is, are not constructible in the sense of Gödel.

To get at the continuum hypothesis, Cohen adds an infinite collection of generic sets that acquire relations among themselves and with the ordinals of the minimal model as they are constructed and ensure that the continuum hypothesis is violated in the expanded model that results.

When talking about highly infinite objects that are not completely accessible, forcing can frequently be used. It is a form of leverage. First you create a stripped-down model and then make a stealth incursion into it, insert a generic set or many generic sets, jigger the books and get out without having disturbed anything. It is a wildly non-constructive existence proof, but it shows that if we assume full Zermelo–Fraenkel set theory, where such proofs are allowed, then we can also show the consistency of the "set theories" we have now created in which the axiom of choice and the continuum hypothesis do not hold.

Cohen had been working almost as an amateur, but in close proximity to one of the most active logical communities in the world. At the edges of the San Francisco Bay were Berkeley under Alfred Tarski (1901–1983), a figure almost comparable to Gödel, and the Stanford logicians. This community had connections to logicians at Princeton University and the Institute. Even before Cohen's article saw print in late 1963, there was an avalanche of work using his new methods. Tarski is reported to have commented, "They have a new method and they'll get everything."[45] It resembled a feeding frenzy. At the same time, there was still skepticism and criticism. According to Moore, Yiannis Moschovakis wrote Cohen asking why he had made such a "ridiculous assumption"[46] as the standard model axiom. Cohen says he took this approach because it made what he was doing easier to explain and that it really didn't affect his argument.[47] He said that originally his thinking was "syntactical," that is, focused on strings of symbols and their manipulation according to rules. Cohen found the rush of activity a little overwhelming and stepped back. People who were truly expert in the field had problems readily at hand that they could apply forcing to. Feferman, Lévy, Solovay, and others published results. Solovay, the other outsider, was particularly strong and proved an important result about measure theory. Along with Scott, he reformulated forcing under the rubric of Boolean-valued models, and so our history turns full circle. One view of the aftermath of Cohen's results is supplied by James Baumgartner:

> Once upon a time, not so very long ago, logicians hardly ever wrote anything down. Wonderful results were being obtained almost weekly, and no one wanted to miss out on the next theorem by spending the time to write up the last one. Fortunately there was a Center [UC Berkeley] where these results were collected and organized, but even for the graduate students at the Center life was hard. They had no textbooks for elementary courses, and for advanced courses they were forced to rely on handwritten proof outlines, which were usually illegible and incomplete; handwritten seminar notes, which were usually wrong; and Ph.D. dissertations, which were usually out of date. Nevertheless, they prospered.[48]

Forcing has been used to settle a wide range of problems, from infinite graphs, to questions in measure theory, to infinite games, to many intricate questions in set theory. The common denominator in all these applications is structures that are infinite and not countable. At various stages Cohen's construction was simplified; at the same time, the subject has become more complex. By the time one gets to someone like Saharon Shelah (Wolf Proze, 2001), whose bibliography has hundreds and hundreds of entries, starting in 1969, one feels that a primitive folk instrument has been so developed that it has entered the concert halls for virtuoso performance. Cohen makes a wonderful remark about the difficulty or simplicity of his original proof.

> One might say in a humorous way that the attitude toward my proof was as follows. When it was first presented, some people thought it

> was wrong. Then it was thought to be extremely complicated. Then it was thought to be easy. But of course it *is* easy in the sense that there is a clear philosophical idea.[49]

This captures much of the history of the entire subject. Gödel could have said it about a number of his results, including incompleteness—though after a short time no one was willing to bet against Gödel being right. Over the 150 years that people have been developing workable notation and thinking about formal symbolic systems, the pace has sometimes seemed slow. Obvious observations were not made; Kreisel uses the expression "blind spots." The mathematical structures created, though of some complexity, do not begin to rival the complexity of structures studied and created in analysis, e.g., string theory. Yet confusion was common; the difficulties in this area are real. They are often philosophical. What do we mean by consistent? Calculable? True? Provable?

In 1966 Cohen was awarded a Fields Medal—the rough equivalent to a Nobel Prize for a mathematician. Davis described to me an incident at the 1966 International Congress in Moscow, where Cohen received his medal. "Paul and I were with a group of people waiting for a cab for an interminable time when he remarked to no one in particular, 'Don't you think that if one of us was in charge for a month, we could make things much more efficient?'"[50] Cohen has been attracted to big problems, and having done the work he had, he could afford to run the risks

Paul Cohen (*courtesy of Paul Cohen*).

that big problems entail. He thought hard about Riemann's conjecture about the zeta function, which we will soon encounter. He thought about hard problems in differential equations. He toyed with ideas about unified field theories. He has not solved any of the other big problems. He has said one of his limitations is that he does not create programs or think about interrelated problems the way that Weyl or Poincaré did (or Hilbert). His aesthetic looked for the tricky or bold simplifying idea. What Cohen did do was have a very good idea, so good that it amounted to a program for a generation of workers in the foundations. What problems are open to forcing? A single original idea, like Cantor's diagonalization argument, can have a long life. Cohen's search for the simple and simplifying idea is one essential style in mathematics.

Cohen has remained at Stanford, surrounded by striking physical beauty. The large campus is entered through groves of eucalyptus trees—there is an Olympic-sized pool, a stable, a golf course. Skies are blue for months at a stretch. The hills behind Stanford are gentle (somewhat like the hills outside Göttingen). Rolling grasslands with oak trees rise to a ridge, then change to redwood forest as the coastal range falls to the ocean. Palo Alto's "Professorville," with its older, redwood-shingled houses, blurs into modernist housing tracts. Hot enough in the summer to grow sweet tomatoes, it cools at night. It's radish soil; nearby Mountain View was "the radish capital of the world" in its agricultural heyday. The seasons change, with a beautiful fall but almost never a freeze or snow. Mathematics has been good to Cohen, and Palo Alto seems a long way from Brooklyn or Chicago. But perhaps it doesn't seem so far to him.

When I was at Stanford, I would go to Tressider Union and sit, somewhat dazzled, under the oaks. Often Paul Cohen occupied a table, passing part of an afternoon, sometimes with a single book or pad of paper, but evidently thinking, thinking about problems.

At Halle, in the 1870s, '80s, and '90s, Georg Cantor conceived of the transfinitum, which he believed was perfectly realized in the mind of God. He believed its essential features could be made transparent to the human mind through reason. The fact that many of its features can be seen clearly—that there are different-sized infinities, that the real numbers are uncountable—is a great human success. Cantor used ordinary language in his investigation. His original theory is now referred to as "naive" set theory.

Logicians and other interested mathematicians realized that a sharper, more precise instrument was necessary to analyze this incursion into the infinite. Set theory was axiomatized. Gödel's constructible sets are a large "natural" model of set theory. (The statement that all sets are constructible is plausible enough so that Gödel at least considered it as an additional axiom—though it was fairly quickly rejected.) Gödel used the Löwenheim–Skolem theorem in this model as a tool to prove the consistency of the axiom of choice and the continuum hypothesis.

Cohen came in from the outside and pushed the technique of model-building to the wall. He accepted the letter of the law—all valid models are equal logically—and constructed a model that is almost entirely "unnatural" but that *proves* not only the independence of the continuum hypothesis but also of the axiom of choice relative to Zermelo–Fraenkel set theory. The method is so powerful it would seem difficult to get around, though there is active work still going on today. The method of models continues to be important. For example, in 2000 a topic of conversation was Hugh Woodin's announcement that he has shown that in a class of theories and models the continuum hypothesis is false—theories he feels answer more questions correctly than Zermelo–Fraenkel set theory and that resist forcing. Another approach to the continuum hypothesis might be found that decides it—a new axiom that seems obvious to all (or almost all) mathematicians. Some consequence of the truth or falseness of the continuum hypothesis might trickle down into some more practical theory, where it might be "empirically" verified. We look for a model so natural that it has to be true. Through all of this we can hear the play of Hilbert's words in his Paris talk: "This conviction of the solvability of every mathematical problem is a powerful incentive to the worker. We hear within us the perpetual call: There is the problem. Seek its solution. You can find it by pure reason, for in mathematics there is no *ignorabimus.*"

Until those efforts are successful, we must say that the Löwenheim–Skolem theorem and Cohen's results suggest that essential features of the transfinitum as basic as the continuum hypothesis are not accessible to investigation by the human mind and its finite language. With each passing decade a decisive new idea would seem less likely. The problem begins in the center of Germany and ends, roughly a hundred years later, on the edge of the Pacific.

Can't We Do This with a Computer?

The Tenth Problem

Matiyasevich, Robinson, Davis, Putnam, et al.

The first and second problems have strong philosophical content, visible in Hilbert's statement of them. In contrast, the tenth problem asks a natural question in arithmetic. It reads in its entirety:

> Given a diophantine equation with any number of unknown quantities and with rational integral numerical coefficients: *To devise a process according to which it can be determined by a finite number of operations whether the equation is solvable in rational integers.*

Yet this problem turns out to be almost as "philosophical" as the first two.

Named after Diophantus of Alexandria (third century A.D.), a diophantine equation is any algebraic polynomial equation with whole number coefficients, for example: $3x^4 + 5x^3 - 123x^2 - 17x + 37 = 0$. Only whole numbers appear; there are variables and + and – signs. There could be more variables and higher powers—the higher powers are just a notational convenience and could be written out as multiplications. With diophantine equations, we are looking only for whole number solutions or sometimes, by choice, only non-negative whole number solutions—though Diophantus himself was looking for any rational solutions. A simple and mathematically famous example is Pell's equation, $x^2 - ay^2 = 1$. More complicated examples could cover a table top or football field. Hilbert asked that these be dealt with as well. Diophantine equations are notoriously difficult to solve. Number theorists were, and continued to be, somewhat amazed that Hilbert posed this problem, because it has so little contact with what they can actually do. It was a siren song: "Look at all diophantine equations at once and the particular complexities will melt away." Hilbert might have had his own bold solution of the problem of algebraic invariants in mind.

Over time, individual mathematicians approached special cases of the problem with some success—Thue in 1908, Siegel in 1929, Alan Baker in 1968—but nobody got anywhere near a general solution. The problem asks us to find a "process." What is a process? If we are shown a purported process, we can check if it actually works. If it can be shown that it works every time, it's a process—an effective process. Mathematicians observed early that a decision process generates a string of yes's and no's to the questions asked it. We may substitute a 1 for yes and a 0 for no, thus making the process generate a function of the natural numbers. Therefore, effective processes can be viewed as a subset of the functions of the natural numbers that can be "effectively calculated." Since any method for effectively calculating a function is a process, the two ideas are identical.

In the 1920s mathematicians working in symbolic logic knew that a large number of functions can be generated using a few simple rules applied again and again—recursively. These functions are now called "primitive recursive." They are generated by a finite number of rules that work in every combination every time, so Cantor's diagonalization method can be used to generate an effectively calculable function that is not primitive recursive. In 1928 Ackermann, Hilbert's collaborator, displayed a reasonable function that could not be generated by these rules, because it grew too fast. (This sort of function was actually discovered earlier, but the result was not widely known.)

In 1931 Jacques Herbrand (1908–1931) suggested how to expand the range of functions recursively definable in arithmetic, in a letter to Gödel. His proposal was more open-ended than the guaranteed-to-work rules for primitive recursive functions. Many candidate definitions would be duds and it wasn't (and isn't) obvious which. At this point Herbrand did not connect the desire for "more functions" recursively definable to finding "all effectively calculable functions." Almost immediately after writing to Gödel, Herbrand fell to his death while climbing in the Alps. Gödel did not immediately follow up the suggestion.

At this point the American logician Alonzo Church (1903–1995) and his graduate students Stephen C. Kleene (1909–1994) and J. Barkley Rosser (1907–1989) enter the story. With his first paper published in 1924, the year he finished his undergraduate work at Princeton, and his last in 1995, the year of his death, Church was active for seventy-two years. He studied in Göttingen when Hilbert was there and in Amsterdam, home of Brouwer—the poles of logical debate. He returned to Princeton and made it a world center for logic, with assistance from von Neumann at the Institute. Church endeavored to make all his lines on the blackboard absolutely parallel when he lectured, and he erased with such thoroughness that it is mentioned in his obituary in the *Bulletin of Symbolic Logic*. He preferred to work in the middle of the night, when it was absolutely quiet, and spent his summers in the Bahamas.

In the early 1930s Church developed a general logical system in the spirit of Frege or Russell and Whitehead. His goal was to have functions of the whole numbers appear naturally in the logical calculus from the beginning. Davis, a later student, describes the next development:

> Church published a pair of substantial papers on the system he developed and set his students Stephen C. Kleene and J. Barkley Rosser to work on it. Their work was extremely effective, if not exactly what one dreams of having one's graduate students accomplish for one: Kleene and Rosser proved that Church's system was inconsistent![1]

A "natural and elegant" part of Church's system, the lambda-calculus, was extracted and proven to be consistent. By 1932 Kleene had discovered that a large class of functions could be defined in this formalism. Church was surprised. By fall of 1933 the lambda-calculus was well worked out "and was circulating among the logicians at Princeton."[2] Church was already speculating that the lambda-definable functions included all effectively calculable functions. They could not be disqualified by a straightforward application of the diagonal argument.

After Gödel arrived in Princeton, Church advanced the idea in conversations with him. Gödel did not like it; it lacked philosophical justification and clarity. Church challenged him to come up with a candidate definition of effective calculability of his own. A month or two later Gödel saw how to modify Herbrand's idea to define what are now called Herbrand–Gödel recursive functions, or simply, recursive functions.

In response, Church and, independently, Kleene proved that all lambda-definable functions were recursive. Then, in a paper submitted July 1, 1935, Kleene proved that all recursive functions were lambda-definable. Despite appearing to be quite different, the lambda-definable functions and the Herbrand–Gödel functions were one and the same. At a meeting of the American Mathematical Society on April 19, 1935, Church—anticipating Kleene's results—proposed that the Herbrand–Gödel recursive functions should be taken by definition to be the class of effectively calculable functions. Only in 1943 did Kleene introduce the word "thesis" into "Church's thesis," adding an appropriate clarification. Church's thesis is the real-world assertion that the Herbrand-Gödel recursive functions are precisely all effectively calculable functions. Recursive functions are definite mathematical objects, whereas "effectively calculable functions" is open-ended. Church's thesis is a quasi-philosophical statement, and belief in it is based partly on intuition and partly on empirical evidence. If someone displayed a method of calculating a function that could not be calculated using general recursion, then Church's thesis would be proven false, but it can never be proven to be true.

What is the evidence for Church's thesis? Every effectively calculable function that anyone has ever thought of is general recursive. The equivalence of lambda-definable and recursive functions, which start from general principles that are quite different, was a second kind of "empirical" confirmation that the essence of computability had been captured. However, as of 1935 only these two different attempts to capture all effectively calculable functions had been shown to be equivalent.

In 1935 Church took Gödel's arguments about incompleteness in formal systems and moved them into the context of recursive functions. All candidates for general recursive functions can be listed. A given candidate may or may not yield an actual recursive function. Many send the calculator off on infinite searches, but we cannot in general tell which ones do. This is where the uncertainty is. Even if we have been calculating for a million years we don't know if the next step will lead to an answer. This uncertainty does not stop us from using this list to make another list that includes all possible deductions from any of the entries on the first list. One keeps doubling back—e.g., first do the all deductions ten steps long from the first ten candidates, then all deductions a hundred steps long from the first hundred candidates, etc. This list would include every successful calculation of the value of any candidate recursive function defined at any natural number.

We use this new list to recursively define a set for which there is no decision process whether a given counting number is included in the set. Again one uses Cantor's diagonalization method. In essence, every time a candidate recursive function succeeds in a calculation, we try to choose a member for the set we are generating that guarantees that the function we have just succeeded in calculating a value for can't be the decision process for the set. Instead of the statement "I am lying," we use "a number n is included in the set if and only if the nth recursive function says it isn't." If n is in the set, then the nth recursive function can't be a decision process for the set. If n is not in the set then either the nth function is a dud and didn't calculate anything or the nth function says it is in the set (which it isn't). Since this is true for all n, there is no decision process for this set.

We are in the same head-spinning state we achieved with Gödel's incompleteness theorem. However, Gödel's first incompleteness theorem is a result about formal axiom systems and what can be deduced from a set of formal axioms. This result about recursive functions is a hard result about the arithmetic and functions of whole numbers, whether expressed formally or informally. This is an essential difference. If we believe Church's thesis that the recursive functions capture all possible effectively calculable functions of whole number arithmetic, then we believe there are unsolvable problems in arithmetic. We have even constructed an example. Might Hilbert's tenth problem be another such example? The miracle here is that (almost) everyone believes Church's thesis. In 1935 many, including Gödel, didn't believe Church's thesis, that computability had been captured.

Alan Turing (1912–1954) changed Gödel's and the other dissenters' minds. Working alone in England in 1935–36, in ignorance of the events in Princeton, Turing set out to capture all computable functions under a single scheme, with the goal of proving that a later problem of Hilbert's, his *Entscheidungsproblem*, was not solvable. In this problem Hilbert asked for an effective procedure for deciding whether a given formula in the predicate calculus could be proved. Turing showed there was not an effective procedure for doing this in "On Computable Numbers, with an Application to the *Entscheidungsproblem*." When Turing, only twenty-four at the time, and his advisers discovered that Church had already used this method and published on this problem, there was discussion of whether this rendered the paper unpublishable. The heart of Turing's paper, though, is in his analysis of what a computable function is and his answer: It is what can be calculated by an idealized, imaginary computing machine (subsequently called a Turing machine). The following gives a feeling for Turing's style of argument:

> Computing is normally done by writing certain symbols on paper. We may suppose this paper is divided into squares like a child's arithmetic book. In elementary arithmetic the two-dimensional character of the paper is sometimes used. But such a use is always avoidable, and I think that it will be agreed that the two-dimensional character of paper is no essential of computation. I assume then that the computation is carried out on one-dimensional paper, i.e., on a tape divided into squares. I shall also suppose that the number of symbols which may be printed is finite. If we were to allow an infinity of symbols, then there would be symbols differing to an arbitrarily small extent.[3]

Where Gödel and Church start from seeing how much performance they can get out of a formal system, Turing starts out by analyzing how a child would be taught to calculate, and the word "computer" drifts from reference to someone who computes to a machine that computes. This was a beautiful moment. Turing asks us to imagine a machine that could do the same simple things a child can, but never tires. He goes on to show that this very simple machine can be programmed to calculate anything one can think of, including all lambda-definable sequences or Herbrand–Gödel recursive functions. His paper convinced Gödel that computability had been captured.

In its simplest form, the Turing machine has an infinite tape divided into squares, with a head or reader that at any given moment lies over a square on the tape. The machine has a finite number of states in which it can be: initial, final (i.e., a state that says the calculation is finished, stop), and interim states. The machine reads a square—in the simplest form, this is simply to see if the square has a mark in it. Based on what it read, according to the state the machine is in, it erases a mark, writes a mark, or leaves the box alone. It moves the tape one square to the right or one square to the left or stays put, depending on the state it is in. It enters a new state. It stops when and if it enters the final state. There is a convention for how to read the answer off the tape when a calculation is complete.

The question that first presents itself is not whether this Turing machine is capable of calculating all possible computable functions, but whether it can calculate anything. With great economy Turing shows how ever-more-complicated functions can be calculated. By the time he is fourteen pages into his paper, he is giving the details of a universal Turing machine so powerful that, given the proper request, it will calculate any function one wants. In fact, one could run Windows or the most sophisticated weather modeling program on some Turing machine—though not at great speed. Today, little more than sixty years removed from Turing's paper, most people would have little doubt that anything calculable could be calculated, at least in principle, on a computer.

Turing was central to breaking Enigma, the code the Germans were using in the Second World War, and building one of the first computers, the Colossus, to decode more rapidly—an important contribution to winning the war. However, Turing was homosexual, which was illegal. In 1952 he brought a nineteen-year-old man home. When Turing's house was later broken into and he reported the burglary, the police arrested Turing. He was tried, convicted of homosexuality, and given hormones in an attempt to "cure" him. He killed himself by eating an apple poisoned with cyanide in 1954. (Like Gödel, he had been fascinated by Disney's "Snow White" and, sadly, as early as 1937 had talked of suicide involving a poison apple.)

Paralleling or anticipating almost all of the developments we have recounted, including Gödel's result for the second Hilbert problem, was an outrider: Emil Post (1897–1954). He illuminates the subject from another side. Born in northeastern Poland in 1897, Post emigrated to New York at age seven with his parents. His father joined a family fur and clothing business in New York. Around the age of twelve Post lost an arm in an accident. He managed well all his life with one arm and resented efforts to help him. Post attended Townsend Harris High School, a school for gifted students occupying the same campus as the City College of New York and sharing the faculty, which allowed him to begin serious study of mathematics. He continued at City College and received his bachelor's degree at twenty. He had already done significant mathematical work (which he was able to dust off in 1930 in a time of need and publish as "Generalized Differentiation"). His first published paper, in 1918, was "The Generalized Gamma Functions." Generalization was to remain a theme.

At Columbia, where he received his Ph.D. in 1920, Post attended a seminar given by his adviser, Cassius Keyser, on the recently published *Principia Mathematica*. Also, in 1918, Clarence I. Lewis pointed out in *A Survey of Symbolic Logic* that all of the new systems of logic that had been coming out embodied different approaches to the generation and manipulation of finite strings of symbols. Post carried this point of view to its limits. His thesis, "Introduction to a General Theory of Propositional Functions" (1921), occupies a place of honor in van Heijenoort's collection of key papers in symbolic logic. In it Post touches on many topics—in an essay on Post, Davis lists three that are of particular importance. Post picked out of *Principia* what is now called the propositional calculus, logic without symbols for quantifiers—"there exists" and "for any." He applied the method of truth tables and proved the propositional calculus complete and consistent. This has been called the birth of proof theory. Post introduced the possibility of truth tables that had more than two (true and false) values. Much subsequent work followed in multi-valued logic. Mathematicians had many motives and strategies for making logic formal. Post saw all of these efforts as capable of being studied from the outside by studying their combinatoric possibilities. His third major innovation was to attempt to treat all the versions of formal logical systems as special cases of one general system, which he called "Canonical Form A." Canonical Form A is so general that *Principia Mathematica* is a special case. This was a bold sally into logic by a young mathematician in America in 1920, where not a lot of logic was being done.

Post's originality was recognized. He went to Princeton for a post-doctoral fellowship, flush with success and high ambitions, and continued to investigate the structure of formal systems. If the combinatorics of a system in Canonical Form A seemed more simple than all the details that *Principia* treated, he introduced Canonical Forms B and C, which seemed simpler still. Post's gambit was similar to Hilbert's in the tenth problem—look at the entirely general process of generating strings of symbols with the hope that the immensely complicated details of number theory, set theory, analysis, etc., would fall away.

Post succeeded in proving a result that first seemed as if it might lie on the path to a decision process for *Principia*—an effective process for deciding which strings of symbols could be proved. This is similar to the *Entscheidungsproblem*, but Hilbert had not posed it yet. Post then spent a lot of time working on a decision process for a simple class of systems called tag systems, but encountered, in his own words:

> . . . an overwhelming confusion of classes of cases, with the solution of the corresponding problem depending more and more on problems of ordinary number theory. Since it had been our hope that the known difficulties of number theory would, as it were, be dissolved in the particularities of this more primitive form of mathematics, the solution of the general problem of "tag" appeared hopeless.[4]

Tag has since been shown to be undecidable, by Marvin Minsky in 1961. Yuri Matiyasevich told me that his teacher S. Yu. Maslov (1939–1982) independently proved this also, but was slow in publishing.

After Post realized he was not going to escape the difficulties of number theory, he wrote that this "led to a reversal of our entire program." He conjectured what is called Post's thesis, essentially Church's thesis in the context of Post's canonical systems. He was quick to see that this, taken with Cantor's diagonalization method, implied that there was no effective general way of deciding whether a string of symbols could be derived in one of his systems. This is essentially the result that the later progress, in the 1930s, led to: There is no process for deciding whether an arbitrary formula is generated by a formal system of any reasonable complexity.

Post had the essential ideas but only an outline of a rigorous and technically realized proof. No one in the world, much less in the United States, was using the terms he was using. Nobody except him was working in Canonical Forms A, B, and C. He returned to Columbia as an instructor for the year 1921–1922. He faced a lot of work and was very excited. Catastrophe struck. Much of the style of Post's work at this time, its all-encompassing nature, its exceptional ambition, the extreme optimism of the early premise and then the turn-on-a-dime change of program, the slightly comical language—*Principia*, i.e., all of mathematics, as a special case of Canonical Form A, B, or C—has some of the feel of a manic episode. And Post had been building to a manic episode. Yet one must make a logical point: It did not make the mathematics wrong. Post landed in a mental hospital, while the mathematics remained correct, if unfinished.

The breakdown, happening before any real recognition of what he had accomplished, greatly damaged Post's employability. He was able to find work in the New York public schools. Imagine encountering the one-armed Post as a teacher, particularly if it was in a year in which he was destined to break down again. He was in possession of results but unable to write them up. Throughout the 1920s he was hardly able to muster an attempt. Briefly in the summer of 1924, during a year in which he had some association with Cornell, Post did some work. He did a little work in 1925 and then some in 1929, but little compared to the project he needed to finish. Davis says he believes Post lectured in the 1920s at Columbia on the incompleteness of *Principia*, but the evidence from this period is like ripples on a pond.[5] In 1930–31 Gödel began publishing in pure, classic form results that were so close to Post's, it must have seemed like the end of the world.

Many people would have subsided into a life of chronic mental illness and uncertain employment, but Post found the courage to turn things around. Gertrude Singer, whom he married in 1929, became a source of strength. This was when Post resurrected his unpublished undergraduate paper on generalized differentiation. Its publication re-established him as a producing mathematician. In 1932 Post received an appointment to City College that didn't last long, due to a new flare-up of illness. However, in 1935 he was appointed again, and from this time, despite bouts of illness, he was employed at City College until his death. He adopted the strategy of always working on two problems at the same time to keep himself from becoming overexcited. He would work on one for two weeks and then, no matter how promising the work was, would switch to the other.[6]

City College required sixteen hours of student contact a week. The mathematics professors had to share one room. No secretarial help was provided. For the next nineteen years Post managed to claw out the time needed for research and contributed specifically to the subject of unsolvable problems. He published a paper in 1936 on calculable functions with an analysis of computability similar to Turing's, which Post was not aware of. Post, an American in the era of Henry Ford, analyzed an assembly line for computation with a "two way infinite sequence of boxes" and a "problem solver or worker."[7]

In 1944 Post published what has been called his most influential paper, "Recursively Enumerable Sets of Positive Integers and Their Decision Problems." Church's 1936 paper on an unsolvable problem in elementary number theory treated a problem cooked up to be unsolvable. For over a decade, the only problems that were shown to be unsolvable were problems originating in symbolic logic or created in arithmetic by logicians to be unsolvable, so that it seemed as if unsolvability might be an anomaly confined in practice to logicians. As a first step in bringing these techniques to bear on natural problems, Post attempted to open up the logicians' techniques to the larger community of mathematicians in a 1944 paper, first delivered as an invited address to the 1943 meeting of the American Mathematical Association. He intentionally relaxed his style to reach this larger audience, thinking that he would go back later and properly write up some of the new results he was giving. Ironically, ever since, logicians working in this area have tended to prefer this more relaxed and easily understood style. Post started by going through the subject, simplifying and pointing out the essential elements of the theory. He offered new results—e.g., degrees of unsolvability. In the paper, Post wrote that Hilbert's tenth problem "begs for an unsolvability proof."

In 1947 Post proved that what is known as Thue's word problem is unsolvable. The word problem, despite its name, did not originate in logic or formal systems, but instead came from efforts by Max Dehn (1878–1952) in 1910 to understand a problem in algebraic topology. Dehn solved his specific problem, but in 1914 the Norwegian mathematician Axel Thue (1863–1922) posed the algebraic problem in more general terms. A. A. Markov (1903–1979) of Russia proved the same result independently, also in 1947. (Markov's father was the A. A. Markov [1856–1922] who gave us Markov chains in probability theory.) This was the first breakout for unsolvability theory from symbolic logic into mathematics at large.[8]

Post suffered his final breakdown in 1954. His wife believed electroshock treatment helped him, but electroshock was beginning to fall out of favor, and the private sanatorium Post went to, one he had used before, wouldn't administer it. He was transferred to a public facility and died of a heart attack shortly after an administration of electroshock therapy, at the age of fifty-seven. Post's collected papers were published in 1994.

Post's pupil Martin Davis takes us into the tenth problem proper. Davis's parents were members of the next generation of emigrant Polish Jews. (Davis's father, after entering the U.S., had a name as un-Polish as Post's.) Though married in the U.S., his parents had known each other in Lodz. Davis says: "My mother as a teenage girl frequented the lending library that my father with other young 'free-thinkers' had set up."[9] Davis's father was an embroiderer of ladies' garments, handbags, and bedspreads. Davis was born in 1928. He had one brother, Jerome, born in 1933, who died of a ruptured appendix in 1941. Davis does not remember the Depression as a time of exceptional hardship, though for a period his family was on Home

Relief. His mother became a corsetiere, turning the front of their ground floor apartment into a corset shop. Like Cohen and Post, Davis attended one of New York's elite public high schools, Bronx Science. At City College (1944–48), Davis gravitated to Post. Post's comment that Hilbert's tenth problem "begs for an unsolvability proof" began what Davis has called "my lifelong obsession with the problem."[10] He continued his education at Princeton under Church and, like Post when he was there, was fond of returning to New York for the weekends. Though now a "visiting scholar" at Berkeley, Davis spent most of his career close to New York City. In conversation, he is happy and enthusiastic; in photographs, smiling. An impressive head of hair and beard grown in the 1960s and 1970s followed him into the 1990s.

At Princeton Davis was not one to be drawn off into a quagmire like trying to solve Hilbert's tenth problem, so he selected a project in an area that had hardly been explored, what is now called the hyperarithmetic hierarchy. It was, in Davis's words, "sure to yield results."[11] However, as he says, "I couldn't stop myself from thinking about Hilbert's Tenth Problem."[12] He refers to efforts at disciplining himself away from the problem. He failed in this effort but succeeded in making progress. His thesis had a third section on the tenth problem far more important than its ostensible subject, introducing what is called the Davis normal form. If one is allowed free use of universal and existential quantifiers (the symbols $\forall$ and $\exists$, "for

Martin Davis (courtesy of Martin Davis).

any" and "there exists"), it is easy to define all recursive functions in the framework of symbolic logic using diophantine equations. Davis showed that all the universal quantifiers could be eliminated except one bounded universal quantifier.* He pointed out that if that one bounded universal quantifier could be eliminated, Hilbert's tenth problem could be proven undecidable. Davis announced these results in December 1949. He published an abstract in 1950 and followed in 1953 with a full paper, "Arithmetical Problems and Recursively Enumerable Predicates." This marked the beginning of a solution for the tenth problem, when the work shifted from a general theory of solvability or effective computability to the details of diophantine equations.

Julia Robinson, in California, approached the problem from a different direction. She was born Julia Bowman in St. Louis, Missouri, on December 8, 1919, two years after her sister, Constance (Reid—the biographer of Hilbert and other mathematicians). Their mother, Helen Hall Bowman, died when Julia was two. The girls were sent to live with their grandmother near Camelback Mountain outside of Phoenix, Arizona, at that time effectively in the middle of the desert. Their father, Ralph Bowers Bowman, owned a successful machine tool and equipment company, but the spirit of enterprise went out of him with his wife's death. He closed the business after a year and followed his daughters into the desert. He brought with him a new wife, the former Edenia Kridelbaugh, a loving and competent woman who was soon viewed by Julia as her mother. A little sister, Billie, was born in 1928.

Julia was an otherworldly child. One of her earliest memories was squinting her eyes in the shade of a saguaro cactus and arranging pebbles in patterns, a prelude to a life devoted to the patterns of the natural numbers. She was also stubborn—her new mother, a former kindergarten and first grade teacher, said Julia was the most stubborn child she ever knew. Julia was slow to speak, and when she did speak she was hard to understand, due to poor pronunciation. People would ask her a question and then look to her sister, the only person who could reliably understand her. Later, Constance wrote Julia's "autobiography," a wonderful document that is the source of much of this section. It is based on an interview done June 30, 1985, a month before Julia's death—though clearly Reid had a great deal of background material as well. In the preface to *Julia: A Life in Mathematics*, Reid says that she read the manuscript to her sister: "She listened attentively and amended or deleted as appropriate, sometimes just a word. She heard and approved all that I wrote."[13]

When Julia was five and Constance seven the family moved to Point Loma in San Diego, California, so that the children could go to school. Besides a small community of Portuguese fishermen, a military base, and a lighthouse, there were

*The Davis normal form can be written: $a \in M \Leftrightarrow \exists z \forall y \leq z \exists x_1, \ldots, x_m \, [P(a, x_1, \ldots, x_m, y, z) = 0]$ where $P(\ldots) = 0$ is a diophantine equation. The bounded universal quantifier is $\forall y \leq z$.

only about fifty families on the point. The children had the same freedom to wander and explore as in the desert. However, Julia contracted a bad case of scarlet fever when she was nine. The household was quarantined and Julia herself quarantined from the family within the house, a sign hung on the front warning of the contagion. Her father took care of her and wore an old duster when he entered her room. Rheumatic fever followed, severe enough that Julia had to spend a year in bed, living in the home of a practical nurse. She wanted a bicycle during this period of recovery, and her father said she could have one when she got well. She thought her father was putting her off because she wouldn't get well.

When Julia was ready to return to her education, she had lost more than two years. A retired teacher was hired three mornings a week to help her catch up. They overshot the mark and in one year completed the material for the fifth, sixth, seventh, and eighth grades, so she was able to enter the ninth grade at the local junior high school ahead of her class. This was a social disaster—she later described it as "Kafka-like." Julia knew no one, and the social life at the school was already well set. "I made many stupid and embarrassing mistakes and ate lunch in a corner as quickly as I could so that no one would notice that I was alone."[14] Julia made a lifelong friend, Virginia Bell, who was her only friend until she left San Diego after college. During her ninth grade year Julia was given an I.Q. test and scored a 98, which followed her even into college, where she was called into the office to explain her high performance, unpredicted by the test.

In high school Julia received the standard mathematics education of the time, much the same material as today, except they had no calculus. Mathematics was her favorite subject and soon she was the only girl studying mathematics or physics. She was interested in some of the boys in her mathematics class, but they never paid attention to her unless they wanted help on their homework. Julia found her adolescence less painful than most people would have. She said that she did not think or worry much about what other people thought about her, attributing this to explicit admonitions by her mother and the unconcerned example set by her father, who was oblivious to public opinion.

Julia received numerous awards for mathematics and science when she graduated from high school. Her mother wondered aloud what would become of such an unusual girl, but her father told her not to worry. Julia "would marry a professor."[15] She was given a beautiful slide rule, which she named "Slippy." She then enrolled at San Diego State College, until 1935 a teachers college and before that a "normal school" training only elementary school teachers. During Julia's tenure, San Diego State didn't even offer a full secondary credential. Neither mathematics professor—there were only two, Mr. Livingston and Mr. Gleason—had a Ph.D. In a course titled Modern Geometry, non-euclidean geometry was not mentioned. Each semester two and only two upper-division mathematics classes were offered.

Julia's father, who had thought he could support his family on his money, conservatively invested, was hard hit by the Depression. At the beginning of Julia's sophomore year in college he ran out of money and killed himself. This sent his widow and three daughters into a modest apartment. They received some help from an aunt in St. Louis and the girls were able to continue in college. Though Julia liked her courses, she hardly knew what mathematics was. Then in a course on the history of mathematics, she read E. T. Bell's *Men of Mathematics*, just published. She particularly liked the number theory that Bell discussed and realized mathematics was created by human beings—mathematicians. Even in the present day there were active mathematicians!

Julia intended to get a junior high teaching credential, the highest offered at San Diego State. Constance graduated and discovered that she couldn't get a job with this limited credential. Their mother dug into what savings the family still had and sent Constance to the University of California at Berkeley, where she obtained a general secondary credential and quickly found a job at San Diego High School, teaching English and journalism. Now Julia "conceived an absolute passion" to go to Berkeley or UCLA.[16] Half of the mathematics faculty, Mr. Livingston, urged her to stay at San Diego State and be the only member of an honors program that was just beginning. The other half, Mr. Gleason, urged her to go to Berkeley. This she did, and was suddenly among people like herself, in Julia's words, an ugly duckling who found herself a swan.

Mathematics, Julia found, was alive at Berkeley in fall of 1939, with the university experiencing a dramatic expansion of mathematical quality under department chairman Griffith C. Evans. Only one detail still prevented Julia from becoming a mathematician. Her mother expected that she, like Constance, would get a secondary credential in one year—and a job. However, there were too many female teachers of mathematics and not enough males, so Julia was discouraged from following this path. Instead, she took five mathematics courses her first and only undergraduate year at Berkeley.

She took a course in number theory taught by a young faculty member, Raphael M. Robinson (1911–1995), then twenty-eight. He hailed from National City, just south of San Diego. It was a small class, only four students in the second semester, and Julia Bowman stood out. Soon Professor Robinson and Julia were taking long walks together, and on those walks Raphael opened up many areas of mathematics to her. He told her of Gödel's work and its connection to number theory, and she found this surprising and fascinating. She was very happy. She was elected to the honorary mathematics fraternity and participated in the active social life of the department. She received her A.B. degree and, a dutiful daughter, attempted to find a job in San Francisco. Her prospective employers inadvertently encouraged her to remain in school, as they were only interested in whether she could type. She applied for a teaching assistant job at Berkeley, with the intention of continuing her studies, but there was no job for her. However, Jerzy Neyman (1894–1981), newly arrived from Europe in 1938, heard of her problem and managed to find a small position for her that paid $35 a month, three more dollars than she told him she needed. She continued at Berkeley and received a master's degree in 1941.

Julia Robinson (courtesy of Constance Reid).

Julia's mother was nevertheless concerned about her daughter's future and urged her to get a "real job," which by definition would pay more than $35 a month. Julia had taken a civil service examination for junior statistician. A job opened up for a night clerk in Washington D.C. at the princessly salary of $100 a month. Her mother wanted her to take it, but Professor Robinson persuaded her to stay, and this time she got a teaching assistantship. At the end of the first semester the Japanese attacked Pearl Harbor, and a few weeks later Julia Bowman married Raphael Robinson.

Julia Robinson genuinely and strongly wanted to have a family. Since Berkeley had a nepotism rule that closed off the possibility of her obtaining a job there and Raphael had a good job, she did not think of having a professional life. In her autobiography she says that though she continued to audit classes she was "really more interested in shopping for furniture."[17] When she finally learned she was pregnant, she was very happy—but after only a few months she miscarried. Shortly thereafter she fell ill with viral pneumonia on a visit home, and when the doctor came, he asked her how long her heart had been bad. She had the mitral valve scarring that was a common aftermath of rheumatic fever. This condition usually progressed, and the doctor told her mother privately that Julia would probably be dead by forty. He only told Julia that she should not have another pregnancy. This came as a shock. She had known that she became winded when going up the stairs in the mathematics department, but on the other hand she had courted on long walks. "For a long time," she says, "I was deeply depressed by the fact that we could not have children."[18]

Raphael encouraged her to go back to mathematics. He had written a paper simplifying the definitions of primitive recursive functions and proposed that she attempt to simplify the definitions for general recursive functions. They spent the academic year 1946–47 at Princeton, and she worked hard and successfully on this problem, achieving results that eventually appeared in 1950. Though still unhappy, she returned to graduate school in fall of 1947 and became a graduate student of Alfred Tarski.

Born in Warsaw in 1901, Tarski had remained in Poland until 1939, when, abroad on a lecture tour, he was stranded in the U.S. by Hitler's invasion. He was well known as a mathematician—Julia said, "In my opinion, and that of many other people, he ranks with Gödel as a logician."[19] He didn't find his permanent position at Berkeley until 1942, a measure of how full of emigre mathematicians the United States was getting. Tarski was unusually broad, personally and mathematically. The luck of Julia's mathematical journey was holding when he washed up in California.

Tarski proposed a problem that didn't interest her, and she made little progress. However, one day at lunch with Raphael Robinson, Tarski talked about whether the whole numbers were definable in a formal system for the rational numbers. Raphael came home and mentioned it to Julia. This problem she found interesting. She didn't tell Tarski she would work on it, but she did and she solved it. It is typical of her best work. She works in the formal system but introduces a clever idea from number theory, in this case a quadratic equation discovered and understood using the Hasse principle from algebraic number theory. Davis described this to me as "an absolutely brilliant piece of work." This made her thesis and she obtained her Ph.D. in 1948. In the same year Tarski mentioned another problem to Raphael, who brought it home. The problem was to show that one could not define the powers of 2 (2, 4, 8, 16, 32, . . .) using only existential quantifiers (i.e., the symbol ∃—"there exists") and a diophantine equation. If this could be accomplished it would have a bearing on Hilbert's tenth problem (in fact, the eventual negative solution depended on showing the opposite).

Julia Robinson became entranced with the problem, but not from the direction Tarski had in mind. At first she didn't know she was working on Hilbert's problem. She quickly decided that she didn't see how to prove that the powers of 2 could not be defined in this way. Instead she decided to see if she could define the powers of 2. This expanded to work on defining other sets of whole numbers. She made rapid progress and on September 4, 1950, delivered a ten-minute paper to the International Congress of Mathematicians at Cambridge, Massachusetts. At the same conference Davis gave a ten-minute talk on his results on the hyperarithmetic hierarchy.[20] He had, however, spoken on his results on the tenth problem the previous winter at a meeting of the Association of Symbolic Logic. The two sized each other up. Robinson said: "Afterwards Martin told me that he didn't see how I could hope to prove anything in general from my examples—I said I did what I could and thought 'How are you going to get rid of that universal quantifier?' "[21] They were both on the right track. The phrase "I did what I could" has an appeal: It characterizes Robinson's whole career.

Davis's paper cast the problem in a modern form, introducing the term "diophantine sets." If this isn't literally the first usage, it brought the term to center stage. A set is diophantine if it can be defined using existential quantifiers and a diophantine equation. For a simple example, look at the set defined: x is in the set if and only if there exists a number a such that $x = a^2$. This equation has a solution for $x = 0, x = 1, x = 4, x = 9, x = 16$, etc. We can view x as a parameter and the equations generated by substituting different numbers for x as a parametric family of equations. A given parameter, that is, x, is included in the diophantine set if and only if the equation it generates is solvable. For a less obvious example, take the equation $x = a^2 + b^2 + c^2 + d^2$. The diophantine set defined here for the parameter x is, surprisingly, the non-negative whole numbers 0, 1, 2, 3, 4, 5, 6, 7, 8, 9, . . ., because a theorem of Lagrange's states that any non-negative whole number can be expressed as the sum of four squares. Investigating the possibilities of these diophantine sets, one quickly sees that many sets are diophantine.

In his 1953 paper, Davis made what Yuri Matiyasevich has called his "daring hypothesis": The diophantine sets are precisely the recursively enumerable sets—the sets that can be generated by recursive functions or, equivalently, by a Turing machine. Since mathematicians already knew there is a set generated by a recursive function that is not decidable, if Davis's hypothesis was true they would know that there is a diophantine set that is undecidable. Specifically, they would not be able to decide which members of the parametric family of equations that generates this undecidable set are solvable—much less the more general question, covering all diophantine equations. What Hilbert asked for in the tenth problem would be impossible to do. However, the idea that the diophantine equations could define all the recursively enumerable sets was just a "daring hypothesis." How did Davis get there? It was a guess, based on his feeling for the problem and the fact that he had already gotten rid of all the universal quantifiers (i.e., "for any") necessary to define recursive functions except for one—and it was "bounded." Davis says that he was motivated by the fact that each class had certain common features as sets—they were both closed under "and," "or," and existential quantification, and neither was closed under "negation." Recognition of similarities or analogies of form, almost the way one might recognize a face, is one of the ways mathematical intuition works.

In attempting to see what kinds of sets of whole numbers she could define using diophantine equations and existential quantifiers, Robinson was going after the same thing as Davis's diophantine sets. In "Existential Definability in Arithmetic," Robinson investigated a variety of functions that grow rapidly, as the powers of 2 do. She discovered that if she could define exponentiation, specifically the relation $x = y^z$, then she could define a lot of other functions—the powers of 2, the factorial function, binomial coefficients, etc. More surprisingly, she found she would be able to define the statement "p is a prime" with a diophantine equation. A truism of number theory had been that there was no formula for the prime numbers. This diophantine equation would be such a formula. She worked hard on exponentiation.

Her major result was to show that if one could define any function that grew sufficiently rapidly, one could take this function and, calling on facts from number theory, use it to define exponentiation itself. She conjectured that finding such a function was possible. (This would then show that exponentiation was diophantine.) This hypothesis became known as J. R. The paper embodies her careful voice. She is looking for patterns of counting numbers, just as she arranged the pebbles in patterns. She had been reading number theory. Classical results about Pell's equation play a crucial role. She refers to Edmund Landau's classic book on number theory.

Davis and Robinson continued to work on the problem through the 1950s. Davis was peripatetic. In ten years he was at the University of Illinois, the Institute for Advanced Study, UC Davis, Ohio State University, the Hartford Graduate Center (a branch of Rensselaer Polytechnic), and New York University. He became a consulting editor of the *Journal of Symbolic Logic* in 1953 and wrote a well-known book, *Computability and Unsolvability*, published in 1958.

Robinson was in the odd position (not so uncommon for women, at the time) of generally not being able to be employed as a mathematician, doubly strange since she eventually served a term as president of the American Mathematical Society. Her official mathematics jobs were at the edges of Raphael's career: a stint at the Rand Corporation, a think tank in Santa Monica, California, while Raphael was on sabbatical; a project at Stanford University on hydrodynamics. Hydrodynamics was a completely different subject for her, and she felt embarrassed and that she had failed. An article about Adlai Stevenson caught her interest. Stevenson was Raphael's first cousin. Senator Joseph McCarthy was going full tilt at his anti-Communist inquisition and the Berkeley department was embroiled in the controversy over loyalty oaths. Julia and Raphael were active in the local resistance to the loyalty oath. After Stevenson was nominated for president, Robinson plunged into politics. From 1951 to 1958 she published only one four-page note—except for her 1952 paper on the tenth problem, which reported work completed earlier.

Hilary Putnam came to the tenth problem from philosophy. He was born July 31, 1926, in Chicago, Illinois. His father was a writer and journalist who was part of what has been called the Chicago Renaissance. In the year of Hilary's birth, Samuel Putnam published a translation of Pietro Aretino and from then on spent much of his time translating from Italian, French, and Portuguese. In the same year the family moved to Paris, where expatriate members of the Chicago circle were in residence. In 1934 the Depression drove them back to the United States, to Philadelphia. Putnam became an active supporter of the Communist Party and wrote for the *Daily Worker*. He broke with the Communist Party in 1945. Samuel Putnam published translations of Pirandello, Ignazio Silone, and the Marquis de Sade, among others. His translations of Rabelais and of Cervantes's *Don Quixote* were widely successful and are still available. His real passion was Brazilian writing. He died in 1950 at sixty-two.

Hilary Putnam became a serious student of philosophy, non-Marxist variety, at the University of Pennsylvania, graduating in 1948. He spent the next year at Harvard and, in addition to nonmathematical philosophy courses, took classes from Wang (whom we met in the chapter on Gödel) and the logician and philosopher W. V. Quine (1908–2000). Quine, born in Akron, Ohio, had traveled to Prague in 1932–33 to study under Carnap and to Warsaw to study under Tarski. He was so at home with symbolic logic he wrote: "Sometimes I have escaped from some boring task into logic, thinking up a logical idea for no other purpose, deep down, than the escape itself."[22] Putnam also took mathematical courses in linear algebra and ideal theory at Harvard. Then he went to UCLA to study under Hans Reichenbach (1891–1953), a founder of the logical positivist movement—the group associated with the Vienna Circle that had played a role in Gödel's development. Reichenbach had fled his position as a professor in Berlin in 1933, first teaching in Istanbul and finally landing at UCLA in 1938. Putnam was going back to the source by studying in Los Angeles. Quine had passed through L.A. in the summer of 1949, working at Rand, although Putnam didn't know it. To make the confluence complete, with Tarski in Berkeley, Carnap himself spent his last years living in Santa Monica and teaching at UCLA (arriving in the mid-'50s). Reichenbach was particularly interested in the philosophy of science, and science in his lifetime had experienced more than one revolution that required philosophical rethinking.

Putnam's dissertation at UCLA was "The Meaning of the Concept of Probability in Application to Finite Sequences." He taught philosophy for a year at Northwestern University, 1952–53, and then moved to Princeton University, where he remained for the rest of the 1950s. Papers by Putnam in the area of mathematical logic begin appearing in 1956. He discussed logic and mathematics with Davis, who was at the Institute 1952–54 and for visits later, and with Kreisel, who was present 1955–57. Putnam was a member of the Princeton mathematics and philosophy departments; Smullyan wrote his thesis under Putnam's supervision. In his collected papers of 1975 Putnam chose to reprint a 1957 paper, "Three-Valued Logic." It displays the approach of a philosopher well. Post had investigated the mathematical structure of multi-valued logic as a class of formal system. In discussing three-valued logic, Putnam is interested in whether this nonstandard logic rather than some form of conventional logic is not, in fact, the correct logic to use when talking about quantum mechanics. (Reichenbach had worked on this earlier.) Perhaps logic is like geometry, and we choose the version we use only after looking at the world. Is there empirical evidence that would cause us to choose a three-valued logic over a modernized version of Aristotle's?

Putnam attended a five-week Institute for Logic at Cornell in the summer of 1957 that was attended by "almost all American logicians." Families attended, and Putnam and his family shared a house with Davis and his family. Davis writes, "He and I began collaborating, almost without thinking about it."[23] Davis had still not succeeded in "disciplining" himself away from Hilbert's tenth problem, so they talked about that. Putnam suggested that they try to use Gödel coding to make that one little bounded universal quantifier in the Davis normal form go away. Davis was "skeptical," but they worked on the problem and made some progress, which resulted in a joint paper, "Reductions of Hilbert's Tenth Problem," received December 17, 1957, and published in 1958. They decided to try to get back together the next summer and to get some funding. They were together in the summers of 1958, 1959, and 1960. Davis writes of these three summers:

> We had a wonderful time. We talked constantly about everything under the sun. Hilary gave me a quick course in classical European philosophy, and I gave him one in functional analysis. We talked about Freudian psychology, about the current political situation, about the foundations of quantum mechanics, but mainly we talked mathematics. It was during the summer of 1959 that we did our main work together on Hilbert's Tenth Problem.[24]

In a letter to Davis, Putnam says he remembers most the intensity of that summer. They would propose an attack, work on it, and if it didn't succeed try another attack. Putnam says that it was the only time in his life when he "regularly stayed up to four in the morning." They felt they were close to an important result. This collaboration was funded by the United States Air Force under Contract No. AF 49(638)-527. If the value to national defense of this work remains obscure, the work was successful.

Now for the Davis normal form and a sketch of Davis and Putnam's key insight:

The difficulty the bounded universal quantifier in the Davis normal form presents is that it asserts the existence of a string of separate solutions to the diophantine equation in question. This string has no specified length, and different equations generated by substituting different, specific values of the parameter will have strings of solutions of different lengths—they could be arbitrarily long. Capturing this in a single diophantine equation would seem to require an infinite number of variables. However, we have to pick a specific equation and so have to commit to a specific number of variables. Gödel coding allows us to assign a code number to any string of symbols, hence to any string of solutions. Equally important, Gödel code numbers behave similarly to the numbers they code when substituted into a polynomial. For example, if we find ten solutions to an equation, the code number assigned them also "solves" the equation in a well-defined sense and vice versa. The bounded universal quantifier in this regime becomes in essence: There exists a Gödel code number that solves a specific diophantine equation. One problem with this strategy is that Gödel code numbers grow very rapidly. Therefore Davis and Putnam analyzed what are called exponential diophantine equations, where variables can occur as exponents. If Robinson's hypothesis that exponential functions are diophantine could be proven true, exponential diophantine equations could be rewritten as simple diophantine equations. Here the two strategies link.

I won't try to suggest the mechanics of how Davis and Putnam did the coding. The original result was conditional. In order to get the coding to work they had to assume that there existed arbitrarily long arithmetical sequences of prime numbers; 3, 5, 7 is such a sequence three long; 5, 11, 17, 23 is four primes long. As the strings get longer they are harder to find, and even as of this writing the general result has not been proved. Davis and Putnam published an abstract in 1959 and sent a copy of their results to Robinson. She replied quickly saying: "I am very pleased, and surprised, and impressed with your results. . . . Quite frankly, I did not think your methods could be pushed further. . . ."[25] She showed how they could simplify the proof and get rid of a gamma function and the analysis. More importantly, she said she saw how to eliminate the assumption of the existence of the sequences of primes, which made the result no longer conditional. Her method was quite complicated itself, but a little later she saw how to make what Davis described as a "drastic simplification of the proof."[26] Davis and Putnam withdrew their original paper and the three together wrote the paper that finally appeared in 1961. This paper is very efficient, eleven pages long, and the main result—that all recursive sets are exponential diophantine—is only four pages! Davis and Putnam were most at home with the logical technique, and Robinson was clever with the constructive number theory. The result has a synergy that parallels the dual nature of the problem. Once this was accomplished, the only thing necessary for a solution of the tenth problem was to prove J. R., which implies that exponentiation is diophantine. That proved difficult.

Putnam was appointed professor of philosophy of science at MIT in 1961, though half his salary was paid by the Research Laboratory of Electronics and for this he taught courses with mathematics department numbers. He moved to Harvard as a professor of philosophy in 1965 and spent an increasing part of his time working as a mathematician. At MIT and Harvard he supervised dissertations by mathematical logicians Ward Henson, Herb Enderton, and George Boolos, among others. Putnam was radicalized by the Vietnam War, joined the Progressive Labor Party, and was faculty sponsor for the Harvard chapter of Students for a Democratic Society. He told a *U.S. News and World Report* correspondent that during this period he was "almost unable to function as a philosopher."[27] He eventually became disillusioned with radical politics. In 1975, when he published two volumes of his collected philosophy papers, *Mathematics, Matter and Method* and

Hilary Putnam (*courtesy of* Hilary Putnam).

Mind, Language and Reality, it became easier to recognize the range, coherence, and clarity of his philosophical work. He was appointed the Walter Beverly Pearson Professor of Modern Mathematics and Mathematical Logic at Harvard in 1976. But now he had largely stopped publishing in mathematics. "Pseudo-questions"—questions of value, of ethics, of what comprises the good life, etc.—were beginning to seem to him to have meaning that could be analyzed by a philosopher. Two decades of increasingly nonmathematical, non-analytic philosophy have brought books including *Reason, Truth and History*, *The Many Faces of Realism*, and *Realism With a Human Face.*

Davis was professor at Yeshiva University from 1960–65. Then he went back to NYU, where he remained, becoming emeritus in 1996. He continued to work on the tenth problem but did not finish solving it. Throughout the 1960s he gave lectures on the problem and was usually asked how he thought it would turn out. "I had my reply ready: 'I think that Julia Robinson's hypothesis is true, and it will be proved by a clever young Russian.'"[28]

Robinson's heart broke down as had been predicted, but a newly developed surgery "to clear out the mitral valve" was so successful that she took up bicycle riding. She also continued to be drawn back to the uncompleted tenth problem and worked hard to prove her hypothesis. Every year at her birthday party she would wish that the tenth problem would be solved—by anyone. "I felt that I couldn't bear to die without knowing the answer."[29] For a period during the 1960s she despaired of proving her hypothesis and even worked to prove it false, but in 1969 she published a paper that made some progress. In fact, she was very close. On

February 15, 1970, she received a call from Davis to tell her that he had just heard that a young Russian had, in fact, proved her conjecture and the problem was solved.

Yuri Matiyasevich was born the year Post and Markov proved Thue's word problem unsolvable—on March 2, 1947, in what was then Leningrad, now St. Petersburg. The war was over, but not the repression by Stalin's secret police. Yuri's father, Vladimir Mikhailovich Matiyasevich, was a construction engineer who designed railroad bridges from a desk in Leningrad. Yuri's grandfather, Mikhail Stepanovich Matiyasevich, had been a member of the gentry and a professional officer in the czar's army. He joined the Bolsheviks after the Revolution and rose to a high position in the Red Army, but eventually was declared a "people's enemy." Yuri's father always had to mention this on any vita, though he was a member of the Communist Party, and Matiyasevich says this added to the stress of his life. The family did not talk about it, but Yuri's grandfather probably died in a camp just before the Second World War.

Yuri's mother Galina (Korotchenko) served during the war as a typist in the army. Her dream had been to become a doctor, but she was sent to study to become an agronomist. She didn't finish this training because she didn't like it. Yuri was a late, only child. His mother was thirty-six when he was born and his father forty-five. His father, fearing that an early death would leave the child unprovided for, stopped smoking the day Yuri was born. He insisted that Yuri's mother stay home to raise the child.

Matiyasevich started school in 1954 at the usual age at the time, seven, and school was important to him from the beginning. It was also easy, except for music. During his early years in school he had to go to the hospital twice for surgery. He had learned to add large numbers. The hospital teacher taught him how to subtract large numbers. However, in this strange setting he suspected her, and checked every answer by adding.

In the fifth year, mathematics was taught by a mathematics teacher, and soon Matiyasevich was excused from normal mathematics classwork as long as he did the homework and passed the tests. He read books on radio for amateurs and was perplexed about how a heterodyne receiver worked, spending hours drawing graphs of sine waves with different frequencies and then adding them to make a third graph.

In January of 1959 a friend of Yuri's received a kit to build a superheterodyne radio receiver with four vacuum tubes—the broadcast frequency is first converted to an intermediate frequency before being amplified and detected. They spent hours carefully assembling the kit, but it never worked. Yuri received a second kit on his birthday in March and this time, a veteran solderer and radio assembler, a week after his birthday he was listening to the radio. He says that this made his father very happy, because his father was purely a theoretician and had no talent for doing anything with his hands. This happiness didn't last. A few days after listening to the

Julia Robinson and Yuri Matiyasevich at the Fifth Congress on Logic, Methodology, and Philosophy of Science in London, Ontario, Canada, 1975 (photograph by Louise Guy).

radio, Yuri's father died without warning. The family became much poorer, as Yuri's mother had no job and no profession beyond typist. However, the next year in school marked the beginning of mathematics competitions. Yuri's success in these became a focus for him and a ticket to new opportunities.

Soviet Russia had mathematical Olympiads that started in the sixth year of school, for Yuri 1959–60. From the beginning he did well. The seventh year brought the *kruzhoks*, or extra evening classes. He was invited to join the best class in Leningrad, held at the Young Pioneer Palace, but chose another that had a more convenient schedule. No one told him that all the *kruzhoks* weren't the same. His first teacher was not experienced. The next year he went to the good class. The *kruzhoks* offered new and richer material and interesting problems, but they were also social events. Sometimes they met in apartments, and there were trips to the forest in the summer.

The Soviet Union was undergoing a tentative liberalization under Khrushchev. Opposing tensions were dealt with in an ad hoc way. There was the belief that "everyone should be in equal conditions,"[30] as Matiyasevich told me, and yet the achievements of individuals under communism reflected well on the communist system. Scientists and mathematicians were valued, just as athletes were. The mathematics Olympiads were dedicated to discovering which young comrades were the most unequal in mathematics. These tensions were visible in the schools. Elite schools were formed, often at the instigation of prominent mathematicians—Yuri attended School 239 in Leningrad—but officially they existed to provide "worker professions." School 239 supposedly trained operators of mainframe computers.

(A Web site lists alumni of this school, many now living the U.S. and Israel.) An extra year of school for everyone before work or university was added, but two days a week were spent on worker education, so that a mathematician experienced a net loss of time on mathematics. The *kruzhoks* were one way around this mandate; they also expressed a classically Russian desire for collaboration. In 1962 a summer boarding school outside Moscow was organized by A. N. Kolmogorov. Also teaching were P. S. Aleksandrov (1896–1982) and V. I. Arnold, among others. Matiyasevich attended. In keeping with these professors' love of vigorous physical culture, the students were encouraged to swim a wide river and go into a forest. Sometimes they returned with mushrooms. Those not up to the swim could row across.

In fall of 1963, Boarding School 18 was opened in Moscow by Kolmogorov et al. for able students from outside Moscow. Matiyasevich had to choose between School 239 and Boarding School 18. An uncle who lived in Moscow offered to pay for the extra costs, so Matiyasevich chose Moscow. These schools were also referred to as *internats*. Leaving Leningrad and his mother was hard on Matiyasevich, and the scientific superiority of the Moscow school was partly lost on him. In the summer he had a good class with Arnold, but Kolmogorov taught a geometry course based on spatial movements rather than lines and points and this was too abstract for him at age sixteen. Matiyasevich was also, by this time, on something of an Olympiad treadmill. In his words:

> In spring 1964 I was rather tired, participating every Sunday in some competition. One of them was the selection of the *internat* team for the all-union Olympiad. I easily passed the selection. During the Olympiad itself I used half of the given time and left, being sure that again I had solved all the problems. I remember that I decided, having saved time at the Olympiad, to walk from the building of the university where the competition took place to the *internat*, situated in the suburbs of Moscow, about two hours walk. I felt that I needed to give myself a bit of rest. I was later disappointed to discover a mistake in a solution of one of the problems.[31]

That same year the International Olympiad was held in Moscow and Matiyasevich was chosen for the Soviet team, despite the fact that he was only a tenth-year student. He was not happy about his performance, but still won a diploma of the first degree. Members of the team were granted admission to the university of their choice. Matiyasevich tried to get permission to enroll at Moscow State University—the most prestigious university—but couldn't make his way through bureaucratic resistance. He still had to get his *attestat* degree from school. Fed up, he boarded a train for Leningrad. There it was worked out that he would take the exams for the *attestat* at School 239 while studying at the university. During this first year, 1964–65, he was busy with exams and, though he attended a few seminars on logic, he and all other first-year students were forbidden to study logic.

He stayed at Leningrad, and at the beginning of his second year, fall 1965, Matiyasevich was introduced to Post's canonical systems and his career as a mathematician proper began. He immediately achieved an elegant result on a difficult problem the professor proposed. This led him to meet Maslov, the local expert on Post canonical systems. The logical community in the Soviet Union had developed along different paths than in the West. The heavily philosophical tradition of Frege, Carnap, Russell, Whitehead, Gödel, and Tarski was at odds with Communist Party doctrine. They had their own logic and it wasn't symbolic. Therefore, after Kolmogorov's early bold work, Russian mathematicians had not been quick to pick up and pursue the work of the 1930s. Sometimes, mathematical names were changed. For example, eventually the Russians had their own version of recursive functions and effective computability, which they called the theory of algorithms. In the United States and England the emergence of electronic computers interacted with symbolic logic, while in the Soviet Union this field lagged. Post's resolutely unphilosophical version of symbolic logic—it is about rules for generating strings of symbols—was mathematical in its perspective and therefore unsubversive, so there were Soviets at work in this field.

Maslov made a number of suggestions for research, which Matiyasevich quickly resolved. In late 1965 Maslov suggested a more difficult question about details of the unsolvability of Thue systems. Matiyasevich solved this problem. Publication of this result was offered, but it had to be entirely rigorous in the style of Markov and Post (when he wasn't compromising). Matiyasevich was given an Underwood typewriter of manufacture predating the Revolution (by the second wife of his grandfather). He told me that it was too expensive for him to have bought himself. He spent a considerable part of the next year typing—only five corrections a page were allowed, and the paper was 100 pages long. He missed lectures in school—particularly, he says, in complex analysis—so his early success had the odd effect of slightly narrowing his education. He was invited to give a talk at the 1966 International Congress of Mathematicians in Moscow, a major honor, and was particularly impressed to meet Kleene. (Later, in 1979, he went with Kleene and others on a pilgrimage to the birthplace of al-Khowarazmi near Baghdad, whose name gave us the word "algorithm.")

Toward the end of 1965, Maslov also suggested Hilbert's tenth problem. He said that "some Americans" had done some work on this problem but that their approach was probably wrong—it should have worked already, if it really was a good idea.[32] Matiyasevich didn't read their work, but like Davis and Robinson he was enchanted—he was drawn to the problem again and again. Once, as an undergraduate, he thought he had solved it and even began a seminar presenting his solution. He soon discovered his error, but became known, with an edge of humor, as the undergraduate who worked on Hilbert's tenth problem. As the years of his undergraduate education passed, like Davis, he too began to think he needed to discipline himself away from this trap of a problem. He did read the work of the Americans and recognized its possible importance. If he could find a diophantine equation whose solutions grew rapidly enough—Julia Robinson's hypothesis—exponentiation could be proved to be diophantine, and since all recursively enumerable sets are diophantine there would exist at least one diophantine set that was not decidable. Hilbert's problem would be solved in one shot. However, he was not immediately successful in this. His undergraduate years were ending. He hadn't done anything better than the early work he had delivered at the International Congress. In his words:

> I was spending almost all my free time trying to find a Diophantine relation of exponential growth. There was nothing wrong when a sophomore tried to tackle a famous problem, but it looked ridiculous when I continued my attempts for years in vain. One professor began to laugh at me. Each time we met he would ask: "Have you proved the unsolvability of Hilbert's tenth problem? Not yet? But then you will not be able to graduate from the university!"[33]

When a colleague told him to rush to the library to read a new paper by Robinson, in fall 1969, he stayed away. However, because he was considered an expert, he was sent the paper to review and so was forced to read it. He delivered a seminar on it on December 11, 1969. Robinson did have a new idea; Matiyasevich was caught again.

Up to this point we have constructed a rather complex narrative, starting in the 1930s and building toward Matiyasevich's closing of the last gap remaining in a proof that the decision process Hilbert asked for does not exist. That gap was a proof of Julia Robinson's hypothesis. We must point out that when Matiyasevich came upon this problem, it was far from obvious that the tenth problem was near solution or that the solution lay in the direction of Robinson's hypothesis. Matiyasevich's adviser had told him to ignore the Americans' work. In this period even Robinson despaired of proving her hypothesis. It was necessary for Matiyasevich to recognize its significance. There were other possible paths. For example, in 1968 Davis had shown that if a certain diophantine equation had no nontrivial solutions then the tenth would be unsolvable. This equation did turn out to have a large nontrivial solution, but it is still unclear today whether this approach can provide an answer for Hilbert's tenth.

Here is a sketch of Robinson's new idea, simplified (though still technical):

A specific relation of exponential growth is the list of statements: 1 is associated with 2^1, 2 is associated with 2^2, 3 is associated with 2^3 . . . n is associated with 2^n, . . ., etc. The key feature is that 2^n is associated uniquely with n. If we could find a polynomial (in a, b, and x's) such that it had a solution if and only if $b = 2^a$, this would be accomplished. But this is devilishly difficult to do—nothing worked for years. In her 1969 paper, Robinson hit on a version of Pell's equation, $x^2 - (c^2 - 1)y^2 = 1$, as an opening. The solutions to this equation grow roughly like $(2c)^n$. However, 2^n is in no way associated directly with n—hence there is no relation. Robinson noticed a peculiar fact. If we divide each of the y solutions by c and take the remainder, we get the sequence 1, 2, 3, 4, . . . until we get to c, at which point it repeats again and again. On the other hand, she noticed that if we take the same y solutions and combine them in a different way with the corresponding x solutions to get another number—then divide by yet another number and take the remainder, we get a set that starts growing roughly $2^1, 2^2, 2^3, 2^4$, etc., until again it starts repeating. However, these two sequences are generally of different lengths and so, though they start out nicely in sync, they don't stay synchronized. If Robinson had been able to find a diophantine condition she could put on the two numbers she divided with, so that the two sequences matched up as they repeated, then the sequences yoked together would associate n and 2^n . These sequences would only make the association up to when they started repeating, so there is more to be done—but this is where the essential difficulty lies. This exact approach still has not been made to work, though.

Matiyasevich spent a feverish December 1969 turning these new ideas over. He attended a New Year's Eve party and was so distracted he walked home wearing his uncle's coat. On the morning of January 3, 1970, he thought he found a solution, but by the end of the day he found an error. The next morning he was able to close it. He was now in possession of a solution in the negative of Hilbert's tenth problem. However, he was afraid there was still an error. After all, he had once gone so far as to start giving a seminar on a solution. He wrote out a full proof and asked both Maslov and Vladimir Lifshits to check it but say nothing until they talked to him again. Then Matiyasevich and his soon-to-be wife, a student in Moscow, went to ski camp (they married in June). For two weeks at the ski camp he worked on refining his paper. He returned to Leningrad to find that the verdict was that he had solved Hilbert's tenth problem and it was no longer a secret. Both D. K. Faddeev and Markov, famous for finding mistakes, had also checked the proof and passed it. Matiyasevich gave his first public talk on the result on January 29, 1970. News of the result and methods moved around the country. Grigorii Tseitin took a copy of the manuscript and, with Matiyasevich's permission, presented it at a conference in Novosibirisk. This is probably why Matiyasevich's original paper in *Doklady*, in its English translation, identifies him as belonging to the Siberian Branch of the Steklov Institute. An American mathematician, John McCarthy, attended this talk and it was through him that information about the result made its way to Davis and Robinson. Using McCarthy's notes, both Davis and Robinson were able to construct proofs of their own.

The details of Matiyasevich's result are complex, the parts fitted together flawlessly. He had earlier tried to approach the tenth through the Fibonacci numbers, 0, 1, 1, 2, 3, 5, 8, 13, 21, 34, . . ., each number the sum of the preceding two. Taking every other Fibonacci number, 1, 3, 8, 21, . . . yields a sequence that grows roughly exponentially. Matiyasevich had already rediscovered the equation $x^2 - xy - y^2 = +$

1 or –1, which generates the Fibonacci numbers. He was familiar with this sequence's properties, which are similar to the properties of the solutions to Pell's equation. But as in Robinson's example of Pell's equation, the solutions just exist without labels. Matiyasevich's breakthrough insight was that one could use an obscure fact from the theory of Fibonacci numbers to accomplish Robinson's new strategy of synchronizing two repeating sequences so that n and 2^n would match up. All of these people, Robinson, Davis, etc., had been reading number theory books full of obscure facts. Matiyasevich had just read the third edition of Nikolai Vorob'ev's "popular" book *Fibonacci Numbers*. In that book, published in 1969, there was a new result about the divisibility of Fibonacci numbers. Matiyasevich used it to prove: If the mth Fibonacci number is divisible by the square of the nth then the nth Fibonacci number divides m. This gave him the ability to put a condition on m while talking only about the Fibonacci numbers. That was enough to allow him to synchronize the sequences, using an intricate argument, and exhibit a diophantine relation of exponential growth. Robinson had read the book on Fibonacci numbers, but not the *third* edition! (Of course, she may or may not have recognized the opening, had she seen it.) Matiyasevich solved the tenth.

Unfortunately, Matiyasevich found that an even younger Soviet mathematician, Gregory Chudnovsky, then seventeen and a first-year student at Kiev University, claimed an independent solution. Because Matiyasevich's results were passed around freely before publication, there will probably never be any certainty about whether Chudnovsky also solved Hilbert's tenth. There is no doubt Matiyasevich

Yuri Matiyasevich (photo provided by Heike Photien, University of Stuttgart).

solved it on his own. I have seen no hard evidence that supports Chudnovsky. He has not produced any over the years, and has moved on in his focus. He did publish a solution after Matiyasevich published his, but his methods were very similar. Chudnovsky later published a working paper that seemed to assert that he had solved the problem by proving Davis's equation had a finite number of solutions. No proof of this by anyone has ever appeared. The fact that Chudnovsky is Jewish and ended up emigrating from Russia after harassment by the KGB means that the issue is also muddied by anti-Semitism. An article appeared in the *New Yorker* (1992) about Chudnovsky and his brother building a "supercomputer" in a very hot New York apartment and calculating "mountains of pi."[34]

Matiyasevich received his *kandidat* degree—equivalent of a Ph.D.—in 1970 for his early work on Post systems. The Soviets had a more-difficult-to-obtain doctoral degree, and Matiyasevich received this for his work on Hilbert's tenth. He has worked at the Steklov Institute in St. Petersburg since then. He tends to pick hard, old problems and work on them for years, whether or not he is successful. Like Cohen, for example, he has worked on the Riemann hypothesis. The economy fell apart in Russia after 1990, but Matiyasevich was able to get supplemental money from a grant made by the financier George Soros. In 1997 Matiyasevich was made a corresponding member of the Russian Academy. He is attached to Russia and St. Petersburg. He makes trips to the West and lectures, makes some money, and returns home.

Robinson was elected to the National Academy of Sciences in 1976. In the same year UC Berkeley found a way to give her the title of professor. She won a MacArthur fellowship in 1983. Julia Robinson died on July 30, 1985, in Oakland, California of leukemia.

The solution of the tenth problem was complex and came from many sources. In Kaplansky's words, "[It] is a remarkable achievement of twentieth century mathematics and deserves the close attention of anyone interested in the fundamental nature of the mathematical enterprise."[35] First there was the problem, both philosophical and practical, of figuring out what it meant for something to be "effectively calculable." This was a great moment of agreement in philosophy—just about the only one I am aware of—and yielded a theoretical basis for all modern computers. Second, the problem of diophantine equations itself turned out to be surprisingly broad. The use of the word "all" in the phrase "all diophantine equations" turned out to be just as powerful as it is in other areas in the foundations of mathematics.

In order to solve Hilbert's tenth problem in the negative, mathematicians learned to code much of mathematics into diophantine equations whose solutions, if they exist, carry the content of the apparently more advanced mathematics. The key direct result obtained along the way is that any set that a computer can be programmed to generate can be generated by a specific diophantine equation. Davis, Matiyasevich, and Robinson collaborated on a large paper for the American Mathematical Society's 1974 conference on Hilbert's problems and gave some examples of questions that are answered affirmatively if and only if a specific diophantine

Martin Davis, Julia Robinson, Yuri Matiyasevich in Calgary, 1982 (photograph by Louise Guy).

equation (which we can find) has a solution. Many of the major questions in mathematics, including Fermat's last theorem, Goldbach's conjecture (undecided), the continuum hypothesis, and the Riemann hypothesis (undecided), are each decided if and only if the diophantine equation that codes them has a solution. This is simply astounding and makes a lifelong obsession with this problem quite understandable. If Hilbert had been right and there was an effective method of deciding whether a given diophantine equation has a solution, then each of these conjectures could have been proved or disproved simply by deciding whether their corresponding equation has a solution. Much of the apparent complexity of mathematics would then disappear into this one decision process for diophantine equations.

Where we found that the liar's paradox, in the form of "this statement is not provable," can be coded into arithmetic and thereby thwarts us in proving mathematics consistent, this other coding, into diophantine equations, in effect ends up stating, "There is a riddle and there will always be a riddle." There is at least one unsolvable diophantine equation (probably many) and no decision process that works on these equations. This is more important for mathematics than that there be no *ignorabimus*. Hilbert did not foresee these consequences, but he certainly conceived a provocative problem, which was his aim. In the end, unsolvability is richer for mathematics than the decision process Hilbert asked for would have been.

The Foundations of Specific Areas

« 3, 4, 5, 6 »

In the Original
The Third Problem

⊰ Dehn ⊱

Geometry was the first branch of mathematics to receive mature expression, and Hilbert had just been immersed in writing *Foundations of Geometry* when he composed his list of problems. The third problem, solved almost immediately by Max Dehn, follows the more general foundation problems and furthers Hilbert's program of investigating the completeness and independence of the axioms of geometry.

No solver of a Hilbert problem has received an encomium the likes of the following, from André Weil's 1992 autobiography, *The Apprenticeship of a Mathematician*:

> I have met two men in my life who make me think of Socrates: Max Dehn and Brice Parain [a French philosopher about nine years older than Weil]. Both of them—like Socrates as we picture him from the accounts of his disciples—possessed a radiance which makes one naturally bow down before their memory: a quality, both intellectual and moral, that is perhaps best conveyed by the word "wisdom"; for holiness is another thing altogether. In comparison with the wise man, the saint is perhaps just a specialist—a specialist in holiness; whereas the wise man has no specialty. This is not to say, far from it, that Dehn was not a mathematician of great talent; he left behind a body of work of very high quality. But for such a man, truth is all one, and mathematics is but one of the mirrors in which it is reflected—perhaps more purely than it is elsewhere. Dehn's all-embracing mind held a profound knowledge of Greek philosophy and mathematics.[1]

Weil was the older brother of the French philosopher and mystic Simone Weil. A sentence more typical of Weil's style, about Dehn and other colleagues, precedes this passage: "It is not without feelings of gratitude and affection that I speak of this group. . . ." He is referring to the participants in the Frankfurt mathematics history seminar, formed to explore the historical foundations of mathematics. Dehn was the seminar's moving spirit.

Dehn was born in Hamburg on November 13, 1878, the fourth of eight children of Maximilian Moses Dehn, a successful doctor. The eldest son became a prominent lawyer in Hamburg, and a sister played violin in the Hamburg Opera orchestra. Two brothers went into business, one moving to Japan and then the U.S., the other to Ecuador. Max Dehn's son Helmut said to me that the extended family "was one of the secularized Jewish families of the second half of the nineteenth century who lived by principles of conduct that some people would call 'good Christian.' "[2] Dehn's younger daughter Eva told me that she had not really realized she was Jewish until Hitler rose to power—they just didn't think in those terms. She said she thought that Max officially joined the Lutheran Church at age nineteen, after his father died. The family had come to Hamburg from Copenhagen in the early part of the nineteenth century. Earlier the family name was Tiktin, after the Polish town whence they had moved to Denmark.

Dehn attended gymnasium in Hamburg and started his university education in Freiburg. He then moved to Göttingen, receiving his doctorate in 1900. Dehn has been described as an "intuitive geometer."[3] He was in Göttingen at exactly the right time for his interests to mesh with Hilbert's. Dehn's first paper, which was never published (though an account of it appeared in 1977), is typical of the kind of work Hilbert was then involved with and that Dehn would remain involved with all his life. It seems obvious that any continuous, closed curve that doesn't cross itself, lying in a plane, like a circle or polygon, has an inside and an outside, but this is not easy to prove. Dehn proved this result for polygons with great economy (2,000 words), using minimal assumptions.[4] The result, without proof, appears in *Foundations of Geometry* (section four of the first chapter, theorem seven). In his paper Dehn recaptured some of the simplicity of our intuition of inside and outside. He also proved that a simple closed polyhedron in three dimensions has a single inside and a single outside.

Dehn was twenty-two when he solved Hilbert's third problem before the full list was out of the print shop[5]—it was one of the problems Hilbert did not have time to deliver in Paris. Dehn's solution was partially anticipated by J. Bricard in 1896, and Dehn refers to this incomplete result in his paper. Tetrahedra with bases of equal areas and equal altitudes have equal volumes. The traditional proofs of this involved a limiting process that is not elementary, making first steps toward calculus, derived from what is called the archimedean axiom: the assertion that if I start walking in equal steps, however tiny, I will eventually reach my destination. These proofs involve a series of steadily more exact approximations. The error disappears as the limit is taken. Hilbert wanted to know if the archimedean axiom was necessary to prove the equal volume result. He suspected it was, so in the third problem he asked for "two tetrahedra of equal bases and equal altitudes which can in no way be split up into congruent tetrahedra . . . ," cut up, as it were, with a perfect knife, making straight planar cuts, into two identical piles of pieces. If the tetrahedra were not, in effect, composed of a finite number of identical pieces, you would have to invoke the archimedean axiom. At bottom Hilbert was asking which axioms are necessary to prove the theorems in geometry about volume.

Dehn proved that a regular tetrahedron (all four sides are equilateral triangles) and a cube with the same volume cannot be cut up into identical piles of pieces. When combined with results that were already known, this led to the desired result. Hilbert's third problem was solved.

Here is an example of two tetrahedra that have bases of equal area and equal altitudes that can't be cut into equal piles:

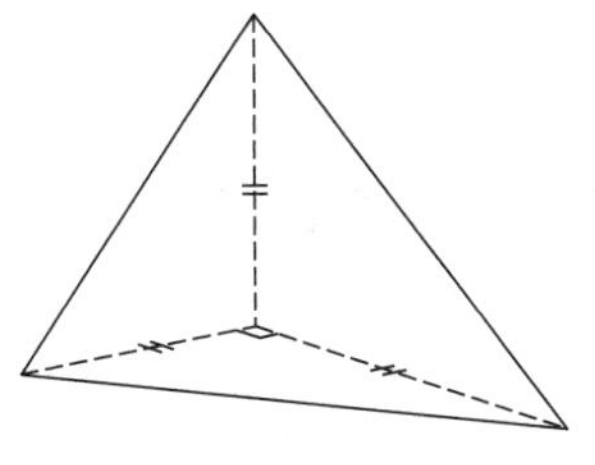
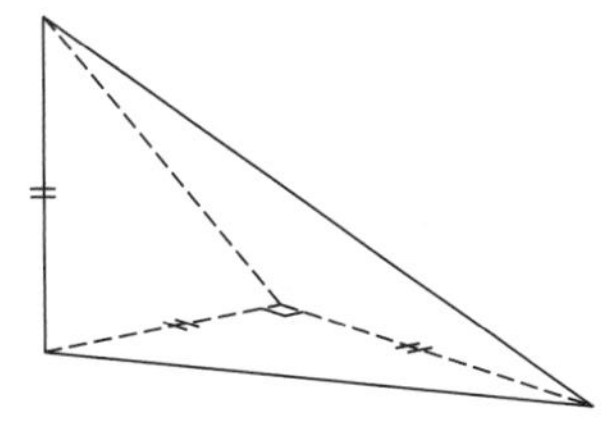

In both, the bases are isosceles right triangles. They have the same altitudes, which are the same lengths as the equal sides of their bases. The only difference is that the vertical edge rises in the first tetrahedron from the right angle in the base, while in the second it rises from one of the 45-degree angles. The first can be obtained by slicing off the corner of a cube. The second is called Hill's tetrahedron.

The reason the archimedean axiom is necessary when using the chopping and comparing strategy in area and volume proofs can be seen with a simple example using parallelograms. We can show that parallelograms with equal bases and equal heights have equal areas, as follows:

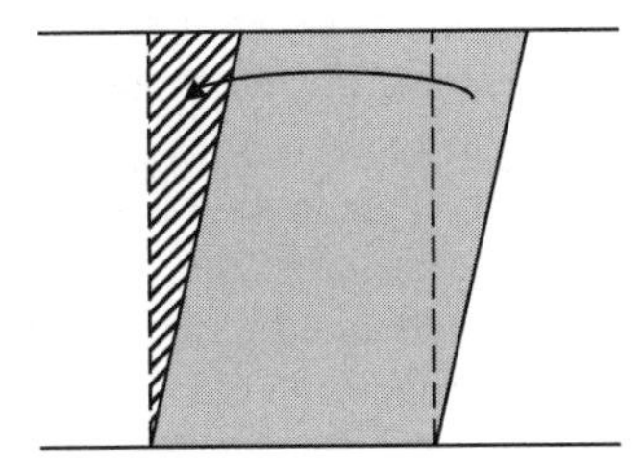

We chop off a triangle in the manner indicated and move it to the other end. This shows that the parallelogram has the same area as the rectangle with the same base and height that we have created. The concept is easily conveyed by the picture. (No wonder Hilbert included diagrams in his book, disagreeing with the axiomatic purists like Pasch and Peano.) However, look at the following:

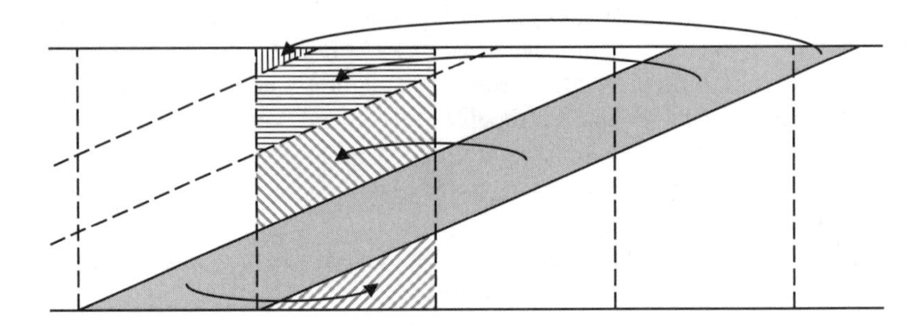

The more the parallelogram tilts, the more pieces have to be moved. We might have thought our first picture covered the situation. But we drew in the dotted vertical line in order to make our argument. Here we have to draw more dotted lines. We have to prove we can draw them. The picture also contains information about where everything is. This example illustrates why the picture can be no more than an aid in geometry. Extending the above argument to prove the general theorem that all parallelograms with equal bases and equal altitudes have equal areas requires the archimedean axiom: If I start this chopping and comparing process, I will eventually come to the end of any parallelogram. Hilbert investigated geometries in which the archimedean axiom is not true.

In *Hilbert's Third Problem*, Vladimir G. Boltianskii writes, "Dehn's own exposition was hard to understand"[6] and, "In 1903 Kagan published a paper, in which Dehn's argument was considerably refined, and presented in a more systematic and readable fashion. This was, so to speak, the rebirth of Dehn's paper." In the 1950s the Swiss geometer Hugo Hadwiger gave a much-simplified version. Boltianskii himself streamlined Hadwiger's proof, and his book provides an accessible account of the problem.

The key tool is the Dehn invariant of a polyhedron, a number built from the lengths of edges where faces meet and the angles between the faces. (The angle is measured in a plane perpendicular to each of the faces and is called the dihedral angle.) The function used to build the Dehn invariant can be explicitly constructed to have certain properties. Each edge, with its angle and length, makes its own contribution. If two angles associated with the same length add up to π or 180°, the total contribution to the Dehn invariant is zero. Likewise, if the same angle is associated with two different edges the contribution to the Dehn invariant is the same as one edge with the combined length and the same angle. Taking a single slice at a face of a tetrahedron will produce a length and two angles that add up to 180°, making a net change to the Dehn invariant of 0.

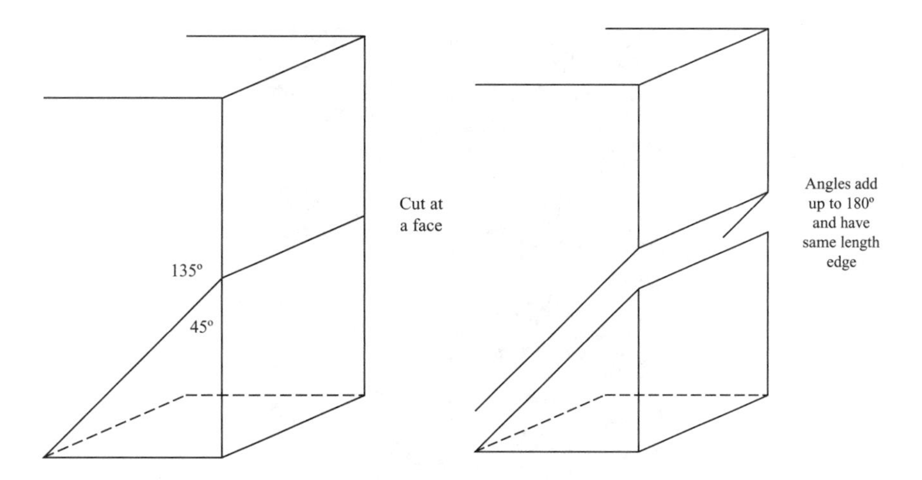

A case-by-case geometric argument that analyzes all possible types of cuts, taking into account how cuts can meet in the interior of the solid, shows that no amount of cutting will change the Dehn invariant of a solid. (An induction on the number of cuts would seem clearest here.) Therefore, tetrahedra would be equidecomposible only if their Dehn invariants are equal.

It is easy to calculate the Dehn invariant for a cube. Every angle is 90° and each edge is exactly the same length. Since there are an even number of them, they can be paired up. The same length is associated with a combined angle of 180° and therefore the contribution of each pair to the Dehn invariant is zero. The two polyhedra we drew earlier have different Dehn invariants and hence can't be cut up into identical pieces, though their volumes are the same.

Dehn was the first to solve a Hilbert problem. His career following this success took a regular course. Assistant at a technical school in Karlsruhe—habilitation—privatdozent in Münster, 1901–11—extraordinarius in Kiel, 1911–1913—ordinarius, or full professor, in 1913 at a technical university (*Technische Hochschule*) in Breslau. While in Kiel, Dehn supervised the dissertation of Jakob Nielsen (1890–1959). Nielsen, at fourteen, had broken with the aunt who was raising him and continued his education on his own, supporting himself by tutoring. He was expelled from school for forming a pupils' club, but was able to enroll at the university in Kiel.

Dehn returned from Kiel to Hamburg for a weekend visit and met an art student from Berlin, Toni Landau, who was boarding in a family member's home. Her father was a newspaper editor and theater critic who also wrote books, including accounts of cruises up the coast of Norway and to the West Indies. Toni's mother translated some of John Galsworthy's novels and Theodore Roosevelt's writings. Max had taken up sailing in Kiel. He asked Toni to come sailing. The courtship was brief. Married at nineteen, Toni was fourteen years younger than Max. Helmut Max was born in 1914, followed by Maria in 1915 and Eva in 1919.

Max Dehn (reproduced by permission of Eva Dehn).

Dehn served in the German army from 1915 to 1918, first on the eastern front as a surveyor and then at western headquarters, where he worked as a coder/decoder. At headquarters it chanced that he was standing at the side of the road when the Kaiser passed. He did not realize who the passing dignitary was, or, perhaps, that a dignitary was passing at all, so he just stood with his hands in his pockets. This was noticed: Dehn was not officer material.[7] He left the army a corporal and after the war returned to Breslau.

Dehn was made full professor at the University of Frankfurt in 1922, when he was forty-three. Frankfurt was a prosperous city with a history of political independence. Dehn was partially insulated from the turmoil of 1920s Germany at its new university, founded largely with contributions from Jewish businessmen. Maria Dehn Peters described the Frankfurt of her youth:

> Founded early in the eighth century by Charlemagne, it has ancient churches and other buildings, the Römer [the old town hall] where emperors were crowned, the concentric rings of old walls and moats showing the perimeter of the city. Early in the nineteenth century the moats were developed into beautiful public gardens, ringing the older city, by Jacob Guiollett. (Frankfurt became the home of many Huguenot families.) My parents used to go down to the Main Kai,

> where the Epsteins then still lived, to listen to the old church bells on New Year's night. Frankfurt was a great cultural center with a beautiful opera house, several theaters and concert halls, art museums—a wonderful place to grow up in.[8]

Dehn's mathematical interest had shifted from geometry to the emerging discipline of algebraic topology. Algebraic topology asks when a surface or, more generally, a manifold is the same as another. We can say two manifolds are the "same" if one can be stretched and pulled, without tearing, into an identical copy of the other. In algebraic topology one attempts to associate "invariants" with surfaces that don't change with stretching, an idea similar to the Dehn invariant in the context of the third problem. An invariant could be a number, like the "genus" of a surface, which is how many holes it has. More often it is an algebraic structure, like a group. Poincaré had created algebraic topology as a modern discipline in a series of papers between 1895 and 1904. In 1907 Dehn had contributed, with Poul Heegaard (1871–1948), a paper on *analysis situs* to the *Encyklopädie der Mathematischen Wissenschaften* that was one of the first systematic treatments of the new subject.

Dehn made important contributions to the algebra used in algebraic topology: to what is called "the word problem"—so named because it involves long strings of "letters" and the question of when two apparently different "words" are the same. Dehn's way of phrasing this problem helped clarify its relevance to mathematical logic. In the chapter on the tenth problem we saw that a generalized version of this, called Thue's word problem, was proven unsolvable in 1947 by Emil Post and A. A. Markov. Even when working in algebra Dehn emphasized visual representation, in this case drawing pictures of the graph of the group:

He also did work applying algebraic topology to the classification of knots. Much of this appeared in his paper "On the Topology of Three-Dimensional Manifolds" in 1910. In this he constructs homology spheres by removing a solid torus from a three-dimensional sphere and then sewing it back in a different fashion. This is now called Dehn surgery. In this same paper he "proved" Dehn's lemma, one of the most important results for which he is known. A serious error was found in the proof, by Helmuth Kneser in 1929.[9] Dehn's lemma was not rigorously proved until 1957, when a complete proof was given by Christos Papakyriakopoulos. Dehn published important papers in 1911, 1912, and 1914.

At Frankfurt in the 1920s Dehn's rate of publication went down. He was giving away ideas to his students now, but the Frankfurt mathematics history seminar was a more interesting, affirmative cause. Others associated closely with the seminar were Paul Epstein (1871–1939), Ernst Hellinger (1883–1950), and Otto Szász (1884–1952). Hellinger and Szász had taught at the university since its opening. Carl Ludwig Siegel (1896–1981) arrived in Frankfurt at age twenty-five, the year the seminar started. Siegel's 1964 address "On the History of the Frankfurt Mathematics Seminar" gives an indelible portrait:

> The fundamental idea of these seminar sessions was to study the sources of the most important mathematical ideas of all time. . . . Thus, we dealt with the ancient authors, Euclid and Archimedes most thoroughly, over several semesters. During another period of several semesters we busied ourselves with the development of algebra and geometry from the Middle Ages to the seventeenth century, in which we especially acquainted ourselves with the works of Leonardo Pisano [Fibonacci], Vieta, Cardano, Descartes, Desargues. Our joint study of the ideas from which infinitesimal calculus developed in the seventeenth century was also rewarding. Here we dealt with the discoveries of Kepler, Huygens, Stevin, Fermat, Gregory, and Barrow among others. . . . [We] did not necessarily strive to publish. The actual meaning of the seminar lay in another direction, that is, to have a fertilizing effect on the student participants, who thereby better understood subject matter already covered in class. The seminar also gave us professors the enjoyment of observing the excellent achievements of long ago.[10]

The seminar never grew large because the texts were read in their original languages: ancient Greek, Dutch, Italian, French, Latin, etc. Even for the comparatively well-prepared students in the German university system this was difficult. Weil sat in on the seminar as a young vagabond mathematician. He recounts how Dehn opened his eyes by showing him how to read a text (by Cavalieri) from the point of view of what the author knew and what was known at the time. In 1926 Weil was already a serious Sanskrit scholar, so Dehn is not being praised for opening the eyes of a neophyte. Weil also recounts meeting Siegel, whom he describes as being "already a legend," rumored to have "drawers full of inspired manuscripts which he kept secret." Dehn explained to Weil "the theory that was circulating in Frankfurt at the time: mathematics was in danger of drowning in the endless streams of publications; but this flood had its source in a small number of original ideas. . . . If the originators of such ideas stopped publishing them, the streams would run dry; then a fresh start could be made."[11]

Siegel in particular had dug in his heels against the attitude of publish or perish, which Weil says came from the U.S., with a depth of obstinacy only he could summon. (In 1924 Siegel was teaching a class with only two students. One day both were late and were surprised to find the class in session and an entire blackboard full when they walked into the room, Siegel busily teaching away.) Weil wonders if some of Dehn's approach was not an attempt to keep Siegel working. If, from the Frankfurt point of view, publication was to be avoided, there was no stricture on working or writing. This explains the 1927 Dehn–Nielsen theorem, though Dehn never published anything on the Dehn–Nielsen theorem himself. Mathematicians gave each other papers for birthday presents and on other special occasions. Weil recounts being allowed to read a manuscript on transcendental numbers that

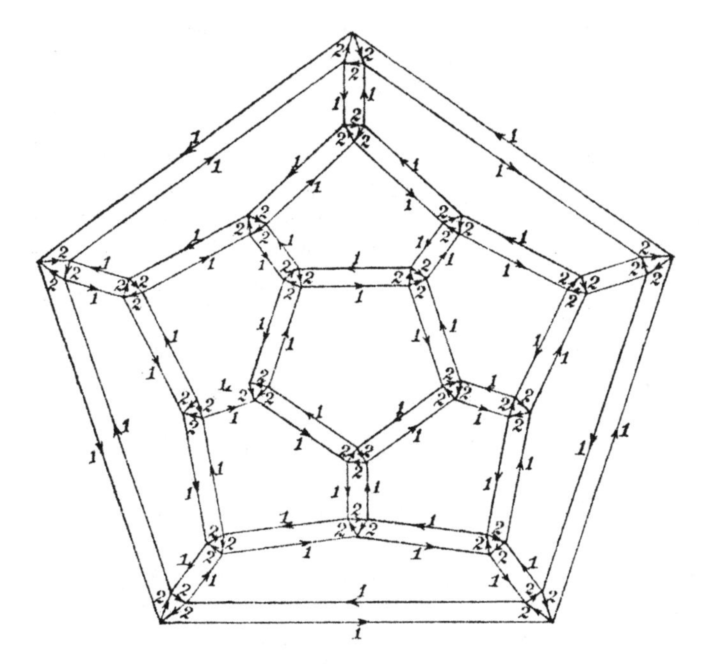

The graph of a group by Max Dehn. From "Ueber die Topologie des dreidimensionalen Raumes," Math. Annalen 69 (1910), 145.

Siegel gave Dehn for his fiftieth birthday in 1928. The manuscript was not allowed to leave Dehn's home, and though Weil was allowed to take notes, he had to promise not to show them to anyone. At the bottom of the manuscript was written: "It's a bourgeois, who still does algebra! Long live the unrestricted individuality of transcendental numbers."[12] This paper did see the light of day the following year, as the first part of Siegel's "A Few Uses of Diophantine Approximation," one of the great papers of the century.

One of the themes we find in Hilbert's life and in the history of his list of problems is the flowering of middle class culture in late nineteenth and early twentieth century Germany, as expressed particularly in the educational system. In prior centuries mathematicians had often been subject to a wandering existence, hostility from family because of their impractical choice of vocation, and in extreme cases, near-starvation and early death. In the most poetic cases (which can be found, with some exaggeration, in Bell's *Men of Mathematics*) they suffered these misfortunes almost gladly for their art. Now the number of stable positions with reasonable pay for mathematicians increased, but with them came a need to "look busy," to produce a product. In an address delivered in 1928 to the university, Dehn said: "The progress of our discipline depends not on mass efforts, not on a flood of papers filled with investigations of insignificant special cases or generalizations, but on individual creative achievements. Such achievements can hardly come about in a factory-like setting."[13] The mathematics history seminar was not a factory-like setting. There were "history coffees" back at Dehn's house. Dehn's children particularly liked the pastries Hellinger brought. There were expeditions into the Taunus, woods-covered hills to the north-northwest of Frankfurt. Every spring Siegel would organize a "great asparagus feast" in the garden of a restaurant.

Eva Dehn remembers the extended Dehn family meeting in Hamburg. She was particularly impressed by her lawyer uncle's large home, which had a wall that could be let down into the basement to make one large room of two. She remembers going to her grandmother's house for an Easter egg hunt and the large yard that sloped down to the Alster River as it ran through the city. She spent two or three summers with an aunt whose husband was a coffee importer. Maria wrote to John Stillwell simply, "Visits to Hamburg were visits to Heaven for us." She writes of the many waterways and the Alster River and lake, before the Alster meets the Elbe:

> Instead of trucks rumbling through the streets, you have barges gliding along canals with their loads. Instead of going to work by streetcar or bus, people take the steamer across the Alster from the residential to the business district, or they walk to work along the water's edge—such a good way to start the day! But for us it was family that made Hamburg our dream town.[14]

She mentions parties, singing, chamber music, dancing, and skits on these visits.

Maria, writing to me, emphasized how much time her father gave to the children. She remembers holding one of her father's hands, her brother holding the other, and running through the woods to kindergarten. Dehn took an active interest his children's education and at bedtime would come and tell them fantastic stories. Eva remembers him reading fairy tales full of monsters. She said her father went to church but was "one of those remarkable people who don't ever get fanatic about any one thing. He saw the good in Buddhism and Mohammedanism and he lived by what he believed. He never had a bad word to say about anybody." Dehn was one of a small group of people who read Greek with Dr. Bölte, a retired gymnasium headmaster, on Saturdays. Maria remembers Dehn studying Chinese with a student. When he was sixteen or seventeen, Helmut read Plato's *Phaedrus* (in the original) with his father.[15] Mathilde Maier, in *All the Gardens of My Life*, describes Dehn coming when her apricots were ripe:

> He would climb up the trellis that reached up to the roof and bring down baskets full of choice orange fruit. It was always an event. Only much later, in California, did I see such beautiful apricots again. When the disaster of Nazism gained ground, he wanted to found a settlement with us, on the other side of the world, far far away from everything.[16]

Maria developed a health problem when she was sixteen and so, around 1931, was sent to a country boarding school in the south of Germany—one of Dehn's former students worked there. Maria quickly improved. This boarding school was a fortunate choice. The headmistress was Jewish, and when Hitler rose to power she immediately moved any students who would follow to Kent in England. Eva went to this school in 1936, so the sisters were safe in England as the situation in Germany deteriorated. Their father traveled to England several times to give advice to the headmistress and taught there between January and April in 1938. Eva eventually worked as a nurse in London at a "casualty receiving hospital" during the Blitz. In that green and pleasant land, Maria went to horticulture college. Helmut stayed in Germany until 1936, when an uncle of his mother's signed an "affidavit of support" and he moved to Cleveland, Ohio. He enrolled in the University of Virginia medical school, supported by his great-uncle, graduated in 1939, and practiced pediatrics for many years outside of Cleveland.

The Frankfurt mathematics history seminar lasted until 1935. At least two remarkable pieces of historical research came out of it. Weil published a book of mathematical history, *Number Theory*, that is very much in the spirit of the seminar. Siegel resurrected a nearly unreadable formula on the zeta function buried in Riemann's notebooks: It is now called the Riemann–Siegel formula.

Siegel wrote the following about his time at Frankfurt with Dehn and the others, and the seminar in particular:

> In retrospect, these hours together with our circle of friends in the seminar, I count among the most beautiful of my life. At the time, I enjoyed this activity, for which we came together every Thursday afternoon from four to six. But it was only later, after we were scattered all over the world, that, through disappointments in other places, I recognized what rare luck it is to find colleagues who disinterestedly and without personal ambition come together to work.[17]

The main seminar participants except for Siegel were Jewish. He continues, "I will also report truthfully how it went for each of the teachers after 1933," and his talk becomes an act of witness. Szász was the first to be touched. Not only was he Jewish, but he was not German, so in 1933 he lost his *venia legendi*, his certificate to teach. In fact, this was lucky because it forced him out early. He went to the United States, where he taught at the University of Cincinnati from 1936 until his death in 1952. Dehn, Epstein, and Hellinger were allowed to continue teaching because of exceptions in the racial proclamations—long civil service, army service in the First World War, etc. However, in 1935 these exclusions were revoked. A few months before this edict Dehn lost his position. Epstein, who had been expelled from Alsace by the French after the First World War, quit on his own. In Siegel's words, "As he said, he wanted to save the German bureaucrats the trouble of what the French had already done to him in 1918."[18] Hellinger was let go with the enforcement of the new edicts.

None of these three left Frankfurt immediately. One thing that kept them there was their books. They were comfortable and had other property they couldn't take with them. These men were in their fifties or sixties, an unusual time to change jobs in any profession, but particularly in mathematics. They could survive, it appeared, where they were. Dehn, though forced into retirement early, was receiving a pension. With money he could help his children. "He couldn't believe they were that bad," Eva told me. In this period, at fifty-nine, Dehn achieved one of his most important results, today called "Dehn twists,"[19] which appeared in a 1938 paper on topological mappings.

However, in Siegel's words: "The actual large-scale terror in Germany began on the tenth of November [1938], with the active persecution of Jews, ordered from the highest positions of the government, in which synagogues were burned, Jewish businesses demolished, and all available concentration camps were overfilled with Jews they had dragged off. At this point, Hitler's minions also came to take Dehn, Epstein, and Hellinger away."[20] Dehn was arrested but let go that same night because the authorities had no place to keep him. Hellinger was not immediately arrested, for the same reason. Epstein was passed over temporarily because he was too sick. Into this chaotic and brutal situation came Siegel, two days later, intent on celebrating Dehn's sixtieth birthday. Siegel had moved on to Göttingen by that time and led, in his words, "a somewhat retiring life."[21] He hadn't witnessed the events of November 10, *Kristallnacht*. Thus, he was surprised when he arrived in Frankfurt. Something was amiss. There was visible evidence of rioting. Dehn's apartment was empty—he and Toni had escaped to Bad Homburg, about twelve miles north of Frankfurt, where they were hiding at Willy Hartner's house. Willy Hartner (1905–1981), an important historian of science and later rector of the University of Frankfurt, was a friend and colleague who had participated in the mathematics history seminar. Siegel went to Hellinger's apartment, where he found that Hellinger hadn't gone into hiding because "he wanted to see just how far, in his case, the state would go against all established law and custom."[22] Siegel continued to Bad Homburg and did succeed in celebrating Dehn's birthday. By the time Siegel returned to Frankfurt, a place had been found for Hellinger in the camps. He was first confined in the city auditorium, then sent to Dachau. After six weeks, he was allowed to emigrate to the U.S., where he had a sister. He obtained a job at Northwestern University and was eventually made full professor. However, he reached mandatory retirement age quickly and was forced to retire with a very small pension. He found another position for two years at the Illinois Institute of Technology, then fell ill with cancer and died in 1950. Toni Dehn traveled to his bedside to help care for him.

Epstein was allowed his freedom, at least temporarily, because of his ill health. Siegel visited him in August 1939. Epstein had seemed calm, sitting in his garden, though a bit sad because he had just had a favorite cat put to sleep because he feared it might antagonize the neighbors by hunting birds. He pointed to the trees and flowers and said, "Isn't it lovely here?" Eight days later he received a summons from the Gestapo and killed himself. He was nearly seventy then, so emigra-

Carl Ludwig Siegel, Max Dehn, Ernst Hellinger. September 26, 1938–forty-two days before Kristallnacht (photo provided by Maria Dehn Peters).

tion would have been difficult. Siegel felt that, under the circumstances, "Epstein acted as wisely as he could."[23]

Dehn decided to go into hiding. With the help of the wife and son of his former student and then colleague, Wilhelm Magnus (1907–1990), he and Toni managed to get to the train station and onto a train. Maria says that "he then spent several days riding all over the railroads till their safe haven"[24]—in Hamburg, where Dehn's four sisters lived. Meanwhile, a way of escaping to Scandinavia was found. Siegel traveled to Hamburg to one of the meetings to plan the escape and was outraged to have to answer questions about his "Aryan descent" before being allowed to check into one of the city's oldest, most established hotels.

Dehn first went to Copenhagen, then found a position teaching at the Institute of Technology in Trondheim, Norway. He had spent time before in Norway and was a member of the Norwegian Academy of Sciences. Siegel visited him in March 1940. There were mysterious German ships in the harbor that the locals called "pirate ships." Siegel himself fled to the U.S.; he couldn't stand the situation in Germany. On April 9, 1940, the Germans invaded Norway, and the pirate ships unloaded their cargo of war materiel. The Dehns were in as much danger as ever. First they hid in a farmer's house and then, when nothing happened, returned to Trondheim. If Dehn could get out of Norway, he had to have someplace to go, or he might be returned to Germany. As a professor, Dehn was exempt from the immigration quota for the United States. However, he had to have a job before he would be admitted. The Dehns had known Clare Haas, a woman dermatologist, in Frankfurt. She was now working as a psychiatrist in Pocatello, Idaho. She managed to arrange a job for him in 1941–42 as an assistant professor at the University of Idaho in Pocatello. After this had been arranged, later in 1940, the Dehns escaped Norway, making much the same journey as Kurt Gödel and his wife, around the same time—across Russia by train to Siberia, then to Japan, then across the Pacific to San Francisco. Reid, in her biography of Richard Courant (1888–1972), recounts how Dehn visited the library in Vladivostock and found the only mathematics books were a shelf of books from the Springer-Verlag "Yellow Series" that Courant had edited. At the ends of the earth, there was Göttingen. Dehn caught pneumonia on the train and almost died. Siegel attributed Dehn's survival to the fact that he refused medical attention. There was also a story about a kind doctor with a wooden stethoscope.

Arriving in the U.S. in 1941, in his sixties, Dehn had a problem. He had been allowed to take out no money. His books had been bought for pfennigs on the deutchmark (and resold to students for many times the amount by "sly Aryan businessmen," to use Siegel's words).[25] He had shipped some of his furniture to London, but in the end it was sold for storage fees. Dehn was not so famous that a position would be found for him no matter what, yet he was too old and well-known to be given a subordinate position at a major university. The result was an odyssey. Pocatello had never been viewed as a permanent appointment. Dehn had a year to find a new job. In roughly the same period Weil was glumly teaching at Lehigh University in Bethlehem, Pennsylvania. Despite the difficulties of his situation, Dehn made the best of things and enjoyed the hiking available in Idaho. In the summer of 1941, Hellinger and Siegel met the Dehns in Idaho and they went hiking.

From Idaho, Dehn moved on to the Illinois Institute of Technology in Chicago. Hellinger was teaching nearby at Northwestern. Dehn's job in Chicago paid better and was more centrally located, but Chicago was noisy, hurried and industrial. Dehn did not like the city. He found himself teaching a class in differential and integral equations to two different sections. The second consisted of students who had failed it the first time, and on the first day of class Dehn received the happy greeting, "Hello, Professor. We're the dumb ones."[26]

He continued in his peregrination, to St. John's College in Annapolis, Maryland. According to Siegel, the college was filled with youths between fifteen and eighteen, because all available older students had disappeared into the armed forces. The idea was that the students would read the 100 *Great Books of the Western World* that had been selected at the University of Chicago. Siegel says they were to be read in the original, but he is wrong in this; only some short sections were to be attempted in the original. However, Dehn was to teach mathematics straight from Euclid, Apollonius, Newton's *Principia*, etc., and finally from Russell and Whitehead's *Principia*! (This last, one of the toughest reads ever written.) There was rhetoric about reading every word in every book. The St. John's program had some of the same volumes as the Frankfurt mathematics history seminar, but these two tetrahedra proved distinct. Under these circumstances, even the even-tempered Dehn was not invariant. He would have been an ideal teacher for this kind of program if the students had been prepared and the program more modest, but as Siegel says, "The whole thing seemed like a malicious parody of the former math history seminar in Frankfurt."[27] Helmut says that his father was particularly rankled when he had to teach Homer's *Odyssey*, read in English, in a week, and that Dehn felt there was a contempt for the sciences at the college.

Dehn arranged to be a guest lecturer at Black Mountain College in North Carolina during spring break of 1944. The correspondence in the Black Mountain archives is with Erwin Strauss, a fellow refugee from Germany. Dehn enjoyed his visit. He was in contact with Black Mountain about coming as a visiting professor, and the job offer was mailed to Dehn in Chicago on December 29, 1944. He went to Black Mountain in early 1945 and stayed there the rest of his life, with occasional forays out.

In operation between 1933 and 1956, Black Mountain is one of the noteworthy experiments in the history of American higher education, in the tradition of Brook Farm or the Fourierist communes. (Mathematicians, this isn't your Fourier.) It never had much more than ninety students (except for summer sessions), but there was a large faculty. Everyone lived at the college. The desire was to make community decisions by consensus (though that wasn't always the way things happened). Black Mountain's artists included, at various times, Willem de Kooning, Franz Kline, Robert Motherwell, and Robert Rauschenberg. Its poets included Robert Creeley, Robert Duncan, Denise Levertov, and Charles Olson, a giant of a man at 6'7" or 6'8" who was rector at the end. In music there were John Cage; the dancer Merce Cunningham; and conductor Heinrich Jalowetz, a disciple of Arnold Schoenberg. There were writer/intellectuals like Paul Goodman and Alfred Kazin. What was Buckminster Fuller? He built an early geodesic dome at Black Mountain, but it collapsed and was renamed the "supine dome." There was the filmmaker Arthur Penn—*The Miracle Worker*, *Bonnie and Clyde*. That is a short list. Mathematicians have sometimes commented that this appointment was below Dehn's level, but, in its own way, Black Mountain was at the highest level. Dehn was flexible enough to appreciate that. On the science side, Nathan Rosen (1909-1995), of the Einstein–Rosen–Podolsky paradox in quantum mechanics, taught at Black Mountain from time to time. Fellow immigrant Josef Albers, coming from the Bauhaus faculty in 1933, had arrived speaking no English, unsure of where North Carolina was. His wife Anni, a noted textile artist and, later, a force in the life of the college in her own right, initially thought it might be in the Philippines. When Albers left Black Mountain after an internal fracas in 1950, it was to head the Department of Design at Yale.

Black Mountain was in the woods, in a beautiful setting. Fuller, who became good friends with Max and Toni, commented that Dehn was "very much of a naturalist." Fuller described the area in a background interview (now coming out with permission) for Martin Duberman's excellent history of Black Mountain College, *Black Mountain: An Exploration in Community*, saying, "It was a very welcoming wilderness, but there were a certain amount of snakes and most of them were non-dangerous. There were quite a number of the big black snakes . . ."[28] Two children of faculty members, David Corkran III and Howard Rondthaler, remembered Dehn

Max Dehn at Black Mountain (reproduced by permission of Eva Dehn).

in their joint interview. Corkran said: "He [was] a little wisp of a fella, and he'd climb. . . . He climbed all over the mountains, and he'd go forever, you know . . . we'd go with him quite often."[29] Both boys, around twelve at the time, remember Dehn saying that he had escaped across Russia in a boxcar.

Dehn was comfortable at Black Mountain. The expatriate German-speakers included Strauss, Albers, Jalowetz, and their wives. Dehn could often help people transcend factional feuding. He enjoyed the bright students and the original, if sometimes abrasive or eccentric, faculty members. Novelist and critic Edward Dahlberg, for example, with a glass eye that had a will of its own, was a passionate enemy of almost all modern literature. Though at first rhapsodic about the wooded setting, he soon started describing the foliage as "homicidal." It would seem that this noted hater of asphalt preferred it to the alternative. Siegel described Dehn as acting as a devil's advocate; Dehn, in fact, got along famously with Dahlberg, despite their differences regarding the foliage. Dahlberg "delighted" in talking about "Strabo, Plutarch and Dionysius of Hallicanassus" with Dehn.[30] Besides mathematics, Dehn taught Greek and philosophy at Black Mountain.

Dehn visited the University of Wisconsin in Madison for the fall term of 1946–47 and went to Madison for the entire school year in 1948-49. The Black Mountain files seem to indicate that he was commuting back on the weekends sometimes in the spring. At Wisconsin he met students after class in the "beer stube." Paul L. Chessin remembers Dehn in the rathskeller at the student union building in Madison: "The table was round and made of a heavy wood, deeply inscribed with student names. We may have numbered ten around the table (it was large). Prof. Dehn, stein in hand, spoke of classic topics obviously dear to him. . . . Greek literature, mythology, some history. When he learned that I had had some Latin classes, he switched to Latin, reciting Horace."[31] He was delighted when Chessin managed to remember a few sentences from Cicero's oration against Cataline.

Though Black Mountain College is not known for having trained mathematicians, Dehn nevertheless had at least two serious students in the few years he taught there. The first was Peter Nemenyi. When Dehn (i.e., the entire mathematics department) left to teach in Wisconsin in 1948–49, Nemenyi followed. When it came time for Nemenyi to graduate, there was a public oral examination at Black Mountain, as in German universities. Dehn asked Emil Artin (1898–1962) to come down from Princeton to give the examination and do some hiking. In the formal examination, Nemenyi remembers questions on elliptic geometry that he didn't do well on. Artin made the arch comment (in light of the small number of mathematics students at Black Mountain) that he must have been asleep during those lectures. However, on a hike afterward, Artin asked other questions that Nemenyi answered well enough, so after graduation he continued in the Princeton mathematics department, where he did eventually obtain a Ph.D. with a thesis about a topic in statistics. Nemenyi particularly remembers Dehn saying "We are building a cathedral"[32] as he explained something in projective geometry. Nemenyi went off into a life of social activism, which included teaching statistics in Tupelo, Mississippi, during the period of the civil rights movement and later in Sandinista Nicaragua.

Trueman MacHenry went on to a more conventional career in mathematics and is now emeritus at the University of York in Canada. He heard about Black Mountain while working in Alaska on a surveying team after having "stopped out" of the University of Wyoming. His friend Andy Fisher, a poet, was still at Wyoming and corresponding regularly with Trueman. Fisher told MacHenry about his new friendship with Ilya Bolotowsky, a painter "who had just landed at Wyoming" after one of the periodic "explosions" at Black Mountain. But he landed with good things to say about Black Mountain.[33] Trueman read the letter in his tent, wrote for information, and went.

At Black Mountain Trueman soaked up everything, studying a bit of philosophy, writing, linguistics, French, German, and Russian. Modern dance was strong at Black Mountain, so he studied under Merce Cunningham and Katherine Litz. Viola Farber, later an original member of Cunningham's company (1953–1965) and then a founder of her own troupe, was a fellow student. He studied physical sciences with Natasha Goldovski. When the composer Stefan Volpe was experimenting with different tempi running concurrently, he asked Trueman to help him figure out when they would come together. And Trueman studied mathematics with Dehn one-on-one. He said he never had another teacher who was as sensitive, deep, or vastly philosophic. They would pick an area that Trueman was interested in and then start talking, sometimes in a room, sometimes while hiking. Trueman would read, Dehn would go to his library and pull down an article and give it to him. Dehn didn't get lost in detail. If he thought Trueman needed to know something or was missing something, he would gently direct him. When Trueman wrote his thesis Dehn arranged for Ruth Moufang (1905–1997), a former student, to read and comment on it. Moufang had been appointed professor at the University of Frankfurt after the war—she was Germany's first female full professor of mathematics. Trueman was also close to Toni. He said she was "very much like a mother to me."[34] Toni was not as free a spirit as Max. Long-haired students would run around without shirts (this in the late 1940s, not the 1960s) and come into her home barefoot. When Siegel visited, he shared some of Toni's view of this. Trueman says that Siegel was Dehn's closest friend and remembers that Siegel and Dehn liked to go down to Florida together. Mildred Harding wrote:

> I asked [Dehn] if it wasn't frustrating to give all that time to just one pupil [Trueman]. Not at all, he assured me, for Trueman was "a real student, a real mathematician"; it was a privilege to be his teacher. "In fact," he went on, "I have been very fortunate. In my sixty years of teaching, I have had at least fifteen real students."[35]

Trueman wrote to me: "Time's metric is very strange: It is curious how some periods in your life contract, while others seem to expand, in some cases to take up more room than would seem possible. Black Mountain and my time with Dehn was one of those periods in my life."[36]

Black Mountain was always precarious financially. For a period the faculty received only room and board and $5 a month. When he came, Dehn was paid the relatively princely sum of $40 a month, which included $15 extra to lure Toni away from her job at Montgomery Ward in Chicago designing displays, where she had remained when Dehn went to St. John's. According to Siegel, Dehn did not regard the money as a serious problem, but he wasn't in a good position financially when he started at Black Mountain and must have badly needed the money his guest lecturing posts brought in—he received $2,750 from Wisconsin for his first semester, in 1946–47.[37] However, as Germany became reorganized after the war, Dehn, now actually past retirement age, began to receive the pension he was entitled to from the University of Frankfurt. The Dehns received a fairly large sum in reparations, part of which they used to buy a house in the village. Dehn retired officially from the Black Mountain faculty in 1952, but stayed on as an advisor. The college was beginning to come apart. (Before it disbanded in 1956 the faculty was being paid in allotments of beef from the college's cattle.)

The Black Mountain community decided to sell off a lovely section of dogwood forest, for lumber. They acquired a sawmill and planned to process the wood themselves. However, everything was done wrong. The fellows of the college were not lumber people. The result was a big pile of lumber with splits in it, cut in the wrong sizes, so that their buyer refused the entire lot. Duberman writes:

> For Max Dehn, the incident assumed the proportions of a personal tragedy. Though he dearly loved the dogwood forest, Dehn had re-

> luctantly agreed to its destruction as a necessary measure to save the college. Soon after the fiasco, he died of a heart attack brought on, some insist, by his despair over the pointless destruction of the trees. He was buried in the woods he so much loved.[38]

Siegel, in his talk, saw a more direct link. He said that when Dehn was hiking on a hot summer day he saw some loggers cutting down trees that were to have been spared and ran up a steep path to try to stop this desecration. Siegel believed this effort caused the "embolism" that killed Dehn the next day. Helmut Dehn says his father spent the hot summer day before he died marking trees that were too special to be cut. He woke the next morning with severe chest or upper abdominal pain. He asked for aspirin—somewhat unusual given Dehn's distrust of medicine. (Helmut remembers his father urging him to prescribe fewer drugs in his pediatric practice.) By the time Toni was back with the aspirin Max was gone. Trueman, who hurried back from New York to help Toni when he heard the news, commented to me that Dehn had survived much greater tragedies.

In Duberman's book we find Olson staying on at Black Mountain alone and overseeing the disposal of the college. Among his tasks was to arrange that proper care be taken of the graves of Dehn and Jalowetz. Toni Dehn lived close to family in Ohio until 1996, but when she died her ashes were brought back to Black Mountain by her children.

Distance

The Fourth Problem

≼ *Busemann, Pogorelov, and . . .* ≽

The fourth problem begins:

> Another problem relating to the foundations of geometry is this: If from among the axioms necessary to establish ordinary euclidean geometry, we exclude the axiom of parallels, or assume it as not satisfied, but retain all other axioms, we obtain, as is well known, the geometry of Lobachevsky (hyperbolic geometry). We may therefore say that this is a geometry standing next to euclidean geometry. If we require further that that axiom be not satisfied whereby, of three points of a straight line, one and only one lies between the other two, we obtain Riemann's (elliptic) geometry, so that this geometry appears to be the next after Lobachevsky's. If we wish to carry out a similar investigation with respect to the axiom of Archimedes, we must look upon this as not satisfied, and we arrive thereby at the non-archimedean geometries which have been investigated by Veronese and myself. The more general question now arises: Whether from other suggestive standpoints geometries may not be devised which, with equal right, stand next to euclidean geometry.

Hilbert proposes an investigation of what happens when one weakens the axiom that states that if we have two triangles with two sides congruent, and the angle between the sides congruent, then the triangles themselves are also congruent. The theorem that a straight line is the shortest distance between two points follows from this axiom, but itself is weaker and does not imply the axiom. Hilbert is interested in geometries in which straight lines are still the shortest distance between two points, but the congruence axiom doesn't hold. He believes that investigations here will "throw a new light upon the idea of distance." He makes technical comments. By the end Hilbert has asked also for investigation of the concept of distance in "the theory of surfaces" (differential geometry) and "the calculus of variations."

In 1903 Georg Hamel (1877–1954), then a student of Hilbert's, found that there were many more of these geometries than Hilbert had imagined, and he proved a few general theorems. He found a way to construct a wide class of these geometries and their metrics (systems of measuring distance) by modifying an already existing technique—the venerable Cayley–Klein model of hyperbolic geometry. This was surprising enough at the time, so that some thought Hamel's results were a solution to the problem. However, he clearly hadn't found all possible geometries of this type or investigated them thoroughly, and so his results today are generally not seen as a solution to the fourth problem.

The following quotation from Herbert Busemann (1905–1994), expresses a later attitude toward the problem:

> . . . Hilbert asks for the construction of all these metrics and the study of the individual geometries. It is clear from Hilbert's comments that he was not aware of the immense number of the metrics, so that the second part of the problem is not a well posed question and has inevitably been replaced by the investigation of special, or special classes of, interesting geometries.[1]

From this standpoint, the problem is a program of research, and Busemann was a major contributor to the enterprise. Investigation of the individual geometries or classes of geometries could go on indefinitely.

Busemann's father was a director of Krupp, the weapons manufacturer, and wanted him to go into business, so Busemann "wasted" several years trying to please his father before he went to Göttingen as a student in 1925.[2] He also studied at the universities of Munich, Paris, and Rome. His rooms in Göttingen became a popular gathering point because he could afford good wine. He learned many languages—eventually he could lecture in seven languages, including Russian, and enjoyed reading the Classics in Latin and Greek. In mathematics he focused on geometric questions, earned his doctorate in 1931 and, at the request of Courant, was supported by his father so that students who needed the money more could be Courant's paid assistants.

One-quarter Jewish, Busemann was not at first directly affected by Hitler's race laws, and Reid reports that Busemann himself begged Hermann Weyl to stay on at Göttingen in 1933.[3] Weyl cleared out, and not long after that Busemann de-

Herbert Busemann (courtesy of the Archives of the Mathematische Forschungsinstitut Oberwolfach).

camped also. He taught first in Denmark and then spent time at the Institute for Advanced Study. In the U.S., he taught at colleges and universities including Swarthmore, Johns Hopkins, and the Illinois Institute of Technology, but his career didn't progress as rapidly as it might have, because of his unpopular choice of subjects. (There are fashions in mathematics. Sometimes a fashion can represent a current enthusiasm in a positive way—something seems new and exciting. Sometimes there is an element of irrational bias. And sometimes a fashion might represent a judgment that most mathematicians in the future might agree with. One place I found information on Busemann was in a book titled *Geometric Tomography*, about the mathematics of putting together a three-dimensional picture from information in shadows and slices—X-rays, MRIs, etc.—not mathematically fashionable in Busemann's era, but useful.) Finally Busemann went to the University of Southern California, where he slowly advanced. He became professor emeritus in 1970, eventually moving to Santa Ynez, California, a little north of Santa Barbara. He had always wanted to be a painter but had never really permitted himself to try, fearing that it would distract him from mathematics. After retiring, he built a studio at his home in the coastal mountains. When reporter Lee Dembart visited him on the occasion of Busemann's winning of the Lobachevsky Prize in 1985—largely for his book *Geometry of Geodesics*—he found the studio "chockablock with dozens of large canvases painted in vibrantly colored geometrical designs."[4]

Building on an idea of Wilhelm Blaschke's (1885–1962), Busemann found a powerful way of constructing metrics in the spaces involved in the fourth problem. He started with a topology—"balls." Any additive function (as in measure theory) defined on these balls and their countable intersections and unions can be used to define a metric. He didn't realize how powerful this method could be, but his 1966 talk at the International Congress of Mathematicians held in Moscow sparked the imagination of A. V. Pogorelov.[5]

Aleksei Vasilevich Pogorelov was born in 1919 on the outskirts of Koroch in the Belgorad region of Russia, at edge of the Ukraine. His parents were peasants who lived almost outside the money economy. Only one ancestor stands out to Pogorelov, a great-grandfather who was a "talented, self-educated" mechanic and designer.[6] He built wind and water mills for grinding wheat and other cereals. Pogorelov spent his childhood in the countryside and his first four school years in the local elementary school. Hunting and fishing have been lifelong activities for him. The family moved to Kharkov, nearby, and in 1931 Pogorelov started attending secondary school there. He became interested in mathematics at thirteen or fourteen and started doing well in the mathematical Olympiads. He says that he was drawn to mathematics, in part, because of his success in solving problems—and because his solutions were clearly recognized—not an uncommon attraction for people who become mathematicians.

Pogorelov pursued mathematics and physics at Kharkov State University with enthusiasm and was then drafted, in 1941, and sent to the Zhukowsky Air Force Academy in Moscow (an excellent technical school). When he was demobilized in 1945, he went to work as an engineer at the Central Aerohydro-Dynamics Institute, also in Moscow. Engineers at this Institute were allowed a free day each week to continue their studies at Moscow State University, so he enrolled as a part-time graduate student, studying under N. V. Efimov (1910–1972) and A. D. Aleksandrov (1912–1999). Pogorelov told me that they put him in contact with compelling problems in global geometry flowing from the work of "such famous mathematicians as Minkowski, Weyl, Hilbert, Blaschke, and Cohn-Vossen. I was lucky enough to solve many of these problems, and so my name became known among mathematicians."[7] Pogorelov defended a *kandidat* thesis in 1947 and a doctoral thesis the following year. He received offers to teach from both Leningrad State University and Moscow State, "but I was not tempted with these propositions and I preferred Kharkov State University." Kharkov, in the Ukraine, has been a center for mathematics since the late nineteenth century, surpassed in the former Soviet Union only by Moscow and Leningrad. It was home, near the countryside where he grew up. Pogorelov was appointed to Kharkov State University in 1947, became chair of geometry in 1950, and was eventually appointed to the Soviet and Ukrainian academies of science.

He has published on many aspects of geometry and on subjects ranging from partial differential equations, to the theory of the stability of elastic shells, to problems in "cryogenic electric machine engineering." Described as an excellent engineer, Pogorelov has been employed since 1960 at the B. Verkin Institute for Low Temperature Physics and Engineering of the Ukrainian Academy of Sciences, where they, interestingly, have a geometry department.

Hilbert's fourth problem can be seen as lying most centrally in the foundations of geometry. After hearing Busemann's 1966 speech and after a period of study of Busemann's *Geometry of Geodesics,* Pogorelov began to believe that Busemann's general method for constructing metrics could be used to construct all metrics, and hence all geometries, of the type asked for in the fourth problem. In this scheme, Busemann's metrics can be found that offer steadily better approximations to any metric in this category, and any metric in this category can, therefore, be seen as the limit of these approximations. In 1973 Pogorelov published a paper titled "A Complete Solution of Hilbert's Fourth Problem" in *Soviet Math. Doklady*. There were some gaps in this—questions about smoothness, about the proof in higher dimensions—but they have largely been answered. Pogorelov wrote to me:

> In my article published in 1973 I have admitted some immodesty when I entitled it as "The Complete Solution of the Fourth Hilbert Problem." In fact, it did not contain a complete solution of the fourth problem, because only the two-dimensional case was examined. Moreover the article contained one incorrect statement. Probably therefore Busemann originally considered my work with some skepticism. In 1975 I published a monograph entitled "The Fourth Hilbert Problem," in which the three-dimensional case was considered. After that the opinion of Busemann changed. I knew about it when Busemann was visiting Moscow because he was awarded the Lobachevsky Prize.[8]

This, from his 1979 English translation, *Hilbert's Fourth Problem*, is Pogorelov's statement of the problem as he solved it, as a question in the foundations of geometry:

> This book is devoted to Hilbert's fourth problem and contains its solution when formulated as follows: Find to within an isomorphism all realizations of the axiom systems of the classical geometries (euclidean, Lobachevskian and elliptic) if, in these systems, we drop the axioms of congruence involving the concept of angle and supplement the systems with the "triangle inequality," regarded as an axiom.[9]

This is pretty close to the spirit of Hilbert, who wanted positive solutions and stated that a good problem ought not be too difficult lest it mock us. Pogorelov's result can be taken as a solution to the fourth problem, even though the problem is somewhat larger and vaguer. In 1986 Z. I. Szabó of Hungary published a paper that extends some of Pogorelov's results.

Aleksei Vasilevich Pogorelov (courtesy of the Institute for Information Transmission Problems of the Russian Academy of Sciences).

Juan Carlos Alvarez told me in response to a letter: "You're right that this problem is very vaguely written. At first I thought this was a weak point, but now I think that the strength of the problem is that it admits a number of different formulations." While many Hilbert problems have been solved and re-solved, the new solutions don't change the nature of what was actually proved. Alvarez wrote to me: "A lot of people forget that Hilbert also asked to study these metrics, not just to characterize them. In this sense, the above solution [Busemann–Pogorelov's] is not sufficient. My own 'solution' in terms of symplectic forms has been a lot more useful for studying the geometry of submanifolds and for extending integral geometry to these spaces."[10] Thus, in this problem, which starts with construction of the metrics but moves on to ask for study of the geometries constructed, a new method of solution brings new methods of investigating what is constructed. Pogorelov's solution shifts part of the difficulty of understanding the geometries constructed to understanding the additive functions that were used to construct them. The "symplectic forms" Alvarez refers to are more obviously fundamentally geometric. I. M. Gelfand and M. Smirnov have approached the problem from the direction of "Crofton densities." So we have a problem that keeps opening up as it investigates "the idea of distance."

Something for Nothing

The Fifth Problem

Gleason, Montgomery, and Zippin

In 1950 André Weil published an essay he had written four years earlier, "The Future of Mathematics." He agreed with Hilbert on the importance of problems for mathematics and wrote, "Even among those of Hilbert, there are still several which stand out as distant, although not inaccessible, goals which will continue to suggest research for perhaps more than a generation; an example is furnished by his fifth problem, on Lie Groups." Weil was not optimistic about American mathematicians making great contributions to mathematical culture. He thought the American education system was inadequate—that "narrow specialization" was required and there was no time allowed for more leisurely exploration, saying, "Beyond this, he [the American mathematician] runs the great risk of not being able to survive the stupefying effects of the purely mechanical teaching which he will have to inflict on others, in order to earn his living, after having undergone it himself for too long a time."[1]

Weil's observation may have had some merit. Nevertheless, in 1952 three Americans, Andrew Gleason, Deane Montgomery, and Leo Zippin, solved Hilbert's fifth problem in two back-to-back papers in *Annals of Mathematics*.

The problem is titled: "Lie's Concept of a Continuous Group of Transformations Without the Assumption of the Differentiability of the Functions Defining the Group." A group is an algebraic structure that satisfies a few simple axioms. There is an operation often called addition that is associative, that is, $(a + b) + c = a + (b + c)$. There is an identity, often called 0. For each element there is an element, its inverse, that we can add to it to get 0. Norwegian mathematician Sophus Lie (1842–1899) had investigated groups of transformations of manifolds that satisfy additional conditions: These groups are continuous and have an analytic structure. The set of all rotations of three-dimensional euclidean space is an example of a Lie group. Two rotations can be done in succession—addition. Doing nothing is the identity. Reversing a rotation takes us back to where we were before. Rotation is associative. Further, there are small rotations, so that one can talk about a "neighborhood" of a rotation. Rotations that are close together remain close together when added to a third rotation, so the group is continuous—neighborhoods are not torn asunder. Better yet—and this is the payoff that comes from it being a Lie group—local cartesian coordinates can be established so that performing successive rotations is analytic. The functions that combine rotations can be expressed, in terms of the local coordinates, as power series (possibly infinite). This is what brings much of the wonderful structure of analysis into play.

Lie created his theory to shed light on differential equations, so it was natural for him to make assumptions about differentiability. Hilbert, instead, wanted to start with geometrical ideas and use Lie groups as a tool to investigate the foundations of geometry. Geometries then come from these groups and can be studied by studying the groups. Since the concept of differentiation comes from calculus, not geometry, Hilbert wanted to avoid the assumption of differentiability in setting up Lie groups.

The text of the fifth problem wanders more than some, and if one interprets Hilbert's request in the most straightforward way, what he is asking for is false—his phrasing is too connected to concrete transformations, and there are some oddball transformations of manifolds. In the modern view of Hilbert's fifth problem, one starts with the group and stays there. This was recognized early on, and the problem was rephrased so that what Hilbert "asked for" was possibly true. Since Hilbert wanted positive results, not technical counterexamples, this rephrasing became by consensus identified as Hilbert's fifth problem. The accepted phrasing of the problem is admirably concise: Is every locally euclidean group a Lie group?

The example of a Lie group we gave above is locally euclidean. All that means is that a cartesian coordinate system can be established and transformations in terms of these coordinates are continuous. But since it is a Lie group, the coordinate system can be chosen so that these functions are also analytic. We are playing the angles here. Are we guaranteed to get a lot of structure even if we only put our money down on a little bit of structure? Can we pay for a coordinate system and get analysis thrown in for free? Another way of putting it comes from Gleason: "[The positive solution of the fifth problem] shows that you can't have a little bit of orderliness without a lot of orderliness."[2] An outsider might observe that an awful lot of work was done trying to get something for nothing, but mathematicians are willing to do this. The "something," once obtained, is forever; the difficulty is left in the past. And it was fun to solve the problem.

This problem is obviously technical in a way that the first four foundation problems and the tenth problem are not. Irving Kaplansky, in his book for mathematicians on Hilbert's problems (it can be found in university libraries as bound xeroxes of typewritten notes for twelve problems), wrote an explanation of the problem, an outline of the history, and offered a few first steps of the proof for "flavor." Then he said: "From this point on, the reasoning becomes lengthy and quite technical, and our sketch will become very sketchy indeed."[3] Montgomery's student and collaborator Chung-Tao Yang wrote in the *Proceedings of the Symposia* of the AMS on Hilbert's problems: "Since the proof of Theorem 1 [Hilbert's problem] is very complicated and technical, it is impossible for us to sketch it here."[4] I will follow their lead.

Gleason, Montgomery, and Zippin's positive solution of the fifth problem—every locally euclidean group *is* a Lie group—was the final step in a chain of partial results on special cases. Understanding of the structure of Lie groups was greatly advanced in the 1920s by Weyl, among others. The (Fritz) Peter–Weyl theorem was particularly important. There then seemed to be two possible approaches to the general result. One division into cases is by dimension. The one-dimensional case is simple. The two-dimensional case was handled by Béla Kerékjártó (1898-1946) in 1931, building on ideas from Brouwer, 1909–10. Montgomery handled the case of three dimensions in 1948, and Montgomery and Zippin handled four dimensions, publishing in early 1952, just before the solution.

This approach through dimension, using sophisticated techniques from topology and geometry, is not what actually panned out. The other approach entered from the direction of function theory and analysis and originally focused on a broader class of groups than Hilbert's problem needed. Alfred Haar (1885–1933), who came from a vineyard-owning family in Hungary, published a paper in the year of his death proving the existence of an essentially unique, well-behaved measure on any locally compact group, which would include all groups relevant to the fifth problem. This allowed the entry of analysis and function theory. In the same year von Neumann, always lightning-fast, published a solution for the case of compact groups. In 1939, L. S. Pontryagin (1908–1960), a blind Russian topologist, solved the fifth for the abelian case, and in 1941 Claude Chevalley (1909–1984) proved the result for solvable topological groups. For all these kinds of topological groups it was proved that you can introduce a local structure that is analytic. This is an elite group of mathematicians. The stage was set for the trio of Americans.

Why was this technical problem so attractive to mathematicians? Deep investigations of algebraic structure and of the links between seemingly unconnected branches of mathematics were resulting in powerful mathematics left and right in this period. Lie groups were becoming increasingly important in the foundations of physics. With this problem, abstract algebra, geometry, topology, and calculus, including differential equations, are united. Deeper understanding of this problem was seen as leading into the foundations of these areas, and from there, possibly to new results of an unexpected kind. (Look at the winners of the Fields Medal for numerous results of this hybrid kind.) Second, since the nineteenth century, mathematicians had been finding strange and highly irregular functions (nowhere differentiable), that were nevertheless continuous. Therefore one couldn't assume much about the differentiability of a function based solely on continuity. Showing a group to be a Lie group shows not only that there is some differentiability, but that it has analytic structure—it is differentiable an infinity of layers deep. Gleason told me that his own intuition in this case was clear. "I don't know how others may have been thinking about the fifth problem, but for myself, I never doubted the result. Without such faith I would never have been able to keep focused on the problem."[5]

Perhaps it is something about endowed chairs at Harvard. We have seen that Hilary Putnam, who contributed to the solution of the tenth problem, did not have a degree in mathematics and mostly published in philosophy. In Andrew M. Gleason we have a Harvard mathematician who never received a single graduate credit toward a Ph.D., much less the degree. Gleason was born on November 4, 1921, in Fresno, in the hot, irrigated farm country of California's San Joaquin Valley. His mother was staying there with her family because Gleason's father, a botanist, was on "an exploring trip" in British Guiana.[6] The Matteis were vintners in southern Switzerland. Gleason's maternal grandfather had come to the U.S. in 1873, starting out as a dairy farmer near Los Angeles, then moving to Malaga, eight miles south of Fresno, to start a vineyard in 1891. Prohibition cut into business, but the vineyard was still there in 1921. When Henry Allan Gleason returned from Guiana, the family moved back to New York, to Bronxville.

On his father's side, Gleason's great-grandfather was a New Englander and graduate of Union College in upstate New York. In 1833, he started an enterprise in Middlebury, Ohio, that would become the Diamond Match Company years later, after he had moved on, himself. Around 1850 he moved to Iowa and literally broke the sod. His son, Gleason's grandfather, "had a peripatetic youth."[7] He attended one of the Lincoln–Douglas debates and was a recruiter during the Civil War. In 1865 he bought a farm in Illinois that is still owned by a family partnership and operated by the great-grandson of the first tenant. Gleason's grandfather didn't go to college, but Gleason says he helped Gleason's father with mathematics. Henry Allan Gleason spent his first nine years on the farm. He graduated from the University of Illinois in 1901 and got a Ph.D. in botany at Columbia. A series of academic posts followed, but he found his permanent job when, in 1918, he went to the New York Botanical Gardens in the Bronx. Gleason's father published botanical books, some of which are still in print, and a prize is given in his name by the gardens.

Henry Allan Gleason had contracted a bad case of malaria in Guiana. As he fought the disease he worked out elaborate arithmetical puzzles in his delirium. Later, he told his son about them, and in his interview in *More Mathematical People*, Gleason recalls the style, if not the details, of one his father liked, called "How the World War Was Fought."[8] In it, each battle of the First World War was fought by following a complicated system of multiplying or dividing numbers with special properties. The battle was decided and the next one popped up. Beyond this fevered number play, Gleason says that his father had mathematical "leanings." His mother was an excellent card player, and the family enjoyed a variety of card games.

Andrew was sent to the public schools, where he did well. In kindergarten he retaliated against another boy by pulling a chair out from under him. The kindergarten teacher was so enraged she went and got the first grade teacher. Both leaned over Gleason admonishing him, so he socked one of them. He said, "Since then I have not been so fierce."[9] When he was nine or ten Gleason became very interested in ancient Egypt. Next he became interested in astronomy. When he was fourteen the family went to Berkeley for the spring semester. Gleason took the first semester of high school geometry and was bored by the teacher, who didn't know much. He helped the other students with their homework, including some in the second semester, so he started following the second half of the class. What this second teacher said made sense to him: "She was an absolute stickler for points."[10] As soon as he figured out what geometry was about he liked it. In Berkeley Gleason's interest in astronomy continued. He would ride the trolley to Chabot Observatory and bought what, at the time, was a very expensive reflecting telescope. Up on the roof of the apartment building, in the mild weather of Berkeley, he looked at the stars. Upon returning to Bronxville he found the seeing was good only when it was very cold and decided he wasn't *that* interested in astronomy.

Gleason went to the local school, Roosevelt High School, in Yonkers. He was dramatically good at taking tests. One semester he took four State Regents Exams and received 100 on each of the mathematics exams and 99 on the physics exam. The high school principal suggested he apply to Yale. He went up for an interview and was sent to William Raymond Longley's office. "He received me cordially and asked what [calculus] book I had been studying from. I didn't know. When I got home I looked at the book; it was Granville, Smith, and Longley."[11] He wrote, apologizing. Longley wrote back that he thought Gleason should take a course he was teaching on mechanics for juniors. Gleason was accepted and received a scholarship. Gleason enrolled in Longley's class, though he hadn't studied the first two years of calculus very well. He rapidly caught up and helped the other freshmen in his dorm and nearby dorms with their calculus problems. He eventually did so many problems that he says he doesn't believe there is a first- or second-year calculus problem in existence that he hasn't seen.

After about a month, Gleason found he had enough time to audit a class for seniors on differential equations. He started turning in the homework, so the professor suggested he sign up. This brought him to the attention of the mathematics faculty. However, the two courses still felt too much like high school to him. The professor of the differential equations course then fell ill, and Einar Hille (1894–1980) came in to teach the course. Though born in New York City, Hille had received his Ph.D. in Stockholm. Gleason felt as if another world had been revealed.

The next year he took the standard course in real analysis for graduate students from Hille. The mathematics faculty suggested that he take the Putnam examination, the national exam on which we have seen that Cohen performed well. Gleason took the test and came out unhappy—he had been able to solve only thirteen out of the fifteen problems. It was probably the first time he couldn't solve all the problems on a mathematics test. If you solve a few of the Putnam problems you've done well, and Gleason finished in the top five, perhaps first—they don't say. He won the Putnam Fellowship to Harvard—that is, admission to Harvard and money. Gleason stayed at Yale and took the Putnam again in his junior and senior years and both times also finished in the top five. As a junior he took a graduate-level course from Nelson Dunford that impressed him. Essentially, he studied as an undergraduate in the graduate program for three years. Since Ph.D. degrees in mathematics were not uncommonly completed in three years, we can assume that Gleason, by the end of his senior year, was operating with a fairly high degree of mathematical sophistication.

When the U.S. entered the war, Gleason signed up with the Navy and was given a commission. He was allowed to complete his undergraduate degree, but in June 1942, at the age of twenty, was sent to Washington D.C. to work in a cryptanalysis group. By this time it was clear to the military that mathematicians and scientists should be used to outthink the enemy rather than be slaughtered, as had happened to a generation of French and English mathematicians in the First World War. The universities emptied. The level of mathematical culture in the cryptanalysis group was high. "It was almost the same as being in graduate school," Gleason said.[12] Donald Menzel, an astronomer from Harvard and an extraordinary applied mathematician, was in the section next door. His specialty was how energy freed in the center of the sun worked its way out to be emitted as sunlight and other radiation. He was working on a "bootleg operation,"[13] trying to explain why planes visible to the naked eye sometimes didn't show up on the radar screen. (The radar didn't always pass smoothly through temperature inversions in the air.) The wartime mission of Gleason's group was to supply mathematical assistance to groups working on specific codes. A code takes a structured message and transforms it into a series of numbers that is supposed to seem random. The mathematical problem is to find a way to recognize the way in which the message is not random. With the computers that were being developed this became increasingly feasible.

The cryptanalysis group was kept on into 1946, part of the time doing puzzles and playing games. Gleason played NIM, a game ancient in its origins that exists in many variants. In one way of setting it up, there are rows of o's of various lengths. On any move a player can take any number of o's in a row that he chooses. The next player then removes o's in the same way. The winner is the player who removes the last o. Gleason worked out a theory of the game. A 1928 theory already existed, but Gleason didn't know about it and his version was more elegant. Menzel recommended Gleason for the Harvard Society of Fellows, and so when he was decommissioned he did finally go to Harvard, as a junior fellow.

Weil had not been the only academic to feel that the narrow requirements of the American Ph.D. program did not promote openness of inquiry. Harvard President Abbot Lawrence Lowell, along with other faculty members, including Whitehead, had founded the fellows program in 1933. Senior fellows were people like Lowell or Whitehead, and junior fellows were bright young scholars from all fields who were to be saved from the burdens of graduate school. Gleason said he had wonderful freedom but was set apart from his generation of graduate students. He wondered if a little bit of hoop-jumping, a habit of writing up answers, might have been good for him. He has drawers full of unfinished papers—something interesting was always drawing him away. Gleason didn't publish a lot as a fellow, but he became interested in the fifth problem from a course George Mackey gave in 1947 or 1948. He began to work on the problem and by the end of his fellowship had begun publishing papers on it. He said he didn't think the Harvard faculty was aware how hard he was working on it and attributed his appointment as assistant professor on July 1, 1950, as partly stemming from a lecture he gave to the math club on the game of NIM.

A few days before Gleason received his appointment, the Korean War started. He was recalled to active duty in a few days and ordered back to Washington D.C. and cryptanalysis. Gleason continued to work on the fifth problem. He was a bit worried that he might not get out of the Navy before his two-year leave from Harvard was up, after which he might have had to start all over again—with no Ph.D. In February of 1952 he had his big breakthrough on the fifth problem and things looked brighter. He lectured on his solution at Princeton, almost getting in a car accident in the snow on the way, and then, contrary to his usual relaxed attitude toward publication, took a week off and wrote the paper in a college guest room.

Gleason's paper is indeed complex. Using the existing topology, build a semigroup. This gets the real numbers in play. Build a "metrically straight" arc. Embed the whole group and the arc in an appropriate Hilbert space, a friendlier place. Show the arc is differentiable in at least one spot . . . etc. . . . end with a "theorem by von Neumann."[14] Montgomery and Zippin polished off the last remaining piece of the fifth problem using a heavily topological method. In 1953 Hidehiko Yamabe (1923–1960) found a way to extend Gleason's methods to accomplish the same thing, as Gleason told me, "putting the whole problem (including much more than the fifth problem itself) squarely in the domain of abstract analysis."[15]

Gleason had been scheduled to return to Harvard in September 1952. However, he got sick and didn't make it until November. Mackey taught his course for him—after all, Gleason had just solved Hilbert's fifth problem. The next year Gleason was given tenure. In this period, a young John Nash at MIT had been designating mathematicians as "geniuses" and adding a "G" at the beginning of their names. In this nomenclature Gleason was "G-squared."[16] In 1969 Gleason was appointed Hollis Professor of Mathematicks and Natural Philosophy, a chair endowed in 1727. The "k" had been removed from mathematicks, but he successfully lobbied to have it restored.

Gleason married Jean Berko in 1959 (Harvard–Radcliffe 1953, Ph.D. 1958, now professor of psychology at Boston University—interested in psycholinguistics, how we acquire language). One thing or another caused Gleason not to travel much on his sabbaticals. After time in Europe and a stay at the Institute in 1959–60, the next time around there were young children they didn't want to uproot—three daugh-

Andrew Gleason (courtesy of Andrew Gleason).

ters. The family stayed home one semester and then Gleason taught a course at MIT. Later, the whole family moved to California to spend two quarters at Stanford. Gleason's home in Cambridge, an easy walk to Harvard, is less than 100 meters from Fresh Pond and its bird sanctuary.

Gleason has published on a variety of topics, including Banach algebras, projective geometry, coding theory, and combinatorics. A 1957 paper on the foundations of quantum mechanics particularly stands out. It turns out the physicists had less choice in how they set up the mathematics of quantum mechanics than they had thought. (When setting up measures on the closed subspaces of a Hilbert space, there is only one general recipe for doing it—using the trace function.) Gleason has enjoyed teaching midlevel classes and has written undergraduate texts on calculus and abstract analysis, as well as *An Elementary Course in Probability for the Cryptologist*. He has continued to be interested in problems and wrote a book containing all the problems and solutions on the Putnam exam from 1938 to 1964. Gleason has said, "Proofs really aren't there to convince you that something is true—they're there to show you why it is true."[17] This easy, pragmatic view is in sharp contrast to the charged debates we heard in the earlier foundation problems.

Deane Montgomery and Leo Zippin contributed the second half of the solution to the fifth problem in a paper that immediately followed Gleason's in the *Annals*.

Montgomery was born in the town of Weaver, Minnesota, on September 2, 1909. Both his parents were born in log cabins in the woods. Weaver had a population of around a hundred in 1909, and it has, according to the post office in Altura, "about a hundred people today."[18] Montgomery's Scotch-Irish grandfather emigrated from Ireland during the potato famine—1845–48. Montgomery told his children: "John Montgomery came to Minnesota, walked into the woods with his ax, chose a plot along the river for his farm, built his log cabin, and started farming."[19] Indians hunted and fished near the farm and, after trust was established, Montgomery's grandfather would store things for them in exchange for part of the catch. They walked quietly into the cabin unannounced, which Montgomery's grandmother, Mary (Finn), also from Ireland, found disturbing. If this sounds familiar to readers of Laura Ingalls Wilder or *Caddie Woodlawn* by Carol Ryrie Brink, it is more than coincidence. Wilder's "little house" in the Big Woods of Wisconsin and the setting of *Caddie Woodlawn* are both within thirty-five miles of Weaver. Montgomery's father, Richard, was born in 1861. An older brother became a farmer like his father, but Richard Montgomery became a merchant.

Deane was an only, and late, child. His father was forty-seven when he was born and his mother, Florence (Hitchcock), thirty-eight. When Montgomery was eleven his father died of a cerebral hemorrhage in his sleep. He must have prospered, though, because enough money was left so that Montgomery could count on continuing his education through college.

Montgomery went to a one-room schoolhouse that covered grades one through eight. The record graduating class, his, had three members. He heard all the lessons and skipped several grades.[20] The first member of his family to attend high school, Montgomery got on a train early each school morning and rode to Wabasha. The high school there had twelve male students, exactly enough for a football team—one boy was disabled. The team scored exactly one touchdown, which Montgomery didn't see, as he was at the bottom of a pileup during the golden moment.

When it came time to go to college, the University of Minnesota would have been an obvious choice. However, Montgomery's mother did not think it "safe for Methodists."[21] Therefore, Hamline University, a Methodist school founded in 1854, was chosen. Montgomery did farm work in the summer to supplement the money his father had left and for the rest of his life was proud of his ability to plant cabbages quickly over a long work day. His mother's devout Methodism left a lifelong impression on him. G. Daniel Mostow wrote:

> From his widowed mother Deane inherited a rigorous egalitarianism. She believed that all people are equal under God and that everyone was as good as anyone else—an attitude Deane kept throughout his career. He also inherited from her the habit of understatement.[22]

Graduating from Hamline in 1929, a few months short of his twentieth birthday, Montgomery went to the University of Iowa. He got a solid background in analysis. His thesis and first four papers were in point set topology with a Polish flavor—the Poles had a strong, original school in point set topology. The questions in this area explore which oddball shapes are possible. Montgomery received a National Research Council Fellowship and studied first at Harvard in 1933–34, where he encountered algebraic topology. He was part of a private study group that included Norman Steenrod (1910–1971) and Garrett Birkhoff (1911–1996). He then moved to Princeton, 1934–35, already a world center in algebraic topology. Before his fellowship was over Deane married Kay Fulton. Montgomery, and then the young couple together, had been frugal with his fellowships. They were able to go to Europe in the summer of 1935, traveling with another couple.

They returned to Smith College, where Montgomery had landed an appointment as assistant professor. He was tall and handsome, with easy good manners. In the words of Atle Selberg, who first met him in 1949 at the Institute for Advanced Study: "My first impression was that he was a remarkably handsome man. . . . As I learned to know him more, it was his character, his honesty and integrity that impressed me."[23] At Smith, Montgomery was a hit. It became clear that he was a very good mathematician. He was appointed associate professor in 1938 and full professor in 1942. He was appointed associate professor at Yale in 1946.

Montgomery's early career was punctuated by visits to the Institute for Advanced Study, and in 1948 he was made a permanent member and took up residence. When Oswald Veblen and Weyl retired in 1950 and 1951, James W. Alexander (1888–1971) withdrew as a professor, and Siegel returned to Germany, there was suddenly a gap in the faculty. Only Marston Morse (1982–1977) and von Neumann were left, and von Neumann really wasn't a mathematician most of the time anymore. Veblen groomed Montgomery as a replacement and later, when he died in 1960, left him not only his office, but his personal archives of Institute business, going back to the earliest years. Montgomery was appointed professor, as was Selberg, in 1951. They worked together and the departed faculty members were replaced by Hassler Whitney (1907–1989), Gödel, and Arne Beurling (1905–1996), in 1952, 1953, and 1954. Armand Borel was added in 1957 and Weil in 1958, so the permanent faculty was again strong.

Algebraic topology was Montgomery's focus, more specifically, transformation groups. His first paper in collaboration with Zippin was published in 1936; eight more came out before the Second World War brought the collaboration to a temporary halt. Their names became so strongly associated that once Zippin appeared at a conference and was handed a nametag that said "Montgomery Zippin." Already in the 1930s they were working on the fifth problem. Montgomery returned to it and began to make progress in the mid-to-late 1940s. He solved the problem for dimension three in 1948 and then published important papers in the following four years. By the end of this process he was back in active collaboration with Zippin and they published their solution for dimension four in early 1952. The pace was picking up. In February 1952, Gleason told Montgomery of his result. Gleason was not a topologist and didn't immediately see how to close the gap.

Montgomery and Zippin worked around the clock. They submitted their paper to the *Annals* on March 28, 1952.

In the words of Borel:

> This was the climax of a major effort and, as I remember it, some people were mildly curious to see where Deane would turn, now that this big problem had been solved. But he did not have to look around at all. . . . He just went back full time to what was really his main interest . . . : Lie groups . . . of transformations on manifolds, so that, in the context of his whole work, the contributions of the fifth problem appear almost as a digression, albeit a most important one.[24]

By the 1960s a lot of energy in algebraic topology was being spent building general machinery that would allow the study of differential structures on manifolds. Montgomery and Yang, in a brilliant series of eight papers appearing 1966–73, explored how to apply this new theory to particular examples.

Montgomery played a big role in the culture and operation of the Institute almost from his arrival, using his position to foster students, in general, as well as the discipline of algebraic topology. In the 1950s, as the French—Jean-Pierre Serre, René Thom—made a series of revolutionary advances, they were invited for visits, with a resulting cross-fertilization for topology. Montgomery believed that only mathematicians could (or would) appreciate mathematics. Borel recalls Montgomery saying in 1948, when a well-known Dutch mathematician was unhappy that a theorem of his wasn't getting the recognition he thought it should have, "If we don't show appreciation, then who will?"[25]

At the Institute Montgomery talked to everyone. He concerned himself with the grounds and maintenance and the Institute budget. An early riser from his small-town days, he was usually the first mathematician to arrive in the morning—after a long walk. (From what I have seen of mathematicians, it may not have been necessary to rise very early to accomplish this.) He welcomed each new visitor, topologist or not, often in his office, and for many years, he and Kay invited the school, including spouses, for a party. As the number of Institute members increased from twenty-five to sixty this became more heroic. Alcohol flowed freely. Raoul Bott recalls one party ending with him and von Neumann on the floor playing marbles, with Montgomery looking on "with great humor."[26] Deane and Kay had two children, Mary and Dick. The family enjoyed vacations, but Montgomery tended to stay away from cities, driving through the countryside and pointing out the crops or birds. (At home he had a rain gauge.)

Morse was set to retire in 1962. Earlier appointments had gone reasonably smoothly, but war broke out on this one. A faction wanted a separate section of the Institute for the natural sciences, including eventually the social sciences. This eventually happened, but the battles were bitterly fought, new ones begun, and enough noise was coming out of the quiet grounds that the dispute was reported in the New York Times. Montgomery was resolute in defense of mathematics, of what he felt

Deane Montgomery (photograph by Herman Landshoff, courtesy of the Archives of the Institute for Advanced Study).

were Institute standards, and fiscal responsibility. He liked to recount that J. Robert Oppenheimer had called him "the meanest sonofabitch he ever met."[27] This conflict dragged on, and Montgomery suffered an early heart attack that many blamed on stress. He had to change his diet and adopted a rigorous exercise program.

Tragedy came into his family's life. Dick Montgomery fell ill with cancer and came home to die. Montgomery slept in a sleeping bag outside his son's door so he could respond more quickly in the night and planted flowers that his son could see out the window from his bed.

In 1988, Deane and Kay moved to Chapel Hill, North Carolina, where their daughter, then Mary Heck, had moved with her daughters. Montgomery felt the loss of daily contact with the Institute. Borel recounts that he, also an early riser, often felt this loss as well, when he arrived early and yet Montgomery was not there. In Chapel Hill, Montgomery asked James Stasheff if he could get library privileges at the University of North Carolina library, and he was given an office in addition. Deane Montgomery died in his sleep on March 15, 1992. Kay followed not long after.

Much of the material I have used for this section appears in a memorial booklet printed by the Institute, and so the tone is elegiac. Even with this caveat, I am struck by this passage from Yang:

> In my mind, Professor Deane Montgomery was much more than a friend. In fact, I always regarded him as a respected teacher as well as a beloved uncle. For this reason, I named my son after him. For the same reason, I never called him by his first name. This is a Chinese tradition I have followed all the time. . . . To me, Professor Deane Montgomery was not only a most outstanding mathematician, but also the greatest human being I have ever met."[28]

Four years older than Montgomery, Leo Zippin was born January 25, 1905, the only child of Bella (Salwen) and Max Zippin. Zippin's daughters, Nina Baym and Vivian Narehood, told me that their grandparents arrived in New York around 1903 from Chernigov in the Ukraine. Max was an intellectual and had two loves, Yiddish culture—particularly theater—and the dream of a workers' revolution. The years 1899–1905 in Russia were full of agitation, conflict and strikes, leading to the inconclusive Revolution of 1905. Violent anti-Semitism gave them another reason to leave. If Max was partly driven by his fancy, Bella provided stability. She helped bring other members of the family to the U.S. Vivian said her grandmother adored her grandfather, though she was jealous of "the CP"—the Communist Party and its antecedents. Nina said that in America what Max really wanted to do most was write plays for the Yiddish theater. He was not successful, and she is not aware that any plays were staged. Instead he wrote for the *Jewish Daily Forward* of Abraham Cahan. The Zippins participated in the immigrant community focused around Yiddish culture and radical politics. Leo met his wife, Frances, for the first time when at age nine he was given charge of the younger girl and took her to the kissing booth in a festival at the Labor Temple.

When Zippin was twelve, the Russian Revolution swept away the czar and the family took off for Russia to join in and witness the big event. Like John Reed, author of *Ten Days That Shook the World*, and despite "unrestricted submarine warfare," they made it. Conditions were chaotic—civil war, famine, typhus. Max was disillusioned with what he found but never officially expressed this. Leo was horrified and later in life wouldn't raise the subject himself but would recount it if pressed.

The family decided to come home. This was not easy. Exits to the West were closed. Instead, as Gödel and Dehn would later do, they found their way out via the Trans-Siberian railway. For the Zippins, the journey involved backtracking and getting stuck. The family story has them as far as Harbin, Manchuria, nearly out of money. Bella marched up to the local potentate's residence and introduced herself as an English teacher to the startled man. She taught members of the household until the family was able to move on. The trip took, perhaps, close to two years.

Zippin didn't let this interlude delay his education. His father found a job in Philadelphia. Zippin studied at Central High School, another public school with an entrance exam. (This high school is still technically allowed to grant Ph.D. degrees.) According to his transcript, Zippin enrolled in September 1919 as a transfer from DeWitt Clinton High School in New York. He was originally placed on academic probation, probably because of the gap produced by the trip to Russia. He received a 90.87 average his first term, taking nine courses. Max Zippin remained involved with the Yiddish press and politics. Leo was frustrated that his obviously talented father remained in what were to him backwaters. He later complained to his daughter Vivian that Max could have been city editor of the Philadelphia Inquirer but instead chose to stay where he was. Leo never accepted his father's revolutionary politics, particularly after the trip to Russia, and when as a boy he had to accompany him to Party events, he would wear an American flag in his lapel.

Zippin graduated from Central at seventeen and entered the University of Pennsylvania in 1922. He graduated in three years and continued as a graduate student in mathematics. Working as an assistant instructor through 1927 and spending the next two years as a Harrison Fellow, Zippin received his Ph.D. on June 19, 1929. Like Montgomery, his first work was in point set topology. He spent 1929–30 as an instructor at Pennsylvania State College.

If the Russian Revolution had been a bad time to visit Russia, 1929 was almost as bad a year to receive a Ph.D. The collapse of the economy led to a tight market for anyone, but in addition, there were quotas limiting Jewish faculty in most departments. Once, in a job interview, Zippin was asked first if he were Jewish. Then he was asked if he were a religious Jew. Having received his answers, yes and no, the interviewer commented that they did not hire Jews, but if they did it would be good if they followed the religion of their fathers. Despite this, Zippin was able to get grants for two years as a National Research Fellow. In 1930–31 he was at the University of Texas (Austin) and then in 1931–32 at Princeton University. Like so many mathematicians, Zippin liked to walk as he thought. In Texas this activity was somewhat atypical. Once, the Texas police found Zippin wandering under the stars between two and three o'clock in the morning and took him in to the police station.

The year at Princeton was important for Zippin, because now he was exposed to the new techniques of algebraic topology. However, the money ran out. In 1933 the job situation became even worse, as the U.S. was flooded with mature, often first-rank, refugees from Hitler's Germany. Zippin spent the next four years as a research assistant to Alexander at the Institute for Advanced Study. They published a paper on topological groups together in 1935. Zippin had addresses on Wilton Street and Nassau Street that appear on documents from the Institute, but it is not clear whether he was paid for his position. His daughter Nina says he traveled to Göttingen, but it is unclear when.

In 1932 Zippin married Frances Levinson, three years his junior. She was a schoolteacher working at James Madison High School in Brooklyn, so there was a lot of commuting back and forth from Princeton by Zippin. He met Montgomery during this period and the two hit it off. In 1936 Zippin got a job as an instructor at NYU. Nina was born that year. Zippin got a job at the newly opened Queens College branch of City University of New York in 1938. This was a position with a future, but it came with the same workload that Post faced during the same period at the City College campus—fifteen to sixteen student contact hours a week. Zippin had to commute to Queens and worked hard to fit research and his collaboration with Montgomery into his schedule.

Zippin went on leave from Queens College in 1942. Too old for the draft, he volunteered to do mathematical work for the war effort and became technical director of the Philadelphia computing unit of the Aberdeen Proving Ground. He was there until late 1945. His visits home were two to four weeks apart, and Vivian, born in 1940, remembers being frightened by this man who kept appearing in the home. Frances was a bit annoyed that her husband had volunteered for the job—they had already been through four years of separation. Leo had been promoted to assistant professor at Queens College and returned there in 1946. He continued to do applied work after the war, at the Institute for Mathematics and Mechanics at NYU. (His 1952 paper with Montgomery solving Hilbert's fifth problem bears a note saying it was partially funded by the Office of Naval Research.)

As soon as he could after the war, Zippin resumed his collaboration with Montgomery. Given the workload at Queens in addition to his applied work, this re-beginning was slow. He didn't publish a paper again until 1950, and his first post-war collaboration with Mongomery appeared in 1951. They were both on the trail of the fifth problem. In early 1952 Montgomery and Zippin proved it for four dimensions. They knew they were on to something. In his memorial reminiscence for Montgomery, Zippin wrote that they were in constant contact. They met at least once a month, sometimes in New York, sometimes Princeton, talked on the phone, and exchanged letters.

Interesting people flowed through the Zippin house, and conversation ranged broadly. There were parties. Zippin was acerbic and fond of word play; Frances was part of the flow. The family would go to the Catskills or Vermont to get away from New York City summers, which Zippin described as "beastly hot." Both parents loved birds. Unfortunately, Zippin usually had to return to the city to work. His schedule allowed mathematics in the summer, and Zippin had a strong desire to do mathematics. Where Weil went wrong in assessing the potential of American mathematicians after the war was that he underestimated how attractive discovering new results was to them. Weil himself was in a French prison awaiting trial for draft-dodging when he did some of his most important work.

Both daughters remember the mismatched pair Zippin and Montgomery made. Montgomery was tall, Midwestern, a prototype WASP. To Vivian he seemed chivalrous, and everyone saw that he was handsome. Zippin was a short New Yorker, culturally Jewish though not observant. Vivian says he was handsome too, but in a different way. "He had a manly, square face and twinkly blue eyes and a wonderful mane of curly hair."[29] Montgomery and Zippin conducted their collaboration while walking. (And when you think of it, what better way to get away from others?) In Princeton the countryside was nearby; in New York they took long walks in Central Park. Vivian says they were "wonderful friends." Post's daughter commented that

Leo Zippin (courtesy of Vivian Narehood).

when her father did mathematics, he needed quiet to concentrate, which was difficult for a child in the household. Zippin's daughters give a softer echo of this. They remember the period of work on the fifth problem, particularly the uproar Gleason's paper caused—the phrase "a little rivalry" was used. Nina describes the work sessions of this period as "extremely strenuous." Montgomery and Zippin did get their results.

Zippin was forty-seven in 1952 when the fifth problem was solved. His research papers stop in 1956. Beginning in 1941 he had worked editing various mathematical publications, including *Annals* and *Transactions of the AMS*. His strong anticommunist credentials allowed him to stand up against the McCarthy hysteria of the early 1950s. He was a political liberal. Zippin helped create CUNY's doctoral program in mathematics in the years 1964 to 1968, and Nina says he was proud of his administrative work. When Russian émigrés started arriving, he helped them. He retired from Queens College in 1971 and quit smoking, though damage to his lungs was already done. Zippin and Montgomery remained in contact. They spoke twice when Montgomery was in the hospital for the last time. Montgomery joked about being discharged, saying that his wife and daughter would have to do the marketing and cooking. His last words to Zippin were, "I am a little tired and I had better lie down to rest."[30] Leo Zippin died about three years later on May 11, 1995.

On Again, Off Again: Physics and Math

The Sixth Problem

The sixth problem asks that physics be axiomatized and the structure of the axioms investigated. It would seem that Hilbert overreached here. In 1900 physicists did not frame their science in terms of axioms, and their methods, though often wonderfully effective, were irregular at best. (There is no manual of uniform procedures for piracy, either.) Physics was riven with contradiction. In physics that is often a good thing: Something might be about to happen. The quantum revolution had begun. Even the great achievements of classical physics, Newtonian mechanics and electricity/magnetism, were on their face inconsistent with each other. Axiomatization would have been premature. Weyl, who made helpful contributions to the mathematics of the new physics, put it well:

> The maze of experimental facts which the physicist has to take into account is too manifold, their expansion too fast, and their aspect and relative weight too changeable for the axiomatic method to find a firm enough foothold, except in the thoroughly consolidated parts of our physical knowledge. Men like Einstein or Niels Bohr grope their way in the dark toward their conceptions of general relativity or atomic structure by another type of experience and imagination than those of the mathematician, although no doubt mathematics is an essential ingredient. Thus Hilbert's vast plans in physics never matured.[1]

The opening paragraph of the sixth problem reads:

> The investigations on the foundations of geometry suggest the problem: *To treat in the same manner, by means of axioms, those physi-*

> *cal sciences in which mathematics plays an important part; in the first rank are the theory of probabilities and mechanics.*

Axiomatizing the theory of probabilities was a realistic goal; Kolmogorov accomplished this in 1933. The word "mechanics" without a qualifier, however, is a Trojan horse. Webster's defines the English word: "a branch of physical science that deals with energy and forces and their effect on bodies." (It is essentially the same in German—*Mechanik.*) That definition could include gravitation, electricity/magnetism, the strong force, the weak force, and any force that is newly hypothesized this week. What exactly is a body? This part of the problem has clearly not been solved. In his article on the sixth problem for the AMS symposium on Hilbert's problems in 1974, A. S. Wightman wrote: "Since physics has gone through two revolutions in the twentieth century, one might think that the original questions posed by Hilbert would be by now primarily of archaeological interest."[2]

However, physics especially holds surprises. As it progressed by leaps and bounds, the new theories often involved a lot of mathematics, or what appeared to be mathematics. Sometimes physicists made up new, not entirely rigorous mathematics, e.g., the Dirac delta function when first used by P. A. M. Dirac around 1930 or, more recently, the Feynman path integral, which remains popular and poorly defined. Sometimes they rediscovered already existing mathematics. Werner Heisenberg's infinite matrices can be understood as a particular case of Hilbert space. Sometimes they actually talked to the mathematicians in real time, as Einstein did when he used Riemannian geometry in formulating general relativity.

Hans Lewy, a prominent mathematician present in Göttingen during the founding of quantum mechanics, was asked if he had anything to do with its development. Lewy's response says much:

> No. I was repelled by it. To my mathematical mind the things that the physicists were doing were too sloppy. In other words, they had obviously some physical intuition which I didn't have, but their mathematics was objectionable.[3]

However, it became increasingly clear that the new physics was right—bombs went off; the transistor was invented. The physics was clearly mathematically complex. Well before the man on the street realized the new physics was important, respectable mathematicians took stabs at it—including Weyl and von Neumann. Despite this early intervention, things got messier before they got cleaner, as quantum mechanics was made consistent with special relativity. (Make a calculation, the answer is infinite—well, make it again for something you think should be 0. Infinite again. Carefully subtract the two infinities—what's left over must be the answer. I must add that the process has been clarified and made rigorous by many people in the years since this method first appeared.)

The cultures of physics and mathematics drifted apart in the 1930s and '40s, and by mid-century mathematics was very pure indeed. For the purest of the pure, mathematics became elaborately rolled-out set theory. Some popular fields, like most of algebraic topology and algebraic geometry, seemed to have no application whatsoever to physics, yet were of obvious, deep interest to mathematicians. There was little good mathematical "action" in physics to pull mathematicians away from their own concerns. Hundreds and hundreds of "fundamental" particles were discovered at ever-more-powerful accelerators, so the frontier of physics began to resemble a bestiary, not even close to a unified mathematical theory. As the number of serious mathematics and physics departments grew across the world, and individual departments grew in size, there was also increasing bureaucratic and physical separation between the two. Freeman Dyson said in 1972 in his Gibbs lecture, sponsored by the AMS, "The marriage between mathematics and physics, which was so enormously fruitful in past centuries, has recently ended in divorce."[4] (This figure of speech came from his friend, Res Jost.) Dyson said communication was already breaking down in the 1860s when mathematicians failed to grasp the *mathematical* importance of Maxwell's equations.

The divorce was not accepted by all the children, and the Russians in particular stubbornly refused to believe that the marriage was not still intact. Major mathematical figures were sometimes drawn to physical problems even in the years of greatest separation. Stephen Smale solved one of the outstanding problems in algebraic topology, then started working on classical mechanics, and thought a great deal about a radio oscillator. It could be argued that Dyson himself is best described as a mathematician. He was trained as a mathematician, did important work as a mathematician—including interesting work on Hilbert's seventh problem—and has never really stopped. His contribution to physics was to increase the mathematical clarity of how relativistic quantum mechanics subtracted infinites and got the right "physical" answer.

With this as a backdrop, Hilbert's sixth problem kept some positive contact between the disciplines. Anything Hilbert had proposed remained worthy of consideration.

The logjam in physics broke slowly. First the theory of how electrically charged particles interact became more mathematically sound. Then came a theory for the weak force treated together with the electrical force. Then a theory for the strong force emerged. The apparent bestiary resolved as each "fundamental" particle was found to be a different combination of a few kinds of quarks. Mathematically flavored and even axiomatic treatments of these theories became more numerous and more mathematically presentable. By the 1970s some of the worst difficulties had been dealt with, and the physicists possessed theories that at least in principle described what was happening in the laboratory. These theories were complex. Obtaining numerical predictions from the equations was very difficult and often impossible. This is the sort of situation, historically, in which mathematicians have made their contributions. If one looks at a textbook like Herbert

Goldstein's *Classical Mechanics*, most of the names attached to the methods used are the names of mathematicians—Lagrange, Euler, Noether, etc. So the question of how to make the new physical theories work mathematically began to generate interesting mathematical work on the side, as an estranged pair may tentatively meet for lunch, then dinner, before spending the weekend together. We will see an odd point of connection in algebraic geometry, in the chapter on the fifteenth problem.

Mathematics and physics officially remarried as the century waned, in a big church wedding in 1990, when Edward Witten, a physicist, received a Fields Medal at the International Congress of Mathematicians in Kyoto. Two of the other three medalists had done work with physical connections. Since that time, more and more theoretical mathematicians have been working on these physical theories. Standards of rigor are looser in these areas, more like eighteenth or nineteenth century mathematics. It's like Euler playing with series that don't converge, e.g., $1 - 1 + 1 - 1 + 1 - 1 \ldots = 1/2$. By the time he was done it was clearer how series do converge. Today, some of the most pure and obscure mathematics is being grabbed from the shelf by "physicists"; conversely, new ideas from physics are being showered on mathematicians. (Isn't it obvious that the topological theory of knots is of central interest to anyone interested in superstrings?) V. I. Arnold and Michael Monastyrsky captured the situation in a low-key way in the preface of the 1993 book *Developments in Mathematics: The Moscow School*. Referring to Arnold's contribution to the collection, they state: "It should be noted that most of the paper deals with very sophisticated and modern mathematics applied to—or motivated by—physics. This is the general trend in science today."[5]

We have come to an odd position. If one of the new theories is right—say, superstring theory—it should be axiomatizable. If that were to prove to be the case, then Hilbert's largest and dreamiest problem would have a clean solution. However, the relationship between mathematics and physics—clearly a case of opposites attracting—continues to be tempestuous. In response to my query about his quote, Dyson wrote to me in 1999, "The problem now is that there is an impending divorce between string theory and the rest of physics."[6] Maybe string theory will go to live with mathematics—at the present time this "physical" theory has no connection to experimental data.

Number Theory

« *7, 8, 9, 11, 12* »

First, State the Tune

In *A Mathematician's Apology*, the English mathematician G. H. Hardy (1877–1947) explicitly brackets mathematics with poetry and art and says that in judging a mathematical theorem, "Beauty is the first test." Hardy was a number theorist.

Another number theorist was Paul Erdős (1913–1996), who has a connection to the eighth Hilbert problem. Though a remarkably productive mathematician (Janos Pach in *The Mathematical Intelligencer* counts more than 1,500 papers), Erdős never held, in the words of Mark Kac, writing in 1985, "anything resembling a regular job."[1] Erdős published his first papers at eighteen. He left his native Hungary in the 1930s and remained an expatriate, wandering from colloquium to temporary appointment all his life and sleeping on the living room couches of friends. He died while attending a conference in Warsaw in 1996. Stanislaw Ulam (1909–1984) met him when Erdős was invited to give a colloquium at the University of Wisconsin at Madison and writes, "His visit to Madison became the beginning of our long, intense—albeit intermittent—friendship. Being hard-up financially—'poor,' as he used to say—he tended to extend his visits to the limits of welcome."[2]

Later in their friendship, Ulam was leaving the hospital in Los Angeles after an acute inflammation of the brain, never clearly diagnosed, that left him temporarily aphasic and partially paralyzed. Erdős appeared, suitcase in hand, and said, "Stan, I am so glad to see you are alive. I thought you were going to die and that I would have to write your obituary and our joint papers." And then, "You are going home? Good, I can go with you."[3]

Ulam welcomed this, though he says his wife was "somewhat more dubious." In the car on the way down to the house on Balboa Island where the Ulams were staying, Erdős started talking mathematics to the trepanned and bandaged Ulam, who was able to respond to his satisfaction. Erdős pronounced: "Stan, you are just like before," which greatly cheered Ulam, who had been entertaining dire thoughts about impairment. When Ulam won two games of chess later, to Erdős's visible disappointment, he was further convinced of the possibility of recovery. Erdős was not one to throw a game.

Erdős stayed through the recuperation and the two took increasingly long walks on the beach. On one of those walks Erdős noticed a particularly sweet little child and said to Ulam, "Look, Stan! What a nice epsilon." (In a private language Erdős referred to children as epsilons.) Ulam writes, "A very beautiful young woman, obviously the child's mother, sat nearby, so I replied, 'But look at the capital epsilon.' This made him blush with embarrassment."[4]

Erdős was so collaborative in his mathematics and so peripatetic that it was said one was not a real mathematician if one didn't know Paul Erdős. The Erdős number was formulated—in Ulam's words, "the number of steps it takes any mathematician to connect with Erdős in a chain of collaborators."[5] Erdős himself was 0. The number 1 signified someone who had collaborated with Erdős—458 mathematicians had this number.[6] The number 2 designated someone who had collaborated with someone who had collaborated with Erdős. Ulam said few mathematicians had an Erdős number higher than 2.

The analogy between music and number theory, particularly with jazz improvisation, is strong. In both disciplines there is a lot of choice in the area explored. Sometimes there are objective criteria that demand an idea be explored, but often a judgment that centers around ideas like attractiveness, beauty, and significance is important in choosing and evaluating work. In all areas of mathematics efficient, clean proofs are valued, but in number theory the character of the proof is even more important. Is it maximally simple? Elementary? By elementary one means, does the proof use methods that are strictly from arithmetic? Is complex analysis required to prove a given result? Is the axiom of choice used? These make a proof less elementary.

Ideas in number theory are explored and re-explored, as jazz standards are played by many musicians, or even the same one, differently on different nights. Gauss's law of quadratic reciprocity was proved in more than fifty different ways in the nineteenth century. Gauss himself published six different proofs.[7] It also attracted other major figures such as Cauchy, Ferdinand Eisenstein (1823–1852), Jacobi, Kronecker, Kummer, and Liouville. When we come to class field theory in Hilbert's ninth and twelfth problems, we will see the same process of reworking and experimentation. The eighth problem is about the Riemann zeta function, which gained prominence in 1896 when it was used in proving the "prime number theorem." One of Erdős's greatest achievements, in 1949, a collaboration with Selberg, was to prove the prime number theorem without the Riemann zeta function, using purely arithmetical methods. The proof is difficult but pure. The emotions excited in number theory are strong. The personalities involved can be outside the conventional run.

Utility is never the goal and rarely the result. Hardy defended all of mathematics on that ground and believed that freed mathematics from the horrible association with war that he saw in science. Unfortunately, writing in 1940, he honored Einstein and Dirac as mathematicians and gave relativity and quantum mechanics as examples of useless mathematics: "Almost as useless as number theory."[8] Hardy was on firmer ground with number theory. As of this writing it remains, despite important uses in encryption and more tangential and speculative uses in physics, remarkably useless. If some of it has turned out to be useful, that is just a happy accident. There are reasonable people who would disagree, but I think they are wrong.

Hilbert proposed six problems in number theory (depending on how we define number theory), more than in any other area except analysis. Some, including Weyl, believe that Hilbert's own greatest work was in number theory.[9] The number theory problems have been among the most productive. Four of the six—the seventh, ninth, tenth (treated with the foundation problems), and eleventh problems—have been solved cleanly. The twelfth problem has seen great progress. The eighth problem has attracted a vast amount of work.

What is number theory? Here is a sampler of classical examples:

Specify a number or class of numbers and various questions can be asked. Start with what numbers are in the class. For instance, we have shown that transcendental numbers exist. Hilbert clarified Lindemann's proof that π was transcendental. Can we exhibit any other such numbers? We have seen that there are vastly more transcendental numbers than the other kinds, so this ought to be easy. This is not the case, and proving other numbers transcendental is what the seventh problem is about.

The prime numbers make up one of the most obvious classes of numbers. How many primes are there? How many prime numbers are smaller than a given number? Can we make a solid estimate? Individual primes are of interest. A lot of effort has been expended finding very large primes—hundreds of digits—and, of course, finding ways of proving that they really are prime. Patterns are searched for—in the eighth problem Hilbert asked about Goldbach's problem, whether every even natural number greater than 2 is expressible as the sum of two positive prime numbers and whether "there are an infinite number of pairs of prime numbers with the difference 2." In 1931 progress was made on Goldbach's conjecture. Russian mathematician L. G. Shnirelman (1905–1938) proved that every even number is expressible as the sum of not more than 300,000 primes.[10] This is the kind of near-miss that can happen in number theory. Shnirelman's result was a major advance. Another near-miss—based on a 1966 result of Alan Baker's that also represents a major advance—is that if the difference between one perfect cube and twice another is 1, then neither can be larger than 1.5×10^{1317}.

Another type of result is Waring's conjecture that every whole number can be expressed as the sum of four squares, nine cubes, nineteen fourth powers, etc., proved by Hilbert himself. This is a generalized version of the result about squares proved by Lagrange. Or there is the problem of partitions. If we take a given positive whole number, how many ways can it be expressed as the sum of other positive whole numbers? The answer for 3 is 1 + 1 + 1, 1 + 2, 3, making a total of three ways. By the time we look at a number like 100, the number is huge and can only be calculated in messy, tedious ways. Can we find an easier way? Or find other facts about how whole numbers can be expressed?

Another type of number theory problem concerns specific identities giving values for infinite series, infinite products, continued fractions, etc. They are daunting to the untrained eye. A simple and elegant example is:

$$\pi = 4 - \frac{4}{3} + \frac{4}{5} - \frac{4}{7} + \ldots$$

(from Leibniz)

A more complicated example is:

$$\frac{1}{1+}\,\frac{e^{-2\pi\sqrt{5}}}{1+}\,\frac{e^{-4\pi\sqrt{5}}}{1+\ldots} = \left[\frac{\sqrt{5}}{1+\sqrt[5]{\left\{5^{\frac{3}{4}}\left(\frac{\sqrt{5}-1}{2}\right)^{\frac{5}{2}}-1\right\}}} - \frac{\sqrt{5}+1}{2}\right] e^{2\pi/\sqrt{5}}.$$

(from Ramanujan)[11]

The tenth problem was about diophantine equations—algebraic polynomial equations with the proviso that we accept only whole number solutions—and we saw that there is no general method for deciding which equations are solvable. Number theorists have always focused on solving specific diophantine equations or specific classes of them—Fermat's last theorem is about diophantine equations. This kind of limited question is hard enough, and number theorists didn't want to waste their time trying to solve everything at once. Baker's estimate, mentioned above, can be used to decide whether $x^3 - 2y^3 = 1$ (or more generally an equation of the form $x^3 + by^3 = c$) is solvable by checking by hand all possible solutions smaller than 1.5×10^{1317}. This is obviously not a very practically effective "effective procedure." Hilbert's tenth problem must have seemed like pie in the sky (rather than π), and this is probably why it was solved by logicians.

The modern methods used in proving results in number theory have become impressively complex. However, current topics of interest can still often be stated in simple language. The ABC conjecture is a good example of this. If we have two integers that don't share a common factor, but each is divisible by a large, high power of an integer, then their sum "tends" to not be divisible by a large, high power of an integer.

The algebraic numbers have become a broad field of research, and the words "class field theory," "*p*-adic numbers," "ideals," "ideles," and "adeles" are in this area. The original prompting came from studying diophantine equations, but clearly the algebraic numbers and the structure of their various subfields are of interest in and of themselves. We can investigate the values polynomials can take if we allow algebraic numbers to be used as coefficients. The eleventh problem is an example of this type of problem. Being a spectator in this area is a little like watching an Olympic figure skating competition. The word "difficult" is used often. The skater leaves the ground spinning in circles and lands successfully, and we are told that we saw a triple Lutz completed. Dyson has used this analogy in reference to mathematics generally, but I think it is most apt with regard to number theory.

Classifying a problem as number theory or algebra is somewhat arbitrary at the edges, particularly when we are investigating numbers that are defined specifically as being the solutions to algebraic equations. I think Hilbert saw the break as coming at the point where specific numbers or classes of numbers still lie at the heart of the problem, no matter how algebraic the phrasing. Purely algebraic questions for Hilbert involve patterns in the algebraic formalism—like his own work in invariant theory—rather than in the character of specific numbers that might be solutions to algebraic equations. In this way of looking at things, the work of Galois and Abel proving fifth-degree polynomial equations cannot be solved by radicals could be classed as number theory, at least to the extent it was ever applied to a specific equation. The general theory, though, often sits at the center of courses in abstract algebra. When I first encountered Galois theory in an abstract algebra class, I had to do research to discover how what we had done actually connected to solving or not solving a polynomial of the fifth degree. I found a book devoted to making that connection, *Classical Galois Theory With Examples* by Lisl Gaal—a close friend of Julia Robinson's, as it turns out. The twelfth problem straddles this line, and Hilbert writes that it "may form a transition to algebra and the theory of functions." The last chapter of this section treats algebraic number theory, combining the ninth, eleventh, and twelfth problems.

Transcending Local Conditions

The Seventh Problem

Gelfond, Schneider, Siegel

The seventh problem, titled by Hilbert "Irrationality and Transcendence of Certain Numbers," asks for a proof that $2^{\sqrt{2}}$ and similarly formed numbers are transcendental. Recall that a number is transcendental if it is not the solution of any polynomial equation with whole numbers as coefficients.

Around 1740, Euler conjectured that such numbers exist and called them "transcendental," because they transcend the realm of algebra. (Hilbert's seventh problem is an extension of one of Euler's conjectures.) There the subject lay for a hundred years, until 1844, when Liouville not only proved transcendental numbers exist in some abundance but provided examples and a recipe for finding more. However, the numbers Liouville produced were constructed to be transcendental. Were they just oddities? In 1873, Hermite proved that e, the base of the natural logarithms and one of the most important numbers in mathematics, is transcendental. In the following year Cantor proved the more startling result that, in modern parlance, "almost all" numbers are transcendental. With Lindemann's 1882 proof that π is transcendental, there were now two important numbers proven to be transcendental. This result about π solved in the negative the classical Greek problem of squaring the circle—using straightedge and compass only and starting with a circle, can you construct a square with the same area? You can't. It is easy to show that any geometric length constructed from another using compass and straightedge can be written in terms of the first using fractions, addition, subtraction, multiplication, and the square root. Such a number is obviously algebraic. The ratio of the side of a square to the radius of a circle of equal area is equal to the square root of π. If the circle could be squared, then there would exist an algebraic relation between π and 1 using nothing more complicated than a square root. Since π is transcendental, there cannot be such a relation. Lindemann actually proved that numbers created by a more general rule were transcendental, with π the most prominent example. Hilbert's problem was a further attempt to flesh out the realm of transcendental numbers.

When Lebesgue introduced measure theory, he added insult to injury vis-à-vis mathematicians' lack of knowledge of the transcendentals. Recall that he extended the idea of length to a more general kind of "measure" that agreed with length when a length could be determined, but could be applied to a wider class of sets of points. The core of his attack was to say the measure of a set is the lower limit of the lengths of line segments that, cut into pieces, can be used to cover, or sit on top of, every point in the set. It made the idea of length more flexible, and it brought the concept of limit to the idea of measure. Earlier we saw that the measure of the algebraic numbers in any line segment turned out to be 0 because they can be listed. Start with an arbitrarily small line segment and tear off pieces—never taking all of what is left—and methodically cover every algebraic number on the list with a piece. The lower limit of the starting segment is 0. Therefore the set of algebraic numbers, or any listable set, has measure 0. The length of any line segment, then, is made up of the transcendental numbers, yet only two common numbers had been proven transcendental. This certainly was surprising.

Here is a number cooked up in the mode of Liouville to be transcendental:

$$\frac{1}{10} + \frac{1}{10^{10}} + \frac{1}{10^{10^{10}}} + \frac{1}{10^{10^{10^{10}}}}$$

This is an unusual number. I chose it because it is both extreme and easily written in decimal notation. Suppose this number is not transcendental. Then it satisfies a specific polynomial equation with integer coefficients. For the sake of argument, let's say it satisfies the following equation:

$$2x^3 + 23x^2 + 3x + 6 = 0$$

I am just picking a candidate equation to show how the argument works. Our number isn't the solution to this, and on reflection, we will be able to see that the same observation applies as a rejection for any possible candidate equation. Hence the number is transcendental.

First let's examine our number and try to write it down in decimal notation:

.1000000001000000000000000000000000000000000000000000 . . .
1st 1 2nd 1 3rd 1→

Before we get to our third 1 we are off the page. How far off the page? Calculation yields—assuming our zeros come ten to the inch—it's about 14,000 miles off the page. What about the next 1, coming from

$$\frac{1}{10^{10^{10^{10}}}} \quad ?$$

This 1 would appear beyond the end of the known universe. The next term, with five tens stacked up, involves an expansion of scale yet more vast. Our number is isolated 1s with ever-vaster deserts of unremitting 0s between them.

Let's look at our number squared, cubed, etc.

$$x = .\mathbf{1}00000000\mathbf{1}000000000000000000000000000000000000000000000 \ldots$$
$$x^2 = .0\mathbf{1}00000000\mathbf{2}00000000\mathbf{1}00000000000000000000000000000000000 \ldots$$
$$x^3 = .00\mathbf{1}00000000\mathbf{3}00000000\mathbf{3}00000000\mathbf{1}0000000000000000000000000 \ldots$$

Squaring or cubing this number results in more 1s, 2s, and 3s, etc., in the deserts of zeros, but the deserts are just as big. In each case the visible part of the number ends with a 1 preceded by eight zeros and followed by many more. The same general pattern occurs around the next 1 that is 14,000 miles to the right. It is pretty clear that our candidate equation with its small coefficients and exponents can't process these numbers and get 0 out of them. It will take an equation with big exponents and/or big coefficients to have a chance at dealing with even the first two clusters of numerals of our number. If we try to imagine an equation that would zero out the first *three* clusters of our number—i.e., including the 1 that is 14,000 miles out—we need even bigger coefficients to have a chance. Further, if we consider the next 1, from

$$\frac{1}{10^{10^{10^{10}}}}$$

there is another radical jump in scale. That is the key—the scale keeps going up. If this number is algebraic it satisfies a polynomial equation. A polynomial equation has a finite number of terms and has a scale—we could say roughly the biggest number to appear in the equation. ("Scale" in this sense is not used in mathematics but accurately conveys the idea.) The cooked-up number we are considering has no smallest scale. It has one scale for the first 1. Another for the second. Another for the third, and so on. No matter how large the numbers we use as coefficients in a polynomial equation, we can easily find an isolated 1 in our number that is so far out and so isolated that even being raised to a stupendous power and multiplied by stupendous numbers it will still be, written in decimal notation, a series of numbers that are lone and very isolated from any lining up and zeroing out that any candidate equation can do. Our oddball number is transcendental.

We can use this number as a pattern to generate an abundant class of tiny transcendental numbers. Throw out the first two terms. This new number, which will be our template, starts out with a string of zeros that goes on for 14,000 miles before a 1 shows up. We can use the set of real numbers between 0 and 1 to generate an uncountable multitude of small numbers based on this template. Take any real number, for example .257 . . . . Use it to make a number of our type as follows:*

$$\frac{2}{10^{10^{10}}} + \frac{5}{10^{10^{10}}} + \frac{7}{10^{10^{10^{10^{10}}}}} + \dots$$

These numbers are tiny, distinct, and all transcendental. Numbers generated in this way can each be added to any fraction and will result in a cluster of transcendental numbers getting close to the value of the fraction very fast, but as numerous as the real numbers. This incredible clustering of transcendental numbers of even this one very narrow type around a fraction suggests how they really can be the black of the sky and the fractions themselves rare stars in this continuum, as E. T. Bell's metaphor had it.

This is how things stood shortly after Hilbert posed his problem. Transcendental numbers exist and, thanks to set theory and measure theory, we have some feeling for their abundance. However, it remains very hard to prove a number transcendental. In a popular lecture delivered in 1920 Hilbert said that he thought no one in the lecture hall would live to see $2^{\sqrt{2}}$ proved to be transcendental—the seventh problem. Siegel was in the hall. In the same lecture Hilbert said he thought he himself might live to see Riemann's hypothesis proven—the eighth problem—and that the youngest members of the audience might see Fermat's last theorem proven. He had the order of solution the wrong way around. Riemann's hypothesis remains unproven (to date). Fermat's last theorem was proved in 1994. Hilbert and Siegel both saw the solution of the seventh problem by A. O. Gelfond (1906–1968) and Theodor Schneider (1911–1988) in 1934.

The work in transcendental numbers at first concentrated on refining and extending Liouville's result from 1844, which was the inspiration for the example we discussed. Gelfond wrote:

> All methods of proof of the transcendence of a number in either the explicit or implicit form depend on the fact that algebraic numbers cannot be very well approximated by rational fractions. . . .*[1]

* For this process to always work, write any number that ends in 0 as ending 9999..., e.g., .50 = .49999....(This is the opposite of the usual convention.)

* "Algebraic numbers cannot be well approximated" needs a little clarification. First, Gelfond means algebraic numbers that are not themselves rational. Second, 1/3 and 2/6 are equally good approximations of a number near 1/3. The second uses a larger denominator to get the same result. It is viewed as less "good"—10/30 is a lot less good. "Good" here carries with it efficient use of relatively small numbers compared to the error. The transcendental numbers we created are all very well approximated by rational fractions.

Axel Thue in 1909 and then Siegel in 1921 achieved important refinements of Liouville's result, pinning down a more precise numerical measure of how poorly attempts at approximation would work.[2]

We met Carl Ludwig Siegel through his essay on the Frankfurt mathematics history seminar. He contributed to the solution of the seventh problem in more than one way and was a leader in pursuing the program of research that the problem suggested. We will run into him everywhere in the number theory problems. Siegel was an only child, born the last day of 1896 in Berlin. His father was a mailman—specifically a "money-order carrier"—who was gregarious and rejected offers of promotion that would have taken him off the street.[3] Young Siegel was close to both parents, especially his mother. Both had moved to Berlin from the Rhineland, and neither parent's first name is recorded in the accounts I have found.

Siegel grew large-boned, tall, and shy. He attended the neighborhood *Volkschule* and then *Realschule* and *Oberrealschule*, an educational track that was more practically oriented and less academic than the gymnasiums. However, study at the universities was allowed for *Oberrealschule* graduates, particularly for a student interested in mathematics. Siegel's relations with his mathematics teachers, however, were strained. He did not do things the way he was supposed to. When taking an exam, he stubbornly started with the most difficult problems, which he solved. But sometimes he didn't have time to do the easy problems. This led to lower scores than he thought fair.[4] Siegel has said his mathematics teachers ranged from average or mediocre to even "malign"[5] and joked (or perhaps he meant it) that bad teaching drove him to study mathematics. He checked an advanced book out of the library as a teenager, somewhat by mistake, Volume 3 of Heinrich Weber's *Lehrbuch der Algebra*, and worked his way through it. Edmund Hlawka quotes Siegel: "I loved drawing and showed early talent in it, as well. Perhaps if the mathematics classes had been good, I would have ended up a painter!"[6]

Much of what we know about Siegel's life comes from the many stories told about him, including those he told himself. The mathematician Hel Braun (1914–1986), a student and later a friend and companion of Siegel's, wrote that Siegel made a distinction between "how it really was" and "how it happened by chance" ("*wie es wirklich war*" and "*wie es sich zufällig ereignet hat*").[7] Perhaps this is meant to convey a belief that happenstance in a life, accidental details of birth or history, could obscure the underlying, or more important, truths. As an adult Siegel painted in an impressionist style. Most of the stories about Siegel have a shifting quality. They were not written down when they were told, and those who heard them are now either old or dead, heard them third- or fourth-hand, or heard them from Siegel when he was fairly old. Nevertheless, I will use versions I have come upon and try to reconcile their elements.

Siegel commenced an individual and unguided self-study of mathematics in *Oberrealschule* before university. Braun, in a section that piles up emotional themes in his life at the time, says:

> No doubt, C. L. Siegel was the unpopular model pupil throughout his school career. He clung tenderly to his mother and had little contact with other schoolchildren. He told a somewhat sentimental story of himself, at sixteen, sitting in a beautiful meadow and experiencing an unutterable sadness for the first time. Subsequently, this sadness to the point of tears would recur often. He drew a lot during this time. . . .[8]

When Siegel was eighteen his mother died, and this was devastating.

After getting his *Abitur* (high school diploma), Siegel enrolled at the University of Berlin in fall of 1915. The war had begun. In a 1968 memoir about his teacher, Ferdinand Frobenius (1849–1917), Siegel wrote:

> Although I had no understanding of the politics behind the events, I instinctively felt a disgust of this violent human enterprise, and decided to dedicate myself to a field as distant as possible from these earthly matters—which made astronomy seem, at that time, just the right subject for me. That I, in spite of this, came to number theory eventually, was a matter of the following coincidence. The person who was to lecture in astronomy had given notice that he would start fourteen days after the beginning of the semester. . . . At the same time, on Wednesdays and Saturdays from 9 to 11, however, Frobenius was to lecture on number theory. Because I hadn't the slightest idea what number theory might be, I attended the lectures for the two weeks out of pure curiosity. This, however, decided my scholarly direction for the entire rest of my life. I did not attend the astronomy lectures once they began. . . .[9]

After his hard years and bad teachers in *Oberrealschule* Siegel found the university liberating. He admired Frobenius and enjoyed his presentation. Frobenius lectured without notes. Siegel wrote:

> He obviously had a great deal of fun rattling off the various algebraic identities . . . with absolute certainty and astounding speed, glancing slightly mischievously every so often at his listeners, who were zealously attempting to write down the massive flow of information. Otherwise he rarely even looked at the students, concentrating mostly on the board.[10]

In the long vacation before the second semester, Siegel received in the mail a request that he appear at the office of the finance officer of the university. One of the stories that mothers told their children during this period in order to scare them into good behavior (at least in Siegel's experience) was that if they were bad the state would take them from their parents. Siegel went to the office specified in the letter with great trepidation. He was told that he had won the Eisenstein scholarship and would receive 144 marks and 50 pfennigs. Siegel was so shy he didn't ask how he had come to be awarded the scholarship or who Eisenstein was (a gifted mathematician whose wealthy family had established a scholarship after his untimely death). Only later did Issai Schur (1875–1941) tell Siegel that Frobenius, who pretended not to notice students, was probably the source of his nomination, based on problem sets he had turned in.

In his third semester Siegel took a class from Schur. After a lecture on Pell's equation, Schur mentioned a similar group of equations and referred to Thue's 1909 paper. Siegel went to this deep paper and:

> When I attempted to read this, I quickly became confused by certain letters c, k, θ, ω, m, n, a, s, whose deeper meaning seemed a riddle to me. In order to better understand, I changed the order of the lemata and used other symbols among which, more by chance than orderly thinking, was a parameter Thue had not used. This, to my astonishment, resulted in a focusing of the approximation theorem. Because I could find no error in my thinking, I wrote it out—about four pages—and gave it to Schur at the next opportunity. My expectations, however, were deeply disappointed, for a couple of weeks later, Schur handed me a manuscript, yellowed by the sun from sitting on his desk, with the brief remark that I had merely calculated identities, from which no conclusion could be drawn.[11]

Siegel had in fact accomplished what he thought he had, but his four-page paper had not conveyed it to Schur. That a raw but extremely talented twenty-year-old could produce this kind of result is not entirely surprising in this area of number theory. The methods are "elementary" and cleverness at calculation can be what takes the problem to the next level.

Siegel, in only his third semester, had entered the history of mathematics in a permanent though not yet public way. There are few simple, general theorems on transcendental numbers. In 1955 Klaus Roth made a final improvement in the Thue–Siegel theorem—any stronger version of the theorem is false—and three years later was awarded a Fields Medal. Siegel, however, in 1917 had convinced only himself, and he possessed a yellowing four-page manuscript that was too terse to be understood by anyone else. Then he was found "fit for military service"[12] and drafted. Siegel seems to have regarded this as akin to kidnapping. In the 1968

Frobenius memoir he captures his attitude in the phrase "before the military tried to abuse me for their purposes." He tried to escape.

We know that Siegel was sent to Strasbourg. From that point on, the story of his time in the military becomes cloudy. Siegel didn't like to talk about this period, and even Schneider in his memoir flies past it. André Weil wrote that Siegel told him he had decided "this war is not my war" and deserted: "He took off, disappearing into thin air near the Haut Koenigsbourg, where he gave himself over to unhappy reflections."[13] He was caught, but it would appear he was not immediately jailed, merely taken back to a barracks. The most compelling account of the aftermath of his "capture" that I have seen is supplied by Braun. She reports, based on what Siegel told her, that Egon Schaffeld saved him. Schaffeld, the very competent son of a wealthy, wheelchair-bound owner of a wool spinning and weaving factory, was spending the war handling recruitment paperwork in Strasbourg. Before the war, Schaffeld's father had wanted him to run the business, but the son badly wanted to study at the university. His father would agree to law, but Schaffeld wanted to study physics. He was two years older than Siegel and had been drafted early, so the conflict was not resolved. In Strasbourg, Schaffeld went to mathematics lectures in his free time, because mathematics was a prerequisite for physics, and was won over.

While processing recruits' paperwork Schaffeld came upon Siegel's documents, which in an attempt to win exemption from service included descriptions of his mathematical abilities. Schaffeld went to the recruits' dormitory and found a "totally depressed" Siegel. Schaffeld was something of an operator. For example, he himself had a private room. He immediately attempted to buoy Siegel's miserable spirits and got him some special treatment. Braun writes, "Then they made plans for Siegel. Siegel's conditions were hard and nearly impossible to fulfill: To quit the military services."[14] Eventually it was arranged that Siegel be sent to a military mental hospital for evaluation.

Siegel was now found unfit for military service and discharged. Hlawka writes, "He was released after five, for him, very hard weeks."[15] It was still 1917, and it would appear that Siegel returned home to stay with his father. Siegel told Braun that he was in Berlin in 1918 during the fall revolution that overthrew Kaiser Wilhelm and that his father kept him off the street. In this period Siegel probably worked as a private tutor. Though his stay in the military hospital may have been somewhat contrived, Siegel described himself as having had a "nervous breakdown," and the period of late 1917 to early 1919 can be thought of as a period of recovery.

In this period Edmund Landau (1877–1938), professor of number theory in Göttingen, enters the picture. Harold Davenport (1907–1969) had many conversations at meals with Siegel in 1966 and made notes. Landau's father, a gynecologist, lived on the same street as Siegel's family in Berlin. Before Siegel was drafted, Landau senior had heard that he was studying mathematics and invited him over. When Siegel told him that he was studying number theory, Landau senior showed Siegel his son's two-volume book on the analytic theory of prime numbers. He told

Siegel he probably couldn't understand it yet, but maybe someday he would be able to. Of course, Siegel saved up money from tutoring and went out and bought a paperback edition, which he tried to understand. He decided Landau's father had been right. Later Siegel had the books rebound in leather. It is difficult to know for certain, but it is possible that Landau senior later collaborated with Schaffeld in getting Siegel into the military sanitarium, treated, and discharged without punishment. He later introduced his son to the promising young mathematician. Siegel presented Professor Landau with his now six-page-long paper. Landau looked at it and pronounced that it had a 10% chance (in other versions 15%) of being correct, but this was enough for him to see Siegel's potential and take him on as his student. In summer of 1919 Siegel moved to Göttingen and recommenced his formal studies.[16]

Siegel was joined in Göttingen by Schaffeld, Erich Bessel-Hagen (1898–1946), and Maria (Siegel's girlfriend from Berlin). Braun says that Siegel and Bessel-Hagen first met as students in Berlin. Bessel-Hagen revered Siegel for his mathematical ability, and Siegel in turn was gratified to be admired by someone he could accept. They were friends all their lives, until Bessel-Hagen's early death in 1946. Bessel-Hagen walked with a slight limp—Braun thought because of polio. Schaffeld, Siegel, and Bessel-Hagen were united in their attitude toward marriage. Braun writes, "I do not know which of the three was more decided against ever marrying, but all three remained bachelors, expressing their views [on marriage] with varying degrees of theatricality."[17] Schaffeld and Bessel-Hagen filled very different roles for Siegel. In later years Siegel would drop everything if Schaffeld appeared to go for a walk or go out to a restaurant. Braun writes, "I had the impression that Schaffeld was the only authority figure Siegel respected besides his father."[18] In contrast, Siegel couldn't resist baiting or playing sometimes cruel and humiliating practical jokes on Bessel-Hagen. He would then repent and do something to make it up to him.

Siegel spent much of this first year in Göttingen working alone, to show what he could do. He would make a revision in his paper and bring it to Landau. Landau would criticize it, and Siegel would revise it again. Eventually the paper grew to forty pages and Landau pronounced that it was 90% likely to be true. The paper was published in 1921, by which time it must have been viewed as 100% likely to be true, and became a milestone in the history of transcendental numbers. Siegel's papers are known for their density and the quality of the writing. His original four pages had been rather too concise, and he was now learning how to communicate his ideas and methods effectively. His proof of the Thue-Siegel theorem is striking partly because it is almost entirely algebraic—and not in the modern sense. It is an inspired application of school algebra. Hans Grauert wrote, "There are theorems that only Siegel could prove; no one knows how he did it."[19]

Siegel received his degree June 2, 1920. During his first year in Göttingen, he had studied with Erich Hecke (1887–1947). When Hecke was appointed full professor at the newly formed University of Hamburg, Siegel followed and taught there the winter semester of 1920–21. However, he remained apart from the mathematical community in Hamburg. Constance Reid says he was "cold, hungry and unhappy."[20]

Courant, who was playing an increasingly active role at Göttingen, heard that Siegel was not happy in Hamburg. He hadn't met Siegel but had heard of his remarkable ability. He knew Siegel was "not easy," but it was clear to him that it was in the interests of Göttingen to bring him back. Though Courant did a completely different kind of mathematics from Siegel and he already had one paid assistant, he managed to get funds for a second assistantship. Housing was scarce in the years after the war, and Courant arranged for Siegel to stay in a room at Felix Klein's house. Sharing a house with Hegel's granddaughter made Siegel nervous, and "he worried constantly that he 'would say the wrong thing.' "[21] Courant often invited Siegel to his own home, where Siegel would sit on the floor and try to get Courant's infant son Ernst to repeat complicated scientific terms. Courant took Siegel to the professors' beach at a local lake and introduced him to Hilbert in the shack where the professors changed into their bathing suits. Hilbert listened and seemed pleased as Courant explained new work that Siegel was doing on the zeta function.

Stories about Siegel's unusual behavior or practical jokes begin to appear during this time. One was the "sinking" or "drowning" of Bessel-Hagen's thesis—Braun calls this an "evil" Siegel story.[22] Siegel had to review Bessel-Hagen's thesis for habilitation—perhaps as a friend, perhaps in some official capacity. However, it came to him at a time when he was deeply involved in his own research, and Bessel-Hagen's thesis was in a different area. Siegel felt burdened, and the longer he neglected the task the more burdened he felt. During this time he was required to travel somewhere by ship. At some point on the voyage Siegel's feelings overwhelmed him. Though he knew he was in possession of the only copy of the thesis, Siegel threw it overboard and sank it, making due note of the coordinates. When he came home, in one version, he told Bessel-Hagen he had done him a favor. It required months of work to reproduce the thesis, and eventually Siegel started feeling bad about what he had done. To make up, he invited Bessel-Hagen on a long vacation to Greece. Braun writes, "This journey made both of them very happy."

Siegel also played practical jokes that were funny and only marginally cruel, usually only on good friends. Hermann Weyl was an irresistible target. Though he was a great mathematician and broadly literate, Weyl was gullible and could be awkward. A big, bespectacled, blond boy from a small town, Weyl for a number of years was not accepted by the in-group at Göttingen—until his mathematical ability redefined "in." Sometimes he and his own friends would converse in a language with the letter p inserted before every syllable. Once he spent an entire party under a chair, answering questions only by barking. Given such a target, Siegel couldn't resist. Natascha Brunswick, who was married to Emil Artin, told me that once at a party Siegel came storming in with urgent news—someone is lost in the woods—we must go find him—making a big fuss. Weyl set out into the woods, earnestly searching. Brunswick said, "He made poor Hermann go into the woods."[23]

Siegel would make dinner plans with someone new to Göttingen and say, "Meet me at the restaurant near the palm trees." If the person didn't have the sense to question the directions, Siegel would let him look indefinitely for palm trees. (In Frankfurt, there *was* a garden with palm trees.) Another time Siegel took an empty baby carriage up to the top of a steep hill and pushed it down, to see what people would do. Once when the Landaus were having a party, Siegel invited a student whose attendance was required to an early dinner. He dragged it out. The student wanted to leave, but Siegel told him to just write a note to Mrs. Landau and tell her to keep dinner waiting. The student said, "Oh no no no," but Siegel prevailed. The student wrote the note. My own note to the reader at this juncture would be that the veracity of any of these stories is open to question, but taken together they convey at the very least some of the feel of the stories that followed Siegel. Telling Siegel stories came to be a popular indoor sport.

Siegel was moody and difficult, but he worked very hard. Hlawka talks about Siegel's "iron energy" in getting through Thue's paper. In 1921 and 1922 Siegel published thirteen very strong, deep papers. He expanded into algebraic number theory and analytic number theory, including his first papers on zeta functions. He used the methods of Hardy, Littlewood, and Ramanujan in his 1922 paper, "On the Additive Theory of Fields," which served as his habilitation. He clearly deserved a professorship, but none was open in Göttingen. Courant's project of keeping Siegel in Göttingen was therefore doomed. On August 1, 1922, Siegel was called to fill Arthur Moritz Schönflies's (1853–1928) position as a full professor in Frankfurt. Siegel was twenty-five years old and had gone from a military mental hospital to full professor in five years.

Hlawka writes that Siegel stopped working after his arrival Frankfurt in 1922 because he suffered from burnout (*uberanstrengung*), beginning again only in 1926.[24] It is certainly true that in his collected works, from after his initial creative outburst until he resumed publication in 1929, only one two-page paper appears, from 1926. This paper was extracted from a letter Siegel had written to L. J. Mordell (1888–1972). It was too important to remain hidden. Weil, who first visited Frankfurt around Christmas in 1926, explained the lack of papers as a consequence of the Frankfurt mathematics history seminar members' belief that too many papers were being published. Whatever the cause, it would be in keeping with Siegel's moods to collapse for a while after a period of great production and then to explode, keeping the emerging results hidden for a time.

In 1929 Siegel published a large paper that expanded the ideas already present in the 1926 paper, and that, among other things, is his most direct contribution to the seventh problem. The first half of the 1929 paper is the one presented to Dehn as a birthday present. Jean Dieudonné (1906–1992) calls this paper "probably his deepest and most original."[25] Schneider, who was Siegel's student, wrote: "I will not allow myself, nor am I able, to compare the value of Siegel's various publications, however, whoever wishes to sense the genius of the great number theorist Siegel need only read these seventy pages."[26] Siegel wrote the final version of the paper in Pontresina, Switzerland,[27] where he often vacationed. Again, he relies heavily—though hardly exclusively—on inspired application of school algebra to achieve his results.

The heart of Hilbert's seventh problem is to prove that $2^{\sqrt{2}}$ is transcendental and, more generally, that any a^b is transcendental when a is any algebraic number and b is both algebraic and irrational. Earlier in the problem's text Hilbert asks for a general study of what happens when algebraic numbers are substituted into transcendental functions. (While it is difficult to prove numbers are transcendental, most functions generated in calculus are transcendental and can easily be proved so. Roughly, a function is transcendental if it can't be defined by applying a finite number of algebraic operations to the variable.) In the 1929 paper Siegel proved that the well-known Bessel function $J_0(x)$ yields transcendental numbers when algebraic numbers are substituted for x. Not since Lindemann proved π transcendental had a naturally occurring number, or in this case, numbers, been proved transcendental. Further, Siegel in his proof found a way to construct certain auxiliary functions that have a lot of zeros where they are desired. Siegel's lemma (an auxiliary proposition used to prove something else) asserting the existence of these functions would prove useful in the solution of the seventh problem.

This paper also contained a proof relevant to Hilbert's tenth problem: Extending another 1909 result of Thue, Siegel proved that any "nonsingular" cubic equation has at most a finite number of integer solutions. (An example of such a cubic equation is $y^2 = x^3 + ax^2 + bx + c$. Nonsingular just means that the graph of the equation doesn't cross itself or have a cusp $\prec$. "Most" cubic equations are nonsingular.)[28] This is a restricted result, relative to the sweep of the tenth problem. However, it was a major event in twentieth century number theory. Siegel used his earlier Thue–Siegel theorem that algebraic numbers cannot be well-approximated by rational numbers. (If one of the equations in question had an infinite number of solutions, they could be used to approximate a certain algebraic number arrived at through a complicated parameterization and a limit process, too well.)

The period in Frankfurt with Dehn and company was Siegel's happiest. First he stayed with the family of a painter named Wucherer. They lived near the Taunus hills, convenient for hiking. Siegel took lessons and did a lot of painting and drawing. Many years later he left the paintings to his cleaning lady, who sold them. Schneider bought many of them.

Siegel eventually moved to a new house downtown, keeping his address secret from his students. He communicated with his cleaning lady only through notes left on the kitchen table. He had a Swedish girlfriend named Betty who lived out of town and visited on weekends. She practiced what were somewhat mysteriously described by Braun as "Swedish healing exercises."[29] Siegel learned Swedish and they enjoyed speaking Swedish together, went on vacations and hiked. Dehn's daughter Eva described Betty as "well-built" and said that Betty introduced her to a kind of cologne that she has used all of her life.[30] Siegel didn't stay in Frankfurt when he wasn't lecturing. His father had remarried, and Siegel had a good relationship with his stepmother. He visited them or traveled.

Siegel didn't have many students, particularly at first. His lectures were flawlessly prepared, but very difficult. His standards were high. Once he had written a paper on a subject, it was like polished marble—no gaps, no unfinished opportunities to apply the new methods to significant problems. If there was a good opportunity, Siegel took it. Graduate students often make their living off the leavings of their teachers. In this respect Siegel was not a bountiful host. Schneider was Siegel's most important student during his period in Frankfurt. He chose the topic of the seventh Hilbert problem on his own initiative.

Hel Braun was also Siegel's student. Besides being a rich source of anecdote concerning Siegel, her narrative *Eine Frau und die Mathematik 1933–1940* recounts her sometimes difficult experiences as a woman in mathematics. Braun was from Frankfurt and studied at the university from 1933 to 1937. At nineteen her impression of Siegel was that he was bald, fat, and pretty old—thirty-six. Siegel would call on his students by their seat numbers on a grid. She was B10. However, only fourteen days into the term she had been noticed, for he told a colleague, "I have two girls in my class who can do the work," then added, "but one is goddess of the underworld."[31] Siegel had scanned the class list to see if B10 had a name and didn't recognize Hel as a name. He looked it up in a dictionary, which supplied him with "goddess of the underworld." Braun wasn't aware that he had noticed her, at the time, and later expressed gratitude that she hadn't known. She enjoyed the way he solved the hard homework problems on the board with lightning speed. His lectures were influenced by Frobenius. Sometimes he could be testy. When a class was forced to meet in a too-small classroom, Siegel would turn around in irritation when someone moved in a seat or wrote too loudly. Sometimes in the summer he would stroll into class wearing a yellow raw silk suit, with a bright yellow bow tie and flashy shoes (clothes from his vacations in the south).

In autumn of 1935 Siegel, after returning from a visit to the Institute for Advanced Study, began to realize the extent of the trouble the Nazis' anti-Semitic policies were bringing to his colleagues. Werner Weber, the leader in a notorious incident in which Landau, who was Jewish, had been prevented from lecturing in Göttingen, had by then come to Frankfurt. Braun was having her own difficulties with Nazi students, and she thinks Siegel had heard about this. He stopped her in a stairwell and suggested that she become his student.

Later, when Braun was having trouble with her thesis topic, Siegel invited her to his flat instead of his office. She rang at the door, which had a sliding wooden peep slot. After a little while the slot slid open and Siegel's head appeared. He demanded, "What are you doing here?" She reminded him of their meeting. He said he had completely forgotten. He let her in, then asked, "Why don't you make some coffee, will you?"[32] and disappeared into his room and closed the door. Braun informs us that she never could make coffee very well and that she had no intention of making coffee at that moment, so she stood in the kitchen holding the coffee pot.

After a while Siegel came back and in a friendly tone said, "Ah, so you can't make coffee either, huh?"[33] He then made coffee and brought out what seemed to her a lot of cake. It appeared that he had remembered the appointment. He wanted to talk. Braun fled, using the excuse that she had to get to a Morse code lesson. Siegel later expressed disappointment that she left so quickly. Doing a sort of dance, they did get to know each other better, but kept a distance. Siegel invited her to his flat for dinner on the occasion of her Ph.D. exam. Braun had passed with honors. She describes this extraordinary dinner in the following passage:

> This was probably typical of an evening he usually had with one young single guest. I know that Schneider has similar memories. I arrived punctually, of course, to find everything happening in the kitchen. A large box stood on the kitchen table. This I was asked to open. Shock, out crawled crabs, so many that we later had trouble eating all of them. Though their claws were bound, they certainly still could walk. I was familiar with crabs, having dissected them . . . in my zoology class, but they had been dead. These crabs, meanwhile, populated the kitchen waiting to be cooked. I did so want to go home. Siegel was happy as a kid, because his surprise had worked so well. Everything else we needed stood ready for us. The champagne was cool. It was a pink one exactly appropriate for crab. Finally the creatures were in the pot and done. . . . As it was, hours passed before these creatures were finally done for, a small portion of them in our stomachs and mountains of debris in the garbage can.[34]

Braun continues, "During the meal I was treated as if I had just failed a test." She was less than happy about her good grade, herself, because she had answered a question wrong. She possessed self-confidence—which allowed her to stand up to Siegel—but doubted (and at the time she wrote, still did doubt) her actual ability. Siegel saw this doubt "and no sooner would I start to smile than a comment from that cool leather chair would reduce me nearly to tears again." He told her how she had come to get the good grade, that part of it had been random luck and department politics. Then she adds two sentences that are odd in their own right: "Siegel surely did not mean to disparage my work or he himself would have given me another grade, but he felt for the sake of my training that I should not become overly confident. Thus it was all to the good that our crab dinner was not solely dedicated to celebration." Braun went on to do work in algebra, including modular forms and Jordan algebras. After many stops she became an ordinarius at the University of Hamburg.

Siegel and Braun began going on hikes together, sometimes longer trips. By day they hiked with backpacks and at night ate expensive, good meals and slept in comfortable hotels. Siegel told Braun he had a funny feeling that things shouldn't be too comfortable, though, so as not to infuriate the gods.

At the end of his time at Frankfurt, between 1935 and 1937, Siegel published three major papers on the theory of quadratic forms. This study goes back directly to Gauss and Lagrange, and flows through Legendre, Eisenstein, Hermite, H. J. S. Smith, and Minkowski. Dieudonné wrote: "The work of Siegel in this domain may be considered the crowning achievement of the theory: but at the same time, he broadened it considerably and prepared its modern versions by connecting it with the theory of Lie groups and automorphic functions."[35] These papers are a major contribution to Hilbert's eleventh problem in its more open-ended form. Much of Siegel's work in number theory after this time was in this area. In his 1950 essay, "The Future of Mathematics," Weil cited Siegel's work many times as opening up new possibilities in number theory.

The Nazi students in Frankfurt were becoming active, vocal, and dangerous. Siegel, alone among the faculty, refused to open class with the Nazi salute. All the professors on a "day of national solidarity" were expected to go out and help collect donations to help poor families survive the winter. Siegel wouldn't do it. When questioned by a local Nazi official he said curtly, "Not in the mood, no time, no money."[36] He remained friendly with his now-dismissed Jewish colleagues and in 1937 dedicated a paper to Landau for his 60th birthday. This last open defiance was probably what made it too dangerous to stay in Frankfurt. However, Siegel had already decided he couldn't stay. Braun says that "earlier in his life Siegel expected a change of scene to solve all his problems. He never considered what awaited him. He only saw the supposedly unbearable present circumstances."[37] In January 1938 Siegel went to Göttingen. Braun took the next assistant position that opened up and followed.

Siegel's problems did not go away in Göttingen. As early as 1921 he had refused to join the German Mathematical Society, the DMV, because he foresaw that Germany might "take a wrong path and the DMV might follow it too."[38] The year 1921 is not a misprint, and it is very early. His premonition was now coming all too true. Braun writes that Siegel's flat in Göttingen remained undecorated. He had pinned some of his own paintings on the wall but had no other personal things except for a chest-of-drawers his stepmother had given him. He was ready to move again. Helmut Hasse (1898–1979), who had become director of the Mathematical Institute in 1934, had offered Siegel the job, and Siegel conducted a joint seminar with him. Hasse was an excellent mathematician and never stopped trying to further the interests of mathematics in Germany. Grauert writes: "With Hasse and Siegel Göttingen became a worthwhile place to be again. One might say, it experienced a sort of 'final burst of summer.' "[39] However, Hasse expressed open loyalty to the Nazi Party and Siegel, confronted with it on a daily basis, soon grew disgusted with Hasse and his politics. Earlier, in the 1920s, Siegel had said to Hasse that his experience in the First World War had almost killed him.[40] Hasse had responded that his experience had been bracing and had greatly improved his health. Now they took walks together, and on one of those walks they had a final quarrel. Braun remarked that Siegel only classified people as enemies or friends, along with a lot of people who didn't count at all. Hasse became an enemy. A few days later Braun delivered a lecture during the joint seminar. Siegel drank champagne and left the room pretending to be drunk as long as Hasse could see him. He said, "I can't stand to be in the same room with Hasse unless I'm drunk, and I want him to know it."[41]

This was a dangerous time for Siegel. He didn't follow the rules. He visited his Jewish friends in Frankfurt. He helped Dehn escape. He was probably protected by the intervention of others motivated to protect his exceptional mathematical abilities—also, everyone knew he was a little crazy anyway. Great "Aryan" mathematicians were becoming a scarce commodity in Germany, and the ministry of education was not eager to lose another—later, when Siegel was in the U.S., he tried to resign by telegram and the ministry wouldn't accept his resignation. Despite the insults he had received and his adherence to the Nazi regime, Hasse followed his pattern with Siegel of not attacking individual mathematicians.

Braun and Siegel continued their unusual relationship. During the week they did not eat lunch together. Siegel wanted to eat well, and, if not paying, refused to eat at the inexpensive places she could afford. On the weekends he would invite her to expensive restaurants or they would cook together at his flat. Siegel continued to do mathematics and sometimes talked to Braun about it. His moods were erratic. After hours staring and brooding, he would suddenly become euphoric, or sentimental, then weary, or full of hate. She recalled that he was often able to overcome a bad mood by walking and getting into a state of rapture with nature (*Naturschwärmerei*).[42] Then he would get back to his mathematics. However, she reports never having seen him on an even keel—merely satisfied or moderately happy. In 1938, every Wednesday afternoon she would go for walks with Siegel, Gustav Herglotz (1881–1953), and Herglotz's German shepherd Alf. When Alf died, Siegel bought Herglotz a dachshund.

In 1939 Siegel and Braun went on a journey from Basel to Rome. Braun was only twenty-four, and the forty-two-year-old Siegel said it was time for him to show her the world and "how to really live."[43] They went hiking over the passes and glaciers of the Alps. They visited the grave of Rilke and recited his poems. Braun comments that Siegel understood their darkness better than she. Every morning they drank hot chocolate made from goat's milk, which made Braun sick. This, Siegel maintained, was to remind them that life is hard. While they were in Venice, Braun lost a tooth. Siegel rose to the occasion and competently found a dentist for her.

On September 1, 1939, the Second World War started, but first came the period of the "phony war," during which little fighting occurred. Invigorated by his vacation, Siegel continued on its momentum. However, he soon realized that the recent events really meant another war. His mood turned black. He stopped painting, never to start again. His father died close to this time. His stepmother asked Braun to help him, knowing how he had reacted to the last war. Sometime in the fall Siegel and Braun vacationed in Wolfgangssee in Austria. A waitress in the village inn was slow in coming. Siegel called twice in a loud voice and scolded her. Then he impetuously blustered out, realizing moments later that he had cut himself off from the only supply of hot food in town. He was too proud to apologize. Still, he was hungry and getting hungrier. He finally persuaded Braun to apologize for him.

In her memoirs Braun wonders why she was so tolerant of Siegel. She concludes that he had a way of making people forget the bad moments quickly. A similar restaurant incident happened to Schneider twenty or thirty years later, and he forgave Siegel also. Siegel had an earnest if difficult soul and did mathematics as an art. Mathematicians with the same love and respect for mathematics could forgive objectionable behavior, particularly when followed by charm and great mental clarity. Siegel dumped Bessel-Hagen's thesis in the drink, but he helped save Dehn's life. He did not bow to the Nazis or even tacitly assent to their presence. In a time of accommodation, averted gazes, and outright cowardice, Siegel was different.

That winter Siegel planned his own escape. Braun writes that they discussed her going along only once. She felt she had to stay with her parents, who depended on her financially. Also, her brother was in the army and she wouldn't have been able to tolerate not knowing his circumstances. She says she blocked remembering any details. In March when Siegel visited Dehn in Norway, he carried with him just the luggage one would take on a short visit. His mathematics was in his head. He left Norway for the U.S. a few days before the Nazis invaded.

As the world avoided and then plunged into another total war, Siegel found his interests turning again to astronomy and celestial mechanics. Dieudonné says that after number theory Siegel's favorite subjects were celestial mechanics and the differential equations that come from a Hamiltonian treatment of mechanics. In 1941–42 Siegel published his greatest insights in this area. Perhaps the central question in celestial mechanics is the three (or more)-body problem. Will the individual planets as they pass gradually jog each other out of their apparently stable orbits, causing one or more to eventually fall into the sun or fly out into space? Is the solar system stable? What happens if it captures a wandering body? Classically it appeared that systems with as few as three bodies might be entirely unstable over the long term. A central difficulty was "the problem of small denominators," which caused series solutions for the problem to diverge and fail, and this was thought to reflect physical instability. This is a difficult and deep subject. Between 1885 and 1912, Poincaré had proved that the series involved did, in general, diverge, though he also proved the existence of many orbit configurations that are perfectly balanced and stable, in the same sense that there is theoretically a position where one can balance a perfectly sharp pencil on its point. If a gnat breathes, the pencil tips over. In 1942 Siegel saw that if certain parameters satisfied certain numerical conditions, the problem of small denominators disappears. This is one of the central insights that went into KAM theory, which emerged 1954–63. Siegel published a number of important papers in 1941–42. There was a paper on what happens to the solutions when three bodies collide, a paper showing that certain fixed points of perfect balance are unstable (starting from number-theoretical equations), and a paper proving that the problem of small denominators cannot be avoided—it is everywhere.

At the Institute for Advanced Study, Siegel entered into a hastily arranged research fellowship, partially supported by a grant from the Emergency Committee in Aid of Displaced Foreign Scholars. Gödel was another of the nine scholars supported.[44] Siegel worked on extending his ideas on automorphic functions, modular forms, representations of automorphic forms in the context of Lie groups, and the central importance of all of this in analytic number theory. Ilya Piatetski-Shapiro has written of studying notes of lectures Siegel gave during the war that made it to Russia at the beginning of his career, in the early 1950s. He translated them into Russian, their only formal publication, earning a small sum.[45]

On October 1, 1945, Siegel was appointed permanent professor at the Institute. Yet this was a period of exile for him. Hlawka says that in the winter of 1946–47 Siegel returned to Göttingen on the first soldier train as a guest.[46] Hasse was gone—Wilhelm Magnus somewhat uncomfortably occupied his position. (Magnus spent most of his later life in the U.S. at NYU.) But then Siegel went back to the Institute. The Institute was eager to keep him, hence his appointment as permanent professor. Siegel became a naturalized American citizen in 1946.

Siegel was beginning to move beyond his period of greatest production, though he would remain productive well into his seventies. Emil and Natascha Artin moved to Princeton University and Siegel was particularly friendly with Natascha. She recounts being sick with the flu and Siegel, fearing infection, not visiting.[47] She said something like, Well, Siegel, you are a nice friend. When you are sick I constantly come to see you. She was still sick a week later, in bed reading a paper by Siegel, which she happened to be editing for a mathematics journal, when twelve pink roses came anonymously. Fifteen minutes later twelve red roses arrived, this time with a note saying, "Get well." She sat on the bed and compared the handwriting of the manuscript and the note—they were the same. Fifteen minutes later twelve yellow roses came and this time Siegel signed the note. Siegel had been going down Nassau Street from store to store on the way to visit her, buying flowers.

In 1951 Gödel was being considered for appointment as a permanent professor, and there was concern that his obsessive attention to incidental detail would interfere with the running of the Institute. Siegel's comment was that "one crazy man on the faculty,"[48] referring to himself, was enough. The same year, Magnus officially left his position in Göttingen (he was already visiting Caltech in 1949). Siegel accepted an offer to fill it, thereby eliminating his own objection to the Institute appointment for Gödel.

Thus, Siegel returned to Göttingen. Many of his graduate students were from this period. German mathematics had to rebuild almost from scratch. Siegel's greater openness to students can be understood partly in this light. Jürgen Moser (1928–1999) was one of the first and probably the most successful of this group. Moser was born in Königsberg and displaced after the war—Königsberg having been swallowed by Russia. He was shot at as he crossed from East to West.

Siegel was ambivalent about taking German citizenship. However, the decision was made by default. After three years in Göttingen, the U.S. consul judged that since he had not returned to the United States he was no longer entitled to a U.S. passport even though naturalized. (This has since been ruled unconstitutional.) Siegel attempted to live without a passport. He couldn't go to the International Congress in Amsterdam in 1954. Then he began to have trouble going on holiday. Though he had been going to Pontresina, Switzerland, for years, he now had to get special documents certifying that he was known, that he could pay his bill, and he had to pay extra taxes to the commune. Then he wanted to visit Weyl in Locarno, Switzerland. Davenport says:

> The natural route is by Swiss postal bus, but this crosses part of Italy. At the Italian frontier the Swiss examination was perfunctory, but then came the Italian passport officer on the Italian side of the bridge. "Where is your permission to enter Italy? You must get out and return to Switzerland." So Siegel was put down, with a good deal of baggage, on the Italian side of the bridge, and ordered to walk back across it. Fortunately he found a boy to carry some of the luggage. On the Swiss side, he had to wait for a bus, travel to the north of Switzerland and then back to the south, and eventually reached Locarno the following afternoon.[49]

Siegel finally applied for German citizenship. He glumly subjected himself to a medical exam and paid the necessary 500 marks.

At Göttingen in 1951–52 one of Siegel's first lecture classes was on celestial mechanics. Moser's notes were published in book form in 1956 and reappeared in 1971, updated and in joint authorship. Simply, it is a lovely book, written in Siegel's careful style. The topological results, early and late, remain rooted in the equations and their transformations. There are harmonic, number-theoretical aspects in how the "good" or "bad" frequencies are recognized and connections made to the tapestry of work going back seventy years to Poincaré, Tullio Levi-Civita (1873–1941), and Karl Sundman (1873–1949). Moser's final chapters, with Siegel's cooperation, maintain this classic style. Though not at all a history book—one has to read the history from the topics chosen and the footnotes at the back—it is a deep, historical picture of how the problem has unfolded. (When I first found I needed to look at it, I tried the Caltech library. They had eight copies in the two editions, all checked out, close to fifty years after the lectures were first given and more than twenty-five after the revised edition. I found a new paperback edition in a general-interest bookstore.)

The lectures put Moser at the cutting edge of this field at the moment it was set to move. That movement occurred publicly in 1954, when Kolmogorov delivered a talk at the International Congress that Siegel missed because he didn't have a passport. Kolmogorov had an idea similar to Siegel's, but he brought in new insights from measure theory and differential geometry. Moser took advantage of the breakthrough (along with Kolmogorov's student, Arnold) and became the "M" of KAM theory, particularly for the Moser twist theorem.

One reason Siegel might have been willing to leave the Institute relates to the style (and content) of modern mathematics. Princeton University and the Institute were a world center for algebraic topology and algebraic geometry. This meant a lot of set theory and complex structure attached to the more primitive spatial and algebraic ideas that were being investigated. Thom and Serre came to the Institute from France in the 1950s with powerful new methods and ideas—fiber bundles, etc.—and a language that was presented by the imaginary (composite) French mathematician, Nicolas Bourbaki. All mathematics was set theory for Bourbaki (and Bourbaki was himself a set, which included in the second generation Thom and Serre). Siegel didn't like this style—abstract, disconnected from simple questions. Sometimes the machinery seemed to become the subject. He favored hand-to-hand combat with numbers and polynomials—or in the event of war, with the equations for the orbits of the planets. Here the connection between what one did and the basic concerns of mathematics was clear. (Anyway, one couldn't play practical jokes on imaginary mathematicians.) In a June 1, 1959, letter to Weil, a founding member of Bourbaki, a major disseminator of the new style in America, and after his period in Chicago, a permanent professor at the Institute, Siegel wrote:

> It is entirely clear to me what circumstances have led to the inexorable decline of mathematics from a very high level, within about 100 years, to its present nadir. The evil began with the ideas of Riemann, Dedekind and Cantor, through which the well-grounded spirit of Euler, Lagrange and Gauss was slowly eroded. Next the textbooks in the style of Hasse, Schreier, and van der Waerden, had further a detrimental effect upon the next generation of scholars. And finally the works of Bourbaki here provided the last fatal shove.[50]

Hans Freudenthal (1905–1990) wrote: "Only such great mathematicians as Siegel could afford to ignore structuralist tendencies completely, even from Dedekind onwards, and nevertheless create works of genius."[51] One of Siegel's heroes was Axel Thue. Siegel wrote in his introduction to Thue's collected works that on his hospital deathbed Thue was trying to prove Fermat's last theorem. Andrew Wiles's 1994 proof of Fermat's last theorem uses just about all the modern technology, some of which flowed from Siegel's work. It is interesting to imagine how Siegel would have seen this.

Siegel retired in 1959 but continued to lecture. He visited the Tata Institute in Bombay four times and stayed for several months each time, his last visit being in 1967. In 1963 he received an honorary doctorate there and was particularly happy because Dirac was the only prior recipient. He traveled and lectured in the U.S. He

Carl Ludwig Siegel taken in Pontresina, August or September 1972 (photo provided by Maria Dehn Peters).

had visitors. During his 1966 visit, Davenport experienced Siegel as cook. The first course was a large trout. The second course was a large trout. Siegel apologized that there wasn't a third. In 1966 Siegel oversaw the publication of his collected works in three volumes. He spent the rest of his life editing (and writing) a fourth volume. As the story goes, he burned everything else, fearing that a historian—as he himself had done with Riemann—would get into his papers.

Natascha Brunswick visited Siegel at his home late in his life. He told her, "At 10:00 I go for a walk." She came down at 9:55 and he said, "Sit down, we are going at 10:00." He was sad that he couldn't travel anymore.[52]

In 1929 Siegel had achieved one of the few positive results on Hilbert's tenth problem—there is not an infinite set of solutions to diophantine equations (exponents 3 or higher). In 1972, two years after Matiyasevich's solution, Siegel proved, among other things, that Hilbert's tenth, restricted to equations of degree 2, was decidable. This is an important and strong result, and Matiyasevich mentions it in his book on the tenth problem. Siegel was seventy-five years old. His apartment was still furnished sparsely, ready for a quick move. Schneider writes, "Increasingly isolated but still spiritually vital, he died on April 4, 1981, in Göttingen."[53] The year 1929 was a watershed for the theory of transcendental numbers. In addition to Siegel's paper, in that year A. O. Gelfond, a young Russian, proved a special case of the seventh problem, that $2^{\sqrt{-2}}$ is transcendental. Siegel's and Gelfond's papers would provide the foundation for the solution of the seventh problem. Yet despite their very different and powerful contributions to the small group of techniques that were available for use in this area, more fundamental innovations were still needed to finish the problem.

Alexander Osipovich Gelfond was born on October 24, 1906, in St. Petersburg. A. P. Yushkevich, a family friend who was later a prominent historian of mathematics, writes in the *Dictionary of Scientific Biography* that Gelfond's father was named Osip Isaacovich Gelfond. In an effort to get a little more information than was available in the *Dictionary* and obituary articles, I contacted Ilya Piatetski-Shapiro, who in turn contacted Andrei Shidlovski in Moscow. Shidlovski wrote, "His daughter Julia keeps his birth certificate, which mentions his father's name and surname: Iosif Ickovich (1868–1942)."[54] (The spelling of Russian names in English is approximate, at best. Much of the biographical detail in this section comes from Shidlovski, either through Piatetski-Shapiro or from published sources.) A rabbi was involved with the registration of the birth, but it was noted the baby was not circumcised. Gelfond's mother, Musia Gershevna (Reishinstein), was the daughter of a well-to-do banker in Kiev.

Both of Gelfond's parents studied at the Sorbonne in Paris, Osip (Iosif) receiving his medical degree in 1896 and Musia in 1898. They married in 1899. Gelfond's parents were part of the liberal intelligentsia that spent time abroad, sometimes by choice and sometimes in exile in the closing quarter-century of czarist rule. Gelfond's father had philosophical interests, and Lenin, in his only philosophical book, explicitly disagreed with Gelfond senior's philosophy. This book of Lenin's became required reading for all students, including the mathematicians that A. O. Gelfond would teach. Lenin and Gelfond's father were friends in private life.[55]

Gelfond senior was also good friends with the playwright Anatoly Vasilyevich Lunacharsky, a radical who was deported in 1898, but jailed when he was back in Russia during the 1905 Revolution. Lunacharsky joined Maxim Gorky and A. Bogdanov on Capri in 1909, where they started an advanced school for an elite of Russian factory workers. Lenin was against the project and it ended. During the Revolution, Lunacharsky was jailed by the Mensheviks and then appointed people's commissar for education by the Bolsheviks. He is credited with saving many old buildings and much artwork and was interested in the place of religion in the new society. Lunacharsky's hard side comes through in Isaac Deutscher's 1967 introduction to *Revolutionary Silhouettes*: "His role in the events of 1917 was quite outstanding, as all eye-witnesses testify. The 'soft' 'God-seeker' with the air of the absent-minded professor, surprised and astonished all who saw him by his indomitable militancy and energy."[56]

I was able to find out little about Gelfond's childhood and early education. Yushkevich perhaps went to grammar school with Gelfond in St. Petersburg; certainly they were friends in university. Yushkevich's father, Pavel Solomonovich, a well known social-democratic philosopher and man of letters, was a friend of Gelfond's father. Shidlovski says that Lev Abramovich Tumarkin, a mathematician, and Lazar I. Lagin were also early friends of Gelfond's. Lagin, a writer well-known in Russia, is the author of *Old Man Hottabytch (Starik Khottabych)*, about a Soviet boy who while swimming in the Moscow River finds a bottle with a genie in it.

The goal of the Gorky/Bogdanov project of exile on Capri, which Lenin had stopped, reemerged transformed in the Soviet universities. Admission requirements changed—the old standards and ways perpetuated the class structure—and activist students worked to press the Revolution forward. Charges of "sabotage"[57] were made. There were a few "red"[58] professors and there was actually an Institute of Red Professors founded in 1921. So in the '20s it was often difficult to become a student in Moscow, particularly for one of Gelfond's social class. Shidlovski says Lunacharsky personally helped Gelfond. In 1924 Gelfond began his undergraduate studies in the department of mathematics and mechanics at the Moscow State University.

Russia had had outstanding mathematicians from the time of N. I. Lobachevsky (1792–1856) and P. L. Chebyshev (1821–1894), but in the 1920s in Moscow a remarkable amount of talent came together. N. N. Luzin (1883–1950) and his students dominated the department, which was sometimes called "Luzitania." Luzin had traveled to Göttingen in 1910–1912. Travel was allowed again starting in 1923, and members of this group traveled to make connections with the French school of analysis and with the mathematicians in Germany. Soon they were contributing as much as they were bringing back.

Luzin's group focused on analysis, specifically the theory of functions and its foundations, starting with point set topology. Many talented students broke away and began to start new schools. They included P. S. Aleksandrov, P. S. Urysohn (1898–1924), A. Ya. Khinchin (1894–1959), and A. N. Kolmogorov. Gelfond arrived a bit later and was not one of Luzin's group. He focused on number theory, using analytical methods. Gelfond finished his undergraduate work in 1927 and continued his postgraduate studies under the direction of Khinchin and V. V. Stepanov (1889–1950). Khinchin, along with Kolmogorov, was revolutionizing probability theory during these years—though he was also publishing in analytic number theory at the same time. Stepanov was not mainly a number theorist either, so Gelfond was not working at the center of his advisers' interests. However, he had found his lifelong love, number theory. His first papers were on the theory of functions of a complex variable, with connections made to number theory. Shidlovski writes, "A. O. Gelfond studied mathematics hard when he was young. I remember how I. G. Petrovsky said that Gelfond is the only Soviet mathematician who solved all the problems in Pólya and Szegö's book [*Problems and Theorems in Analysis*]."[59] Pólya and Szegö proved useful to him.

Complex analysis was Gelfond's road into number theory. Any complex number is a real number combined with a purely imaginary number ($\sqrt{-1}$, often called i), written as $a + b\sqrt{-1}$ (or $a + bi$); a is called the real part and b the imaginary part. Just as the real numbers are identified with the points of a line, the complex numbers can be identified with the points of a plane. If one generalizes the real number definition of the derivative (how steep the graph of a curve is) to apply to functions on the complex plane, one finds that for functions that have this derivative many calculations become possible. (The derivative in complex analysis corresponds to a local measure of twist and stretch rather than simple steepness.) One can calculate many hitherto unapproachable integrals with complex analysis.

Analytic functions on the complex plane can always be expressed locally as a power series (polynomials of infinite degree). The reappearance of the polynomial here in this expanded and (usually infinite) version is where the connection to algebra and number theory is made. It's as if the ordinary algebraic polynomial stepped through the looking glass. In one view, complex analysis is merely the extension of the idea of the polynomials of algebra into the realm of the infinite. Sometimes these polynomials allow the collection of an infinite class of numbers into one analytic function. For example, all the prime numbers and only the prime numbers are captured in the Riemann zeta function. Families of algebraic numbers can be likewise captured in Dirichlet L-functions—see the ninth, eleventh and twelfth problems. Complex analysis or calculus can then be used to extract information about the structure of the whole class of numbers.

In 1914 Georg Pólya (1887–1985), originally from Hungary, proved a theorem stating that an "entire function"—an analytic function well-defined over the entire complex plane—that assumed whole number values for positive whole number inputs and didn't grow too fast, was, in fact, a garden-variety algebraic polynomial and not an infinite series like most entire functions. Gelfond realized that he could exploit this connection. Pólya's result allows us to take an analytic function, prove that it has certain properties, and then return back through the looking glass with it. Based on the assumption that $2^{\sqrt{-2}}$ is an algebraic number, Gelfond in his 1929 paper proved the function $2^{\sqrt{-2}z}$ must be a polynomial. But it isn't. Therefore, $2^{\sqrt{-2}}$ must be transcendental. (The first thing proved by Gelfond was that e^{π} is transcendental.)

Siegel was excited by Gelfond's result when he saw it published in *Comptes Rendu*. From this, he saw how to extend it to the case $2^{\sqrt{2}}$. Schneider says this in his 1934 paper on his own solution of the full seventh problem. Reid writes that Siegel told Hilbert, who wanted him to publish this result. Siegel felt that the important work was done by Gelfond and that he would soon see how to extend his own result, and so Siegel never published. R. O. Kuzman published this result first in 1930 in Russia.

His 1929 result made Gelfond's career. Sometime in the '20s he married his first wife, Morozova, and in 1930 their daughter Julia was born. In the academic year 1929–1930 Gelfond taught at Moscow Technological College. He traveled to Germany for four months in 1930.[60] He was then appointed to the faculty of Moscow State University, where he occupied at various times chairs of number theory, analysis, number theory and analysis, and finally also history of mathematics. In 1933 he was appointed to the faculty of the Steklov Institute, the research institute of the Academy of Sciences in the USSR, in addition to his university position.

In 1934 Gelfond solved Hilbert's seventh problem in full generality. Building on the method of his 1929 result and using an important lemma from Siegel's 1929 paper, he added several new ideas and created a new method that was used to solve many other problems in the years to come. Hille wrote, "The proof of Gelfond, though more advanced [than Schneider's], has quite a simple basis and gives a beautiful example of teamwork between algebraical and analytical ideas."[61] (Assuming the number in question is algebraic, Gelfond, in Hille's version, used several results from nineteenth century number theory and complex analysis to construct a transcendental function based on the number that has a lot of spots where it is 0. Alternating between algebraic and analytic arguments, he proves such a function has more and more 0s. Finally he proves this function has so many 0s that it is 0 all the time. However, it isn't—a contradiction. So the number is transcendental.) Gelfond was awarded the Russian doctoral degree in 1935 without dissertation.

Political conditions worsened. Ernest Kolman, a Czech who wound up in Russia during the First World War and stayed to participate in the Revolution, had moved into the Communist Party power structure and, unfortunately, took an interest in mathematics and science. S. S. Demidov describes him as "one of the most sinister personalities of the science vs. ideology 'dichotomy' of the 1920s and 1930s. All tragic events in the history of mathematics of the period were accompanied by his exposing addresses and appeals to enhance witchhunting."[62] In 1931 Kolman published an article called "Sabotage in Science" in *Bolshevik.*

Kolman and others singled out D. F. Egorov (1859–1931), who along with Luzin was a founder of the Moscow school of the theory of functions, as an early victim. They had already been pursuing him—"Proletarian students 'declared war' on Professor Egorov in the second half of the 1920s."[63] First he lost power in the department and then was removed altogether. He did not back down. In a letter to the director of the newly reconstituted Mathematical Institute he wrote, "It is the enforcement of a standard outlook upon scientists which is the real sabotage." Soon he was arrested and sent into exile to Kazan, where he died in 1931.[64]

In the 1930s Russian mathematics came perilously close to being destroyed by Stalin—Soviet biology *was* destroyed. The most dangerous moment was a *Pravda*-driven campaign and what amounted to a show trial against Luzin. The campaign was started in earnest in 1936 by someone in the party apparatus, perhaps Kolman. A hearing at the Academy of Sciences was scheduled. Coinciding with the hearing, *Pravda* ran a series of articles between July 9 and 15, leveling a variety of charges, including sabotage. Demidov writes:

> The tone of the ideological accusations against Luzin set up by the newspaper reports was ever-present at these sessions. The most aggressive mathematicians were Aleksandrov, [O. Yu.] Schmidt, Sobolev, Gelfond, and Lusternik. (Khinchin, Kolmogorov, and Shnirelman demonstrated more reserve.) Only two people—Bernstein and Krylov—tried to defend Luzin.[65]

A number of factors were at work in these denunciations of Luzin. Information is still coming out. There was an intergenerational power struggle in progress in the Moscow mathematical community, independent of Stalin. For example, Gelfond and others, including Aleksandrov and L. A. Lusternik (1899–1982), had endorsed the action against Egorov after the fact, when it would do no harm, and had gained some power by it according to Demidov (though perhaps just the power to survive).[66] Luzin felt betrayed whenever a student struck out on his own and often broke off relations with the student. Aleksandrov believed that Luzin had contributed to the death of M. Ya. Suslin (1894–1919) in the aftermath of the First World War, by not giving him deserved credit and recommendations. Suslin had been Luzin's student, and when Suslin didn't get a position, he went home to his family in the provinces where, like millions of others in that chaotic period, he died of typhus. The extent of Luzin's fault in this is open to question, but Aleksandrov believed Luzin could have more actively protected Suslin. Such antagonisms may partially explain why many colleagues turned against Luzin, but these hearings were an unfortunate venue in which to express a grievance.

This seemed to be leading to Luzin's expulsion from the Academy when suddenly, between July 11 and 13, "The campaign was abruptly stopped by someone high up."[67] The tone of the hearing shifted, and in the end Luzin only received a reprimand. Larger and more important trials were in the offing for August, and perhaps Luzin's hearing was seen as a distraction. Distraction or not, no one involved in this event ever forgot about it.

Luzin was allowed to continue his career. He blocked Aleksandrov's appointment as a full member of the Academy of Sciences until he died. Kolman, the "black angel"[68] in this affair, disappeared from view in 1937 and was, in fact, arrested himself. (He failed to mention the arrest in his 1982 autobiography.) Soviet mathematicians felt under intense ideological scrutiny, but mathematics did continue—and this period is genuinely part of "the golden age of Russian mathematics." (Demidov's article appears in a book of that title.)

It is difficult, from the outside, to understand any nuances of Gelfond's participation in the ideological upheaval of the time. Soviet citizens were under extreme pressure. Shidlovski says that Gelfond joined the Communist Party in 1940, but what is the significance? He died early, long before honest conversation was possible, even between friends, and before he had time to write any general memoirs.

Gelfond spent most of his career in analytic number theory, proving new results and extending old ones. Shidlovski says that for many years he tried hard to prove the Riemann hypothesis about the zeta function. He worked applying results about transcendence to theorems about diophantine equations (diophantine approximation), as Siegel and Thue had done earlier. In 1947–1949 Gelfond and Dyson, working separately, found a way to extend the result on the approximation of algebraic numbers further and opened the door to a new wave of results. Gelfond had many students and ran a seminar on number theory, coordinating the work of mathematicians from Moscow and regional universities in Tashkent, Tiflis, Minsk, and Yerevan, among others. In 1938 he was elected a corresponding member of the Academy of Sciences. During the war he did applied work. Gelfond's father was evacuated to Kazan during the Second World War, died in 1942, and was buried there.

Gelfond had a second wife. He then married Lilia Vladimirovna Lukashevich. They had two children together in the early '50s, Olga and Sergei. Piatetski-Shapiro supplies a different image of Gelfond than the one refracted through the political events of the '30s. In 1951 Piatetski-Shapiro had just graduated from Moscow State University and wanted to continue there to do his graduate work. This was during a high tide of anti-Semitism in Russia, which Piatetski-Shapiro oddly says Stalin "inherited" from Hitler.[69] (Gelfond's Jewish background does not seem to have been as important in the earlier era.) Jewish students, or even students who might have been partly Jewish, were asked a more difficult set of questions in entrance exams. If they still did well, they were given bad grades in military training or in Marxism and Leninism. Gelfond was Piatetski-Shapiro's adviser. Piatetski-Shapiro writes, "He was a very warm person, very humane and sensitive to me and to the other students."[70] He says that Gelfond thought that he might be able to help him because of his membership in the Communist Party. But he failed in getting Piatetski-Shapiro admitted to the university. The rejection, true to form, was based on a poor grade in military training. Gelfond did succeed in getting Piatetski-Shapiro accepted into the Moscow Pedagogical Institute.

Alexander Osipovich Gelfond (courtesy of the Archives of the Mathematische Forschungsinstitut Oberwolfach).

It is said that Gelfond played chess at a professional or almost professional level—"first category." He met Emmanuel Lasker (1868–1941) early in his life and played chess with him. (Lasker was world chess champion from 1894–1920 and also a professional mathematician who at one point had studied under Hilbert. Lasker had to flee Hitler penniless and spent 1936–37 in Moscow after a chess championship.) Gelfond was expert in literature and mineralogy. Later in life, he was interested in the history of science, became an expert on Euler, who had spent many years in different periods of his life in St. Petersburg, and was elected a corresponding member of the International Academy of the History of Science in 1968, the year of his death. Gelfond was still active when he was felled by a stroke at sixty-two. Piatetski-Shapiro writes: "I was present in the hospital. I remember that he was trying to write some formula and tell me something which was clearly related to the zeta function. He could not because he was already paralyzed."[71]

The two following quotes, each from a group of Gelfond's students, evoke a strong image of his mathematical personality.

> His mathematical ability was esteemed, above all, for its originality. Many eminent mathematicians think roughly along the same lines as less eminent ones, though more rapidly and in a more organized way: Gelfond always thought in his own way, one that was unconventional and quite original.[72]

and

> When one studies Gelfond's arithmetic works one is involuntarily amazed by the power and force of his analytical methods. Problems for which no approach to a solution was found for many years were fully and completely solved by his methods.[73]

Gelfond published an outline of his proof on April 1, 1934. Theodor Schneider obtained an independent solution about the same time, or shortly thereafter. He submitted his paper on May 28, 1934. It is widely accepted that Schneider's paper was independent, and the work was done so close in time to Gelfond's that Schneider is given credit as one of two solvers of the seventh problem. His paper followed Siegel's general methods more closely than Gelfond's did and, though similar in its fundamental methods, is different from Gelfond's on many points. Schneider learned of Gelfond's solution the day he submitted his paper.

Born May 7, 1911, in Frankfurt, Schneider was the son of Josephine (Breidenbach) and Joseph Schneider, who owned a fabric store in an outer section of the city. Theodor attended the Helmholtz Gymnasium, which concentrated on science. He also studied piano and was good enough to be accepted in the master class of the Frankfurt Conservatorium. He had to choose whether to train to become a concert pianist or to enroll in the university. He enrolled at Frankfurt in the summer semester of 1929 to study mathematics, physics, and chemistry, but soon the lectures of Dehn, Epstein, Hellinger, Szász, and Siegel won him to mathematics. Schneider attended a lecture by Siegel on transcendental numbers in his seventh semester and became Siegel's student. To be accepted into Siegel's seminar one had to pass a difficult exam. Siegel recognized Schneider's talents quickly and urged him to write a dissertation, suggesting a variety of possible topics. Schneider kept returning to the hauntingly beautiful lecture he had heard on transcendental numbers. With youthful energy and optimism, he chose to tackle one of those questions. Schneider later said, "After a few months I gave him a work of six pages and then was told by Siegel that the work contained the solution of Hilbert's seventh problem."[74] Apparently Schneider didn't realize that he had been working on Hilbert's seventh problem! Here we have another great result in this area from a naive but powerful mathematician.

Schneider's paper was so short that Siegel thought he should write a bit more to make the thesis acceptable to the science faculty. Schneider followed with a five-page paper on transcendence results about elliptic functions, which appeared in the same issue of the same journal as the result on the seventh problem. When he was done Siegel told him, "You are the eleventh who started writing a dissertation and the fifth who finished."[75] Yet, in spite of the importance of his thesis, Schneider had trouble finding a job. In the words of a group of Schneider's students:

> For, as of autumn 1933 success depended no longer only upon mathematical achievement, rather also upon having membership in a suitable Nazi organization and views in line with Nazi ideology. If Schneider wanted to continue his studies or later find employment at the university, he would be compelled to join a Nazi organization. The alternative was giving up his university career.[76]

Frankfurt was now a Nazi university, and the Jews were gone or going. Schneider was not Jewish, but everyone not Nazi was suspect. In addition, Schneider was associated with Siegel, whose politics overshadowed, in this setting, his renown. This twenty-two-year-old man who only a few years before had had to choose between music and science, and then had gravitated to the transcendental numbers, was placed in an intolerable position. Schneider joined the SA, the original storm troopers or brownshirts, which by 1935 was a large and weakened organization. This allowed him to get a job as assistant at Frankfurt.

Schneider's compromise didn't make him acceptable to the Nazis. He was not allowed to go to the 1936 International Congress of Mathematicians in Oslo, where he would have talked about his results on the seventh problem. In 1936 he published three papers, one refining the Thue–Siegel theorem on the approximation of algebraic numbers. More generally his research focused on the behavior of certain transcendental functions—elliptic, modular, and abelian functions—as generators of transcendental numbers. This is the more general program that Hilbert mentioned in the seventh problem and that Siegel initiated in 1929. Schneider's work on elliptic and abelian functions was important.

Schneider's habilitation was rejected without explanation in 1938, possibly without being read—it was a historic paper on abelian functions. He followed Siegel to Göttingen and got a job, still only as an assistant. He habilitated in Göttingen on November 9, 1939, with the same work that had been rejected in Frankfurt and in 1940 got a job teaching. However, he was soon drafted into the German army. He spent the war as a meteorologist in France. In 1941, Schneider's habilitation on abelian functions was finally published. Between 1936 and 1949, when he was between the ages of twenty-five and thirty-eight, this was his only publication.

By 1944 it was clear that Germany was going to lose the war, and people in the ministry of education as well as the army wanted to preserve something for the period afterwards. In 1944, Wilhelm Süss managed to start a research institute at Oberwolfach (near Freiberg). He tried to pull mathematicians back from the front lines so they wouldn't be slaughtered in a final desperate defense. Schneider was now in a camp for parachutists—though there were not many airplanes left. Süss, with the help of Theodor Kaluza, got Schneider sent to Göttingen to fill in for Braun, who had caught diphtheria. This done, Schneider went to Oberwolfach in March 1945, where he waited out the end of the war.[77]

As in towns all over Germany, Oberwolfach was short on food. Thus, Schneider had to concern himself more with getting enough bread than with beginning again to do mathematics. Göttingen reopened. In autumn of 1945, Schneider rode back to Göttingen (260 miles as the crow flies—much farther on country roads) on a rattletrap bicycle put together from junked parts. He stayed at Göttingen until 1953, except for the academic year 1947–48, when he was at Münster. He was an assistant first, then started moving up the ranks. In 1950 Siegel invited Schneider to Princeton. He couldn't get a visa, probably because he showed up on U.S. State Department documents as a former member of the SA.

Schneider married Maria Urbach in 1950—finally, unqualified good fortune. He called her "Mieke" and, while he is described as calm, she was full of energy and expression. Schneider's students, writing about her, make it clear how much they appreciated her. Schneider was thirty-nine and Mieke thirty-five when they married. They had one child, Bernard, who became a doctor.

Schneider received a call to Erlangen in 1953 as full professor, replacing Otto Haupt, and was head of the scientific faculty there from 1955–57. He was thought a good judge of people. In 1957 he published an influential 150-page monograph on transcendental numbers, gathering and organizing old and new ideas. It was rapidly translated into French. Schneider rejected a call to the Free University in Berlin but in 1959 accepted a call to Freiberg im Breisgau, replacing Süss, who had died. He also replaced Süss as director of the Institute in Oberwolfach and kept the cash-strapped Institute together, setting the stage for the later expansion under Martin

Theodor Schneider (courtesy of the Archives of the Mathematische Forschungsinstitut Oberwolfach).

Darner. Schneider held meetings on the theory of numbers at Oberwolfach from 1955 to 1972. After that, he organized the meetings on diophantine approximations (and transcendental numbers). In 1969, following Siegel's lead, he visited the Tata Institute.

Some of his students, Luise-Charlotte Kappe, H. P. Schlickewerei, and W. Schwartz, write:

> Saying goodbye to the active university life was not that hard for Schneider, since the years between his sixtieth and sixty-fifth birthdays were in a time of disorder and unsafety, when the scientific mind was asked less while the political part was asked more. Student representatives requested at that time a voice in academic decisions. This was contrary to what for Schneider was the university's task and goal. This and his changing health led him to request emeritus status at the earliest possible time in 1976.[78]

Kappe et al. say, "In his last years of life Theodor Schneider lived a retired life in his beautiful house in Freiburg-Zahringen, enjoying, with his wife, the garden, his sports car, vacation trips, and freedom from all obligations." Siegel's death in 1981 came as a blow to him, "a hurting cut."[79] The two had continued to hike in the Black Forest and in Switzerland, and though Schneider had experienced some of the same troubles with waitresses and crabs, etc., as Braun, he loved Siegel for his great mathematics and his moral courage. Schneider saw to it that Siegel's ashes were put in a place of honor in the Göttingen cemetery, and he wrote the memorial piece that has been a primary resource on Siegel in this book. Schneider died on October 31, 1988. He had been ill for some time, but his death was sudden and came as a surprise.

The Inordinate Allure of the Prime Numbers

The Eighth Problem

A prime number is a natural number greater than 1 with no factors other than 1 and itself. How many prime numbers are there?

When we search through the natural numbers, at first we find there are many small primes. Somewhat larger primes are reasonably common, but scarcer. As we look for larger and larger primes, they appear to become increasingly scarce. But the frequency of appearance of prime numbers is clearly irregular. We find long stretches with few primes; then we find (relatively speaking) a cluster of them. Can we prove anything about the frequency of appearance of prime numbers? How irregular is the distribution of primes? Sometimes two primes differ only by 2; for example, 3 and 5, and 17 and 19. Is there a largest such pair?

We saw Julia Robinson arranging pebbles into patterns in the Arizona desert as a child. She went on to spend her life playing with numbers. In "First, State the Tune," I said musicians play with notes like this. Patterns are discovered, arranged, understood. Anyone who responds to music ought to be able to understand the intense response to patterns in numbers that some people have. Consider a Bach fugue. The fugue is nothing more than a regular collection of notes—but not entirely regular. Pythagoras pointed out that musical notes have a precise mathematical relationship with each other. Someone who loves a Bach fugue responds and recognizes its aesthetic wholeness intuitively—and then understands its pattern. A passionate response to patterns of numbers is more centrally human that many other passions.

"Problems of Prime Numbers" is the title of the eighth problem. Several distinct questions are mentioned. Christian Goldbach (1690–1764) conjectured that any even natural number greater than 2 is expressible as the sum of two primes.

Was he right? Is there a largest pair of primes differing by only two? This has not been decided. The heart of eighth problem is the request for a proof of the Riemann hypothesis about the zeros of the Riemann zeta function. Though on the surface this appears to be a somewhat technical question, the goal is to achieve an accurate estimate of "the density of primes." Many results in number theory begin: "If the Riemann hypothesis is true..."

A mathematician wanting to study the zeta function would do well to start with H. M. Edwards's remarkable book, *Riemann's Zeta Function* (1974). It is hard to say if this work is primarily history of mathematics or mathematics proper. It is both. Edwards believes it is important for mathematicians to read the classics, for much the same reason as the members of the Frankfurt mathematics history seminar did: The picture of the moment of deep mathematical insight the classic papers present is often lost in the more efficient refinements that follow. Riemann's paper is a window into Riemann's imagination.

The zeta function goes back to Euler's product formula, which is:

$$\sum_n \frac{1}{n^s} = \prod_p \frac{1}{1 - \frac{1}{p^s}}$$

The important detail is that the n refers to natural numbers, while the p across from it refers to prime numbers. The left side involves summing over all the natural numbers, while the right side involves an infinite multiplication with prime numbers. The number of terms added on the left is roughly the number of natural numbers below n, while the number of terms multiplied on the right is the number of primes below n. (Both sides of the equality are absolutely convergent and so can be rearranged.) Since multiplication is just repeated addition, there is hope the two can be connected.

Behind the product formula is the fact that if one methodically were to take all possible combinations of prime numbers multiplied together (allowing any prime number to be used any number of times), one would have produced a list (not in order) of the natural numbers (except for 1).* Each natural number has a unique factorization in terms of primes. The Euler product formula is sometimes called the analytic form of the "fundamental theorem of arithmetic." Euler didn't have the language or theory to actually prove much. Nevertheless, in 1737, reasoning from his product formula, he published the following enigmatic equation:

$$1/2 + 1/3 + 1/5 + 1/7 + \ldots = \text{Log}\ (\log \infty)$$

* $1/1 - x = 1 + x + x^2 + x^3 + x^4 + \ldots$ Substituting for each term on the right-hand side of the Euler product formula, multiplying and rearranging, yields the left.

It contains the germ of "the prime number theorem," which, when developed, would give a function that approximates the number of primes less than a given x. Log (log ∞) says take the logarithm of the logarithm of infinity, which is nonsensical.

Gauss reported in a letter written in 1849 that in 1792 or 1793, when he was still a boy, he had noticed that "the density of prime numbers appears on the average to be $1/\log n$," a more precise statement of Euler's idea.[1] In the intervening fifty-six years Gauss checked this guess against every new (and larger) table of prime numbers that came out, and with each larger sample, agreement with the approximation $1/\log n$ improved. This trying to discern a pattern is "empirical" and is essentially the same thing that physicists do when they try to recognize mathematical shapes in data for which there is no underlying theory. Adrien Marie Legendre (1752–1833) published a similar formula, likewise an empirical guess, in 1808 in the second edition of *Théorie des Nombres.*

If the relative frequency of appearance of primes, in fact, is approximately $1/\log n$, then the natural thing to do is to sum (integrate) over the rate of appearance—it's like integrating over velocity to find distanced traveled. The first concrete proof in this area was accomplished in 1850 by P. L. Chebyshev, founder of the St. Petersburg school of Russian mathematics and one of the major figures in the history of probability. Chebyshev, reasoning from Euler's product formula, proved that the number of primes less than n is equal to:

$$\int_2^n \frac{dt}{\log t}$$

to within an accuracy of 11% for sufficiently large n. This expression is usually called the logarithmic integral. The logarithmic integral is the area under the graph of $1/\log t$ between 2 and n.

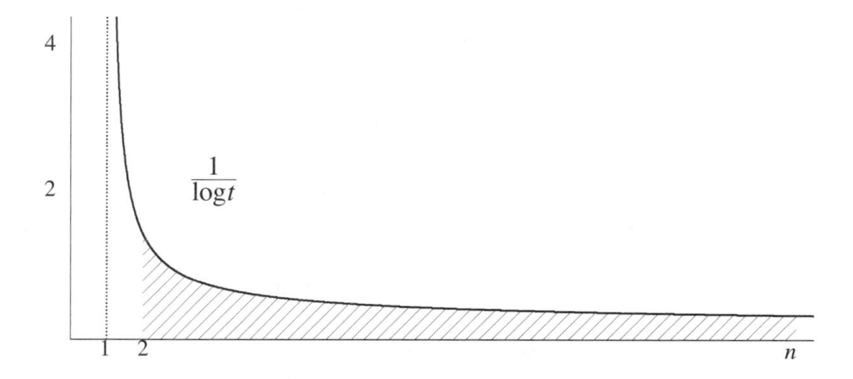

The prime number theorem says that the number of primes less than n is approximated by the shaded area on the graph. This has the surprising "feel" of a great result.

In 1859 Bernhard Riemann published "On the Number of Primes Less Than a Given Magnitude." Much of Edwards's 300-page book is an attempt to read, with full insight, its eight pages. Though a major theme of German mathematics during Riemann's lifetime was to increase the rigor of mathematical proof, Riemann did not sign on to that program. Nor did he spend an inordinate amount of time polishing his published results. Edwards wrote:

> When Riemann makes an assertion, it may be something which the reader can verify himself, it may be something which Riemann has proved or intends to prove, it may be something which was not proved rigorously until years later, it may be something which is still unproved, and, alas, it may be something which is not true unless the hypotheses are strengthened. This is especially distressing for a modern reader. . . .[2]

Riemann's work on complex analysis, Riemannian surfaces, and algebraic geometry was revolutionary. A good deal of the complex analysis that is the meat and potatoes of modern physics goes back to Riemann. He tried to remake physics itself and didn't succeed—his version of electricity and magnetism did not prevail. Even where he didn't succeed, however, Riemann's goals often lived on. He wanted unified theories of all the forces and looked to non-euclidean geometries to supply the mathematics. When Einstein needed to work out the geometry of general relativity, Riemann's geometry, well worked out during the intervening fifty years, was available.

In his famous paper, Riemann starts with the term that appears on the left side of the Euler product formula, $\Sigma\ 1/_n s$. Over time this had come to be treated as a function of a real variable—the letter s in the formula ranging over values greater than 1. Riemann constructed an analytic function of a complex variable that for real number inputs agrees with Euler's function but is defined for all complex number inputs (except for a pole at 1). The function Riemann created is called the Riemann zeta function and is written $\zeta(z)$, consisting of the Greek letter zeta and the variable z. The information about the number of primes that seems visible but unreachable in the Euler product formula is moved into the zeta function, where complex analysis can be used to extract it. Riemann was a master at calculation and waded into the zeta function. After an impressive set-to, with much wandering in the complex plane, Riemann managed to extract a formula for the number of prime numbers below a given number. The steps are giant, but comprehensible—with Edwards's help 140 years later. But how Riemann managed to choose the strategy that would allow him to get where he wanted is more impressive and mysterious. How does a great calculator arrive at an effective strategy?

The formula Riemann arrives at for the number of primes has more than one term that needs to be evaluated. Some disappear without much struggle. One is the logarithmic integral mentioned above—which is not 0. However, the value of the term most difficult to calculate depends on exactly those points (or complex numbers) where the zeta function equals zero. These are what are called "the zeros of the zeta function." Each such point makes a contribution. If the real part of all the zeros of the zeta function is exactly 1/2 (with some well-understood exceptions), this term goes to zero rapidly. Riemann states that most of the zeros have a real part of 1/2 and conjectures that they all do. This is the Riemann hypothesis. A proven Riemann hypothesis would give an extremely strong estimate for the number of primes below a given number.

On almost all of his points Riemann was right. The paper seemed obscure to his contemporaries, and no one made progress toward filling the gaps for over thirty years. Mathematicians had discovered that Riemann's lack of rigor had led him astray in demonstrable ways at other times in his career, and this made them suspicious of his unproved observations. One was his assertion and use of Dirichlet's principle. Weierstrass showed concretely that Dirichlet's principle didn't always hold; only later did Hilbert "salvage" it by showing that it held in most of the cases where one might really want to use it.

In 1893 the French mathematician Jacques Hadamard (1865–1963) proved "the product formula" for the xi function from Riemann's paper. In 1894 Hans Carl Friedrich von Mangoldt (1854–1925), building on the product formula, proved Riemann's "main formula." Von Mangoldt did not follow Riemann's sketched method, but Landau proved in 1908 that Riemann's method could, in fact, be made to work. In 1896 Hadamard and the Belgian mathematician Charles-Jean de la Valée Poussin (1866–1962), building on von Mangoldt's results, separately proved a weakened version of the Riemann hypothesis—each proved that none of the zeros had a real part as large as 1. This weaker restriction on the zeros of the zeta function was enough to show that the term with the zeros in it became arbitrarily small for large enough numbers. This proved what is called the prime number theorem.

The prime number theorem can be stated in two ways. The more precise statement is that the number of primes less than a number n is arbitrarily well approximated for large enough n by the logarithmic integral:

$$\int_2^n \frac{dt}{\log t}$$

This integral can in turn be well approximated by $n/\log n$. Stated in this form, the prime number theorem is more obviously astounding, if slightly weaker. If we want a good estimate of how many primes less than a number n there are, we simply have to divide n by the (natural) logarithm of n. If mathematicians were soccer fans, this is the sort of thing that would cause celebratory rioting in the streets. (The *Chicago Tribune* ran a tongue-in-cheek article by Eric Zorn describing such riots after Wiles proved Fermat's last theorem.)

So the next step toward a new and improved prime number theorem, a more accurate estimate of how fast the relative error in the approximation disappears, is either a proof of the Riemann hypothesis, or, barring that, more information about where the zeros are. This is the point at which Hilbert posed the eighth problem.

As the twentieth century began, it was obvious to calculate where the first few zeros lie. Jorgen P. Gram (1850–1916) published a list of fifteen roots in 1903 that was later proved to be quite accurate. In the next ten years R. Backland and J. I. Hutchinson made more calculations. All were consistent with the Riemann hypothesis. Soon after he completed his pathbreaking 1936 paper on computers, Turing set out to build a computer to calculate the zeros of the zeta function, but the war intervened. A recent tally asserts the total of zeros calculated exceeds 1.5 billion.

Ernst Lindelöf (1870–1946), Harald Bohr, Landau, and J. E. Littlewood (1885–1977) all made progress. In 1914, G. H. Hardy, at Cambridge, proved that there were an infinite number of zeros of the zeta function with real part 1/2. This did not prove that they all have a real part 1/2, but it was a start. Hardy was passionate on the subject of the zeta function and when planning a journey by sea, would always cable ahead that he had just proved the Riemann hypothesis. God had it in for him, he believed. If the boat sank, Hardy would always be famous for his "lost proof." Therefore God couldn't sink the boat.[3] (This might put Gelfond's attempted deathbed communication with Piatetski-Shapiro in an interesting light.)

In 1932 the Frankfurt mathematics history seminar inspired its most remarkable result. When Siegel went back and studied Riemann's papers, including scratch papers, to see if he could figure out how much Riemann had really known and how much he had been guessing, Riemann's eight-page paper was in some disrepute. Hardy had said in 1915 that Riemann "could not prove" the statements he made about the zeta function. Landau called them "conjectures." In the words of Edwards:

> The difficulty of Siegel's undertaking could scarcely be exaggerated. Several first-rate mathematicians before him had tried to decipher Riemann's disconnected jottings, but all had been discouraged either by the complete lack of any explanation of the formulas, or by the apparent chaos in their arrangement, or by the analytical skill needed to understand them. One wonders whether anyone else would ever have unearthed this treasure if Siegel had not.[4]

Riemann's papers are full of disconnected pieces of mathematics. In one place he breaks into a computation of $\sqrt{2}$ to thirty-eight decimal places, apparently from sheer exuberance.

Siegel discovered an approximation formula that made it clear that Riemann knew a great deal more about the zeros of the zeta function and estimates of error in the prime number theorem than anyone had imagined. Edwards reproduces a page of the manuscript where one can make out part of the formula. What is now called the Riemann–Siegel formula allowed Riemann to calculate a surprising number of zeros and they all fit the hypothesis. Riemann probably made his conjecture based on the Riemann–Siegel formula.

Twenty years earlier, on a different continent, someone else had been thinking about prime numbers, Srinivasa Aiyangar Ramanujan (1887–1920). A poor South Indian from a town near Madras, Ramanujan encountered *A Synopsis of Elementary Results in Pure and Applied Mathematics*, by G. S. Carr, a few months before he finished high school. It was a compilation of 6,165 equations and other mathematical facts (Hardy's count), listed without proof, first published in 1880 as a cram book for the undergraduate tripos exam at Cambridge. This second-rate book had filtered down through the educational system of the British Empire to his village, where it was Ramanujan's first contact with something even close to modern mathematics. As a boy, Enrico Fermi found a half-century-old treatise on spinning tops written in Latin in the flea market in Rome in a similar way and became a physicist. Ramanujan worked through the propositions in the cram book, convincing himself of their truth, and kept going, writing his original results in notebooks. (I almost wonder if the detective novelist F. R. Keating knew of Ramanujan, for his Inspector Ghote makes constant reference to a book called *Gross's Criminal Procedures*, purchased at a flea market in Bombay. Every time he encounters a visiting detective from Europe, Ghote inquires of his relationship to the great book. More often than not the detective has never heard of it, though Ghote's own effectiveness at his craft invariably argues that the European detectives ought to have paid more attention to this book. When things are not going well on a case, Ghote invokes Gross and "the correct procedure tirelessly applied.")

Ramanujan's application after he discovered Carr was from that time forward indeed tireless. He wouldn't stop to eat or sleep. His mother put food in his mouth as he worked.[5] Finding enough paper was a problem. Somehow, though his family was close to poverty, Ramanujan managed to work until he was twenty-five. He flunked out of college twice because he wouldn't work on any of the other subjects. None of his teachers or the mathematicians of the fledgling Indian Mathematical Society were in any position to evaluate the quality of the work Ramanujan was doing, though some did recognize something unusual was going on.

Ramanujan wrote to mathematicians. His epistolary style was not elegant; his letters were lists of formulas, stylistically similar to Carr, many written in strange forms. Two of the mathematicians who received letters in 1912-13 returned them. The third, Hardy, at Cambridge, glanced at Ramanujan's letter, dismissed it, and went off about his day. The letter contained identities of the sort:

$$\frac{1}{1^3}\cdot\frac{1}{2}+\frac{1}{2^3}\cdot\frac{1}{2^2}+\frac{1}{3^3}\cdot\frac{1}{2^3}+\frac{1}{4^3}\cdot\frac{1}{2^4}+\text{ etc. } = \frac{1}{6}(\log 2)^3-\frac{\pi^2}{12}\log 2+\left(\frac{1}{1^3}+\frac{1}{3^3}+\frac{1}{5^3}+\text{etc.}\right)$$

$$1+9\cdot(\frac{1}{4})^4+17\cdot\left(\frac{1\cdot 5}{4\cdot 8}\right)^4+25\cdot\left(\frac{1\cdot 5\cdot 9}{4\cdot 8\cdot 12}\right)^4+\text{ etc. } = \frac{2\sqrt{2}}{\sqrt{\pi}\cdot\{\Gamma(\frac{3}{4})\}^2}$$ [6]

The text of the letter contained the statement: "Very recently I came across a tract published by you styled 'Orders of Infinity,' in p. 36 of which I find a statement that no definite expression has been as yet found for the no of prime nos less than any given number. I have found an expression which very nearly approximates to the real result, the error being negligible."[7] Hardy must have been particularly skeptical of this statement.

Hardy kept returning in his thoughts throughout the day to the strange letter. Something bothered him. Some of the identities, though oddly stated, were known results, and hence correct. He had never seen anything like some of the others. They were strange, compelling. Hardy began to think they were too strange and compelling to have been made up by a real crank. In his experience—he had received strange letters before—cranks were not that creative. After dinner, he went back to his rooms with his collaborator Littlewood. Before they went to bed, they had concluded that Ramanujan must be brought to Cambridge. For a detailed account of Ramanujan's story from its beginning, to his election to the Royal Society, to his premature death, I recommend Robert Kanigel's excellent full-length biography, Hardy's accounts, and C. P. Snow's preface to *A Mathematician's Apology*.

As Hardy continued to think about the letter, not the least intriguing part was what Ramanujan said about prime numbers. We can imagine how startled he must have been. What does a natural scientist or mathematician do if the society in which he or she is living does not supply an outlet? Ramanujan had no such outlet. The line "Some mute inglorious Milton here may rest," from Thomas Gray's "Elegy Written in a Country Churchyard," has been criticized as sentimental. H. L. Mencken said, "There are no mute, inglorious Miltons, save in the hallucinations of poets. The one sound test of a Milton is that he functions as a Milton." Hardy had just received a letter from a marginally educated Indian, whose English was poor, who was in a mathematical sense functioning as a Milton. This is, perhaps, easier to accept when we think of other number theorists, like Siegel and Schneider, who did early great work from a naive perspective. Siegel used mainly algebra in proving the Thue–Siegel theorem and for some time couldn't convince either Schur or Landau he had done it. Schneider solved Hilbert's seventh problem using fairly elementary methods, apparently without being aware he was working on it.

It turned out that the work on prime numbers Ramanujan had done in India was wrong. In characterizing this work on primes Hardy, in the obituary, stated (though he took it back years later):

> This may be said to have been Ramanujan's one great failure. And yet I am not sure that, in some ways, his failure was not more wonderful than any of his triumphs . . . this result was first obtained by Landau in 1908 . . . Ramanujan, however, implies much more, and what he implies is definitely false. . . . However, Ramanujan had none of Landau's weapons at his command; he had never seen a French or German book; his knowledge even of English was insufficient to enable him to qualify for a degree. It is sufficiently marvelous that he should have even dreamt of problems such as these,

> problems which it has taken the finest mathematicians in Europe a hundred years to solve, and of which the solution is incomplete to the present day.[8]

The Riemann hypothesis remains a conjecture. We saw that Cohen worked hard on it, and that Matiyasevich has also. Many other mathematicians have. Stephen Smale, in 1998, answering a call by V. I. Arnold, composed his own list of problems for the twenty-first century. The first on the list is the Riemann hypothesis. The Clay Mathematics Institute in 2000 offered $1 million for a solution.

There seems to be a bit of a buzz around the current status of the Riemann hypothesis. Certainly people want something to happen. When Dyson decried the "divorce" between physics and mathematics, the title of his talk was "Missed Opportunities." In the same year as the talk, 1972, Hugh Montgomery, in tune with earlier ideas of Hilbert and Pólya, investigated whether somehow the zeros of the zeta function could turn out to be understood as related to eigenvalues of some hermitian operator in Hilbert space. People have long tried to find patterns by listing the lengths of the gaps between prime numbers as we search ever higher. Nothing has emerged. Montgomery was looking at the pattern the gaps between the zeros of the zeta function made. He showed his data to Dyson, who was trained as a number theorist, though his day job was as a physicist. Dyson said to Montgomery that the pattern the zeros made looked a lot like the energy spectrums of the nuclei of large atoms in quantum mechanics. This was like a shot in the Alps after a huge storm. Where everything was breathtakingly quiet there comes an avalanche. The game here becomes finding a quantum system that generates the zeros of the zeta function as its energy spectrum. (All quantum energy levels are real numbers, hence all nontrivial zeros of the zeta function would lie on the same line!) If physics and math did remarry in a big church wedding in 1990, this has to count as a hot flirtation.

Hilbert's problem also suggests the investigation of an analogous zeta function and analogous Riemann hypothesis constructed for algebraic extension fields. This problem was more tractable and was solved by Erich Hecke (1887–1947) in 1917.

A number of analogous zeta functions for fields of functions were investigated by Artin, Hasse, Weil, and others—and have likewise proven to be more tractable, despite their seemingly more complex origins. The prime number zeta function remains the most simple, compelling, mysterious, and ultimately difficult, attracting not only someone as far from the technical mainstream as Ramanujan but, at the least, a simple majority of the great mathematicians of our time, at one moment or another.

Castles of Air

The Ninth, Eleventh, and Twelfth Problems

Takagi, Artin, Hasse

Go to a university library and take the books on class field theory off the shelf: the Artin–Tate notes, Hasse's book, books by Lang, Iyanaga, Weil—QA241 call numbers. Stack them on a table. There really aren't many. The unfolding of class field theory was not unfocussed or fragmented by jargon, though it is difficult. Language and approach place each mathematician in time, yet each one was outside of time with the numbers. Their biographies will eventually be reduced to their names, dates, and a theorem or two or three. Hardy said, "Archimedes will be remembered when Aeschylus is forgotten, because languages die and mathematical ideas do not."[1]

Teiji Takagi did his work in Japan during the First World War, in the words of his biographer, in "utter"[2] isolation. Emil Artin would do his work under the Weimar Republic, flee Nazi Germany, leave his family in America, and return to Hamburg alone a few years before his death. Helmut Hasse, a German of conservative nationalistic views, would collaborate with the Nazis, undergo "denazification," and go back to being seen mainly as a mathematician before he died, shedding part of the biographical skin even in his lifetime.

The ninth, eleventh, and twelfth problems are about the algebraic numbers, myriad and sometimes nearly unnamable, the solutions of polynomial equations with whole number coefficients. They are not studied one at a time, but as parts of "fields" and "extension fields." A field allows the standard arithmetical operations and is "closed" under them—i.e., multiplication, addition, subtraction, or division (not including 0) involving numbers in the field always leads to a number also in the field. The fields studied the most usually start with the rational numbers and have additional algebraic numbers added. The "extension field" is the set of all the numbers one can get by adding, subtracting, multiplying, and dividing the numbers we started with. (It also includes the rules for addition, multiplication, etc., for the elements of the field.) In the extension field of the rationals generated by $\sqrt{2}$, 2 is no longer obviously prime because $2 = \sqrt{2}\sqrt{2}$. Even in this most simple extension field, new definitions of "integer" and "unit" are required, and the details of factoring are surprisingly complicated. In this case $(2 + \sqrt{2})(2 - \sqrt{2})$ also equals 2—what are $2 + \sqrt{2}$ and $2 - \sqrt{2}$? Is $\sqrt{2}$ an integer?

The early arguments in number theory all turned around "unique factorization," the fundamental theorem of arithmetic—any natural number can be factored in one and only one way into a product of prime numbers. Pierre de la Fermat (1601–1665) focused on results about whole numbers, for example: Every prime of the form $4n + 1$ can be expressed as the sum of two squares in exactly one way, and if $2^n - 1$ is prime, then n is prime. The methods that Fermat had available required great dexterity. All the classical number theorists played with numbers, and their ideas and intuitions flowed from experiments with calculation. When the complex numbers (any number with a real part and an imaginary part) were added to the growing body of mathematics and the algebraic numbers better understood, mathematicians thought unique factorization could be extended. With the idea of "integer" defined in this more general context (which leads to different primes), they tried to factor in finite algebraic extension fields of the rational numbers. They weren't called "fields" at the time, but the ideas that would resolve into class field theory begin here.

At first, calculation seemed to confirm that, in general, unique factorization held. Euler used unique factorization in a proof of Fermat's last theorem for $n = 3$ in the mid-1700s. Fermat's theorem states that there are no whole number solutions to any equation of the form $x^n + y^n = z^n$ for $n \geq 3$. In his proof, Euler assumed (without proving it) that unique factorization held when $\sqrt{-3}$ was added to the rational numbers.

In 1801, when he was only twenty-four, Gauss published *Disquisitiones Arithmeticae* and changed the discourse in number theory. He had written it at twenty but had trouble finding a publisher and finally published it himself. Gauss used imaginary numbers systematically. In *Disquisitiones* he proved the law of quadratic reciprocity and investigated the structure of the nth roots of 1.* He developed a theory of congruences (statements of the form $2 = 9 \bmod 7$) and was adept at using it. He investigated quadratic forms. Gauss used these second-order theories to prove theorems about whole numbers, but they are interesting in their own right and inspired Hilbert's ninth and eleventh problems. Several of the pathways to modern class field theory start with Gauss and are interrelated. In later work Gauss sometimes worked in an extension of the counting numbers now called the

* One is always an nth root of 1, because $1^n = 1$. Minus one is an nth root of 1, if n is even, e.g., $(-1)^n = 1$. It can be shown that there are n nth roots of 1 or "unity," and all the others have imaginary parts. They all lie on the unit circle. These fields are called cyclotomic.

Gaussian integers. These are the counting numbers with $\sqrt{-1}$ thrown in—numbers of the form $a + bi$ where a and b are counting numbers. Unlike Euler, Gauss felt it necessary to prove that unique factorization worked in this case, and did so.

Gauss, Euler, and others had been at play in specific extension fields; Galois gave birth to a more fully realized concept of the "field." In 1829 he invented the Galois group when proving that there was no general algebraic method of solving a fifth-degree (or higher) polynomial equation. A finite extension field is generated by a finite collection of generators. The Galois group is the group of all the ways we can swap the generators and still get the same structure in the extension field (the "self-isomorphisms" of the field). He could use the Galois group to study the structure of any extension field he chose. But Galois died in a duel before his twenty-first birthday, in 1832, and his writings were difficult. Further, Abel had proved the unsolvability of fifth-degree polynomials in 1826, using what seemed at the time more straightforward methods. (Abel also died young, of consumption at twenty-six.) It took the next forty years for Galois's ideas to enter mathematics. When they finally did, they became fundamental.

Fermat's last theorem continued to be of interest, and the developments here contributed to the partial understanding of unique factorization. Mathematicians continued to grind out yardage with the same type of argument used by Fermat and Euler. Sophie Germain (1776–1831) proved a theorem about Fermat's equation for $n = 5$ that bears her name. The following excerpt from *Journey Through Genius* by William Dunham makes one reason plain why we haven't seen more women mathematicians in this book:

> As a child, Germain had been fascinated by the mathematical works she found in her father's library. She was especially intrigued by Plutarch's description of the death of Archimedes, for whom mathematics was more vital than life itself. When she expressed an interest in studying the subject more formally, her parents responded in horror. Forbidden to explore mathematics, Sophie Germain was forced to smuggle books into her room and read them by the candlelight. Her family, discovering these clandestine activities, removed her candles and, for good measure, removed her clothes as well in an attempt to discourage these nocturnal wanderings in a cold and dark room. It is a testament to Germain's love of mathematics, and perhaps to her physical endurance, that not even these extreme measures could keep her down.[3]

Dunham also writes of Germain lurking about the doorways to classes she was not allowed to attend because of her sex, in the hope she might hear a bit of the lecture or succeed in borrowing a set of notes.

In 1825 P. G. Lejeune Dirichlet (1805–1859), then barely twenty, building on Germain's theorem, proved that Fermat's last theorem for $n = 5$ reduces to two cases and proved for the first case that there were no whole number solutions. Legendre quickly supplied a proof for the second case. In *Fermat's Last Theorem: A Genetic Introduction to Algebraic Number Theory,* Edwards says, "Legendre's proof of the second case was rather artificial and involved a great deal of unmotivated manipulation, perhaps a symptom of his great age and long experience."[4] Dirichlet offered a more natural proof of the second case a few months later and returned to Fermat's last theorem in 1832 and proved it for $n = 14$. In 1839 Gabriel Lamé (1795–1870) proved the case $n = 7$. (There is some doubt whether his proof was complete.) Though similar in style and technique, these proofs offered no common thread in their particulars.

Lamé appeared at the March 1, 1847, meeting of the Paris Academy and excitedly announced that he had proved Fermat's last theorem using arguments based on unique factorization. He would use the nth roots of 1 to help factor the left side of Fermat's equation. So far so good. That can be accomplished using algebra. Lamé graciously gave credit to Liouville, who got up and gave it back. Liouville said many others had used similar ideas and, in Edwards's words, "practically said that Lamé's brainchild was among the first ideas that would suggest themselves to a competent mathematician approaching the problem for the first time."[5] Liouville pointed out that the existing proofs worked because they dealt with counting numbers (or extensions of them) where unique factorization held. He cast doubt on whether unique factorization was true of the nth roots of 1 Lamé was proposing to use.

Cauchy spoke next and was much more sanguine, stating that he himself had had a similar idea, which he had given the Academy only half a year earlier. However, he had not had time to pursue it. There followed a period of great excitement and activity. At the meeting of March 15, 1847, Pierre Wantzel (1814–1848) stated he had proved unique factorization for the nth roots of 1. On March 22 Cauchy stated he hadn't and started publishing papers attempting to do so himself. At the same meeting both Cauchy and Lamé presented "secret packets," a method the Academy had of deciding priority when the possessors of an idea were not ready to publish. Nevertheless, both protagonists kept publishing notices of their work as it progressed.

This went on until May 24, when Liouville read a letter from a then-little-known German mathematician named Ernst Kummer, a professor at the University of Breslau (now Wroclaw in Poland). Until five years before, Kummer had been teaching at a gymnasium in Leignitz (now Legnica)—one of his students was Kronecker. Kummer's letter announced that unique factorization fails for the 23rd roots of 1. He pointed out that he had published a paper on this subject three years earlier. It was published obscurely. Kummer also wrote about an idea that could save this method of attack on Fermat's last theorem and said he was far advanced in accomplishing this and that more details would be forthcoming in the proceedings of the Berlin Academy. This letter stopped Lamé in his tracks, but Cauchy continued his publications for a number of months.

Kummer had done a large number of calculations with the *n*th roots of 1, exploring their structure. Like most mathematicians of the time, he had an intuition that told him that unique factorization worked. He knew many cases where it did work, and through his calculations he knew exactly how it worked. When Kummer discovered that the system broke down for the 23rd roots of unity, his "intuition" was deeply enough entrenched that he tried to save the situation. Kummer was deeply conservative. He was not interested in speculative forays that would open up new ground, but instead wanted to extend and elaborate the tried-and-true results of the masters. His interest in the *n*th roots of 1 flowed directly from Gauss through Jacobi, and what he was specifically trying to do was extend Gauss's law of quadratic reciprocity into higher orders.

But confronted with trying to save unique factorization, which was as fundamental to algebraic number theory as it was to whole-number number theory, Kummer did a radical thing. He said, if we add some "ideal" numbers to the pot we can make this work. It is unclear what these ideal numbers were supposed to be—a kind of complex number drafted for new duty? Something pulled in from an entirely different astral plane? In order to make the theory work mathematically one was not required to be specific. The ideal numbers tended to disappear before the end of proofs in which they were used. (They were ideal for the job in that sense.) They could be viewed as useful fictions that did not damage the logic of the proofs, somewhat the way an imaginary friend can contribute to a developing child, yet be gone for the adult. With hindsight one can even think of ways to avoid the ideal numbers completely in the proofs that Kummer used them in.

Kummer was probably aware early on that his results had bearing on Fermat's last theorem, but he placed extending the reciprocity laws on a higher plane, referring to Fermat's theorem as "a curiosity of number theory rather than a major item."[6] However, the fuss in Paris got him going and he finished his results by April 11, 1847. He did not prove Fermat's last theorem, but he radically increased the number of cases for which it was proven.

Kummer was responding to work done by Dirichlet, who had proved that any arithmetic progression, $a + b$, $a + 2b$, $a + 3b$, $a + 4b$, . . ., where a and b are "relatively" prime (not sharing any common factors, e.g., 6 and 35), includes an infinite number of primes. He also proved the Dirichlet units theorem, which gave the structure of the units of an arbitrary field extension. Units are like doppelgangers of 1 that appear in extension fields. If you can't tell the units from the primes, unique factorization disappears.

As they began to understand that unique factorization broke down (without attempts to save it, like Kummer's), mathematicians introduced the concepts of "class group" and "class number" to measure the breakdown. The class group is a group constructed from all the different ways factorization can be done, and the class number is the "rank"—a measure of size—of the class group. If the class number is 1, then unique factorization holds. This is the "class" in class field theory. Number fields, particularly finite extensions of the rationals, had emerged as mathematical objects to be studied as existing separately from either algebraic or complex numbers in general. Unique factorization, the key tool, can be partially saved by the introduction of ideal numbers. However, as of 1870 there was still no organized language for these things. Also, finite extensions, though finite in their generating instructions, were clearly infinite aggregations, just as the rational numbers that were being extended were infinite. Set theory or something like it was needed to talk about concepts like fields, extensions, class groups, and ideals.

At this point Richard Dedekind (1831–1916) appeared. He received his Ph.D. at Göttingen under Gauss in 1852. As a student he struck up a close friendship with Riemann and the two continued together as lecturers in 1854. In the interim Dedekind studied in Berlin under Jacob Steiner (1796–1863), Jacobi, and Dirichlet. (Steiner didn't learn to read or go to school until he was fourteen.) In 1855 Dirichlet succeeded Gauss at Göttingen, and the young Dedekind started following his lectures. Dedekind couldn't have been better placed to ask, "What exactly are you talking about?" As early as 1858 he possessed a model for the real numbers built from sets of rational numbers, the Dedekind cuts. These ideas came to him as he was preparing lectures in analysis for a teaching post at the Polytechnikum in Zurich. However, for years he did nothing to disseminate them. He was talking about completed infinites and knew he would get into trouble. He waited until Cantor came along, drew all the fire, and made him look like a moderate.

Dedekind assumed responsibility for editing Dirichlet's lectures, which naturally connected to the concept of ideals, and he added a supplement trying to clarify Dirichlet's theory. In 1871 he published a second supplement containing what has become the modern definition of ideal, and in 1879 and 1894 in updated supplements supplied the modern treatment of extension fields and unique factorization, using ideals in "rings" of integers. Dedekind introduced the modern concepts of ring and unit so that the details of how unique factorization was supposed to work could be clearly stated. In this regime an algebraic object—a field, or a ring—is a set, with operations of addition and/or multiplication defined on it that satisfy certain rules. Dedekind moved this idea—ideals—further toward abstractness, that is, logical independence from the complex numbers themselves or any other parent. An ideal was not a number but a certain kind of subset of the field or the ring. Ideals could be "multiplied" and an ideal could be factored as the product of other irreducible or "prime" ideals. In many cases this factorization was unique! In 1880 Dedekind, with Weber, Hilbert's first university teacher in Königsberg, published a long paper that rigorously proved the Riemann–Roch theorem in a general form, for the first time using the new algebraic number theory and no analysis. The appeal, to a number theorist, is similar to the appeal of climbing Mt. Everest without oxygen, with the proviso that the number theorist would like to be barefoot and naked, as well. Hilbert probably first became aware of the power of set theory through Weber. In this work Dedekind clarified existing ideas and dramatically forwarded what would become fully realized as modern abstract algebra under the stewardship of Noether and others.

If unique factorization was finally understood (to some extent) in terms of class number and class group, mathematicians were soon asking what would happen if they looked at extension fields of extension fields, and they guessed that the class group of the field being extended might tell them something about the extension. This idea first appeared (under a different name) in the work of Kronecker in 1882, and he commented on the similarity of the class group to the Galois group. Weber used the term "class field" in 1891 and in 1896 enlarged the concept to cover a wider territory, enlarging it still further in 1908.[7] (It was this "algebra" book of Weber's that Siegel picked up by mistake at an early age.) Hilbert, in 1897, came up with his own definition and made a more explicit connection between the Galois group of an extension and the class group of the field being extended. Hilbert wrote his problems.

Hilbert's contribution stood out for those who followed, because in the *Zahlbericht* (1897) he had organized the whole theory of algebraic numbers and number fields. In papers in 1898 and 1899 he had also produced a prescient series of conjectures about the structure of his version of class fields. He conjectured that if one started with an algebraic number field, there existed a unique "abelian extension" of it such that the structure of this extension as reflected in its Galois group was identical to the structure of the class group of the original field. Hilbert's conjectures were proved by Furtwängler in 1907 and by Furtwängler and Artin in 1930, though this version of the theory didn't seem the most general possible.

Class Field Theory

Only with the work of Teiji Takagi (1875–1960) did class field theory emerge as a well-realized theory. He started with Weber's more general, 1908 definition of ideal class, a broader canvas than Hilbert's ideal class and class group, and proved a more general form of the existence theorem in Hilbert's conjecture. If one takes a class group—one of the many possible in Weber's theory—in the starting or base field, then there exists a unique extension, the structure of which is a mirror of that class group. The more startling result, however, was that any abelian extension of the base field was a class field generated by one of Weber's generalized class groups. This told us that anything we can know about the general structure of an abelian extension (at least as reflected in its Galois group) was already contained in one of the class groups of the original field. Takagi at first could not believe this result. We can think of extensions of fields as representing nests of numbers, building ever larger. At each step there are many possible directions to go. Attempts to understand how unique factorization broke down in any one nest (expressed in the concept of class group) led to an understanding of where to go for the next nest. Class fields and class field theory became the ground of investigation of ever more complex structures in algebraic number theory and algebraic geometry.

Takagi did not solve any Hilbert problem outright, but his class field theory was central to Artin's solution of the ninth problem and is part of the context of

Hasse's solution of the eleventh. Both of these results followed the emergence of Takagi's theory by only a few years. Takagi's class field theory is fundamental to progress made on the twelfth problem.

Teiji Takagi was born into a family of farmers on April 21, 1875, in what is now Kazuya village, a few miles to the west of Gifu in central Japan. His birthplace was a dirt-floored farmhouse shared with a horse. Even in the 1970s the setting was rural. Takagi's biographer, Kin-ya Honda, was impressed with the ripe persimmons, abundant birds and clean, refreshing wind. My treatment is based on a biographical essay in English by Professor Honda.

Takagi's great-grandfather, known as Kansuke I, had become wealthy through brewing and died with "an unusually vast farm."[8] It was said that he could make a journey to a place called Tokino without leaving his own land. Unfortunately, Kansuke II, Takagi's grandfather, reversed the process of accumulation, ending in the loss of all the property. Later, he left his family and went to Osaka, where he took refuge in a Buddhist temple, working as a clerk until he died in 1853. His wife and children stayed in the rural area. His daughter Tsune was Takagi's mother. His son, Kansuke III, at age six witnessed the family house being torn down by a creditor. The little boy challenged the destroyer, with no result, and Takagi's grandmother explained that the house no longer belonged to them. Takagi's uncle became "an extremely austere and hardworking man."[9]

Young Kansuke III worked "in another person's home."[10] He managed to start buying back what had been the family property. Takagi's mother, who was eight years younger than her brother, worked on a farm. In 1874, at thirty-one, perhaps conventionally past the time of marriage, she married Mitsuzo Kinomura, a fifty-five-year-old farmer from the town of Kitagata. He had two sons by a previous marriage and also worked for the town office. Tsune was put to work in Kinomura's rice fields, which were full of leeches. She was not averse to farm work but did not like the leeches. She became pregnant quickly and returned to her family's home to bear the child. Kansuke III, now wealthy and also the treasurer of Kazuya, had married a cousin named Iwo, who had a daughter by a former marriage. They were unable to have children themselves, so there was no heir. The birth of a male child to a sister who never wanted to go back to her husband was a stroke of good luck. Teiji was recorded as the child of Kansuke III and Iwo, his wife. This was not unusual in Japan at the time. Tsune remained at the family home in the role of Teiji's mother.

Takagi was a small child. He was not allowed to swim in the nearby river or, in general, to play with other children, particularly if any danger could be imagined. Since his mother was a devout Buddhist, he visited many temples to pray. When he was about five, he recited the Godensho Scriptures from memory to a startled group of adults. Takagi was enrolled in the private school of a highly cultured neighbor, Kyohei Nogawa. In addition to being a medical doctor and running the school,

Nogawa was a writer and expert in "Chinese literature, calligraphy, Chinese painting, flower arrangement, and tea ceremony."[11] The main subject of study at Nogawa's school was Chinese literature—Confucius, Mencius, etc. Honda believes Takagi's later writing style was influenced by this: "succinct, with a lofty tone."[12]

Takagi entered the village primary school in 1882. He did exceptionally well, completing the six-year course in three years. As village treasurer, Kansuke III had an office next door to the school. Instead of playing with the other children after school, Takagi would study in the office. Parents held this up as a good example, and one is tempted to imagine the reaction of the other children in simplistic terms. However, one little girl in the class said years later:

> Since he was such a lovely little boy, I wished to touch him. However, on the other hand, since he had such a distinguished record, and rarely talked to anyone, he was indeed one boy who was difficult to approach.[13]

The Gifu Daily News ran a story about Takagi's achievements when he was only ten, stating, "He is indeed a genius, for whose future we have great expectations."[14] Even more remarkable, though, is that Takagi had competition for top in his class. It came from Tazuko, the daughter of Dr. Nogawa, the neighbor whose school Takagi first attended. She was a pretty girl, but she was also "strong-minded and romped as if she were a boy. She was skillful in walking on walls and in tree climbing."[15] When the Nogawa family wanted persimmons, it was Tazuko who climbed the trees to pick them. Sometimes Teiji was first in the class and sometimes Tazuko. When Teiji was only second, Kansuke III made the small boy stand outside with a heavy desk on his back. This did not turn Teiji against Tazuko, though.

In 1886 Takagi was enrolled in the Gifu-Prefectural Ordinary Middle School, some miles away, but close enough so he probably continued to live at home. (Honda does not say.) Takagi himself wrote about the curriculum of this school. After the Meiji Restoration in 1868, European education was introduced to Japan. When Takagi attended the Gifu school, there were still only imported books for texts. Takagi recounts the particular difficulty of the English spelling text—the students were taught the spelling of words for which they did not know the meaning.

Takagi impressed the mathematics teacher, who is said to have caused dissension in the ranks of his students by diligently trying to find examination problems that Takagi couldn't solve. Takagi graduated first in his class on March 31, 1891, and in September entered the Third Higher Middle School in Kyoto. When he left on the train, Kansuke III and his mother carried big loads of baggage to the train station on their backs. My inference is that Takagi traveled unencumbered. However, little more than a month after his arrival in Kyoto, on October 28, the "Great Earthquake of Nobi"[16] occurred. Also called the Mino, Owari quake, it caused the destruction of more than 220,000 dwellings and killed more than 7,000. Gifu was hard hit—Honda says there were 230 deaths, 906 houses collapsed, and 2,017 houses were destroyed by fire. Takagi rushed home to his family, taking the train as far as it went and then walking the rest of the way. He was relieved to find them safe, though shaken up. Kansuke III was angry that Takagi had left his studies and wouldn't let him enter the house. Takagi's mother tried to persuade Kansuke III to let Takagi spend the night, to no avail, and Takagi left immediately to return to school. He is reported to have told a friend fifty years later that he had been able to succeed in mathematics because Kansuke III had not let him enter the house.

At the Third Higher Middle School Takagi decided to become a mathematician. His professor, Jutaro Kawai, had studied in Germany and attended lectures by Klein and Weber. Kawai was a function theorist and the first practicing mathematician that Takagi had encountered. Kawai's firm focus inspired stories. In one he is walking down a road reading a mathematics book, runs into a light pole, but instead of finding a way around the obstacle simply stops where he is until he gets to a stopping point in the book.

Takagi entered Tokyo University in July 1894, when the university was itself only twenty-five years old.[17] Two central figures of modern Japanese mathematics, Dairoku Kikuchi and Rikitaro Fujisawa, taught there. Prior to the Meiji Restoration there was a Japanese mathematics called *wazan*. Kikuchi had gone to England to study at the early age of eleven, and it was he who had introduced European mathematics to the Japanese school curriculum. Fujisawa had studied in Germany and had written original papers on elliptic functions. Honda says he was the first research mathematician in modern Japan. Fujisawa gave the incoming students a calculus exam. Takagi did so well on this exam that he was awarded 140 out of a possible 100. (I am amazed at the cross-cultural, almost universal, tendency of mathematicians, of all people, to award exam scores that exceed 100%.)

Also at the university was Tazuko Nogawa, the pretty little girl who had often bested him in primary school. The logic of the narrative demands they meet again and fall in love. They did meet each other and we know that Takagi, at least, did fall in love, for he proposed marriage. The proposal probably went through Kansuke III and his mother to Tazuko's parents. It was refused, probably because of Kansuke II's ruin and disgrace. It is not certain that Tazuko ever knew she had been proposed to. There is a remarkable contrast here between her being, as a woman, at the national university and the fact that she might have never known Takagi had proposed. She married a scholar of English literature. He died young, so she resourcefully started a beauty parlor, which became "one of the most famous parlors in Tokyo." As to Tazuko's possible feelings about Takagi, Honda writes:

> One day, after the Second World War, when Takagi once happened to have an interview with a radio announcer, Tazuko heard him, and joyfully, she cried to her family. "Teiji is now on the radio!" She listened to it quite longingly.[18]

At university Takagi acquired the nickname "Buddha."[19] It is not known if it was applied because of his intelligence (as is sometimes said) or because he would sit entirely motionless, studying, for long periods of time. He graduated in July 1897 and continued in graduate school. He wrote two books in his first year as a graduate student—*New Arithmetic*, which treats the real numbers in the manner of Heine and Dedekind; and *New Algebra*, which largely follows Weber's *Lehrbuch der Algebra* (1895, the first volume). Takagi was already thinking about algebraic number theory and had managed to find one of the most significant current expressions of the subject.

In May of 1898 Takagi was selected for study abroad by the Minister of Education, the highest honor for a student. The order as reproduced by Honda is admirably terse: "You are ordered to go to Germany in order to study mathematics for three years."[20] Takagi started the long journey to Berlin on August 31. His professor, Fujisawa, accompanied him to Yokohama, as did Kansuke III and his mother. Fujisawa advised Takagi to sample the mackerel in Marseilles. But when he landed, Takagi hurried off to begin his studies. It was a journey of considerable magnitude. Kansuke III would be dead by the time Takagi returned. The voyage had taken about forty days, first on a Japanese steamer to Shanghai, then a French ship that stopped in many ports.

Berlin was in its heyday when Takagi arrived, capital of the German Empire, full of noisy horse-drawn carriages, cabs, and trolleys and even larger in population than Tokyo, with 1.8 million inhabitants. Sausage and beer, German being spoken everywhere, Prussian military officers. There was at least one theater in the part of the city where Takagi lived, located across the street from a military barracks, something that Honda identifies as particularly strange to a Japanese person. He writes, "Perhaps it must have been Prussian taste."[21]

Takagi took lodgings at Schumannstrasse 18, about fifteen minutes walk from the University of Berlin. He enjoyed his stay and often went to the theater. However, mathematically the university was a disappointment. Honda says, "To the eyes of Japanese young men of that time, professors of European universities were something like gods!"[22] The mathematics faculty at Berlin was not in one of its heydays. Fuchs and Hermann Schwarz were sixty-five and fifty-five years old. Frobenius was nearing fifty. A young professor at Tokyo University who had just returned from Germany warned Takagi before he left to be wary of Frobenius. According to him, Frobenius had delivered an inaugural address on being appointed dean of the faculty in which he said, "In order to study science in Germany, there come to this country many foreigners—Americans, and so on, and lately even Japanese! In the future, no doubt apes will be coming. . . ."[23] Takagi didn't know whether to believe this report, but it colored his first impressions. As it turned out, he enjoyed the liveliness of Frobenius's lectures, though he found their content not much more advanced than what he had already studied in Japan. When Takagi approached Frobenius with questions he received a reasonable reception.

Takagi stayed three semesters in Berlin. In the meantime, a friend from Japan, Takuji Yoshiye, went to study in Göttingen. In a surviving postcard Takagi writes: "It is just one year today since I arrived in Berlin. How fast time flies! I am very much ashamed of the lack of progress in my study."[24] In the spring of 1900, Takagi left to go to Strasbourg, where Weber was teaching, but on the way stopped in Göttingen, came under its spell, and never went on. The small university city must have seemed an attractive change from Berlin. Takagi found lodgings a mere seven minutes from the university. Honda comments on the many birds, reminiscent of Takagi's youth, and on the bell ringing at dawn. Göttingen then had a population of 30,000. The duels among the students that Reid mentions in *Hilbert* did not go unnoticed by Takagi and his friends. Yet it was the mathematics of Göttingen that was significant. Takagi wrote:

> I was much astonished by the striking contrast in the atmospheres of the mathematics departments of Göttingen and Berlin. In the former, once a week a meeting was held, and in attendance was a group of brilliant youths from all over the world, as if here were the center of the mathematical world. I was already at the age of twenty-five, and felt painfully that my knowledge was fifty years behind the then stage of mathematics. To recover that fifty years' lag seemed to be very difficult. But, within three semesters in the atmosphere of Göttingen, I began to feel I had almost recovered the lag. Thus I was made to realize how important atmosphere is in pursuing scientific truth.[25]

Takagi attended Klein's lectures for six weeks only, because after the six weeks free tryout period he would have had to start paying the lecture fee. Takagi wrote:

> That was quite enough. Such attendance for six weeks helped me so much to recover the fifty years' lag. . . . Klein often used the expression "Three Large A's"; they were *Arithmetic*, *Algebra*, and *Analysis*. I interpreted his implicit intention as follows: "Ordinarily, one studies mathematics, dividing it into several fields, while I unify them by my geometry."[26]

Takagi was both lucky and unlucky in the timing of his visit. Hilbert had finished his work on algebraic number theory and left the field. He was now two subjects past class field theory, was thinking about analysis, and gave Takagi no direct help. When Takagi met Hilbert, the master (*sensei*, as Takagi viewed him) was suspicious of Takagi's level of preparation. Hilbert asked a question that Takagi failed to answer properly. Walking Hilbert home as they talked, Takagi announced his intention of tackling Kronecker's *Jugendtraum* or "youthful dream," which is part of the twelfth problem and concerned how extension fields could be generated. Hilbert said only, "That's fine."[27]

If Hilbert had personally turned his attention to other things, however, his spirit was still fresh in algebraic number theory in Göttingen. Takagi responded.

He probably had already read Hilbert's book on algebraic number fields, the *Zahlbericht*, when he was in Berlin, but he would turn to it again and again. Honda says, "This book played the role of the Bible in Takagi's scientific life."[28] Hilbert's ability to find simple ideas that explain and unify complexity was particularly attractive to Takagi. Honda describes this point of view or ability as "high essentialism" and finds it present in Takagi's work. Takagi thrived for a year and a half and actually did make progress on Kronecker's *Jugendtraum*, proving it for a special case. This was no mean achievement. Then it was time to return to Japan.

Takagi arrived in Japan on December 4, 1901, and returned to his home village of Kazuya. He came riding on a *jinrikisha* and by tradition was welcomed under five tall pine trees near Okua Temple. The whole village turned out, as did 300 current students of the village primary school where Takagi had excelled. His study abroad conferred great honor on him and tangible benefits. He was appointed assistant professor at Tokyo University and started to lecture. He rented a house and his mother, Tsune, came to run the household. He also came back with a high-quality wardrobe and wore a pince-nez and grew a mustache in the style of Kaiser Wilhelm II. He was referred to by his neighbors as "that dandy gentleman in the house over there."[29]

Takagi was highly marriageable on his return and a little over a month after his landing was formally introduced by his landlord, a professor at the Tokyo High School of Commerce, to the woman who would become his wife, Toshi Tani. She was a bit taken aback by Takagi's near silence at their first meeting and was herself somewhat preoccupied in trying to repress a sneezing fit. Nevertheless, the meeting went well enough so that they were married on April 6. Takagi's mother stayed on for three months evaluating the situation and then returned home.

Takagi rose rapidly at the university, largely based on his studies in Germany and the significant result toward Kronecker's *Jugendtraum*. He published five short papers of no great importance and then, in 1903, published the paper that he had actually written in Germany, reporting his results on the *Jugendtraum*. He received the degree of doctor of science that year and the next year was made full professor. He then didn't publish a single paper until 1914, eleven years later. He lectured, was an honored man. He was a dandy. He and Toshi had eight children. He was an austere head of the household. Toshi is described as "a cheerful, sociable woman. She talked with visitors very loudly."[30] With visitors Takagi was quiet, except when his wife got a date wrong. He is said to have been very perceptive about his visitors' characters, particularly about their faults. Takagi's father-in-law had been a samurai before the Meiji Restoration. If he saw one of his daughters sitting in anything other than the perfect Japanese posture he would hit her with his pipe so she would straighten up. But he is reported to have said, "When I am talking with Teiji, I feel quite stiff."[31] One wonders if a certain amount of pain was involved in Takagi's behavior. Takagi wrote:

> Generally speaking, I am a man who cannot do anything without stimulation. Different from the circumstances of today, there were

> then almost no persons in Japan who were studying the same specialty as mine. Therefore, I could not get stimulation from others. It was an era when one might spend his days vacantly.
>
> Now in 1914, the World War broke out. That was a good stimulation for me—a negative stimulation; European books stopped coming to Japan. A certain scholar was reported with ridicule in the newspapers to have grumbled that, since German books were not to be obtained, science could not be studied in Japan. In my own case, since European books did not reach me, I had to study by myself, if ever I wished to study. If the World War had not broken out, I might have achieved nothing.[32]

Honda uses the expression "utter (scientific) solitude"[33] to describe Takagi's situation. We have the picture of Takagi trapped in a materially pleasant situation, but uncreative. Deprived now of even imported publications, the war caused him to go deep within himself and to create what is described by Honda as "the magnificent edifice of class field theory."[34] Weyl writes:

> There is nothing in any other science that, in subtlety and complexity, could compare even remotely with such mathematical theories as for instance that of algebraic class fields.[35]

The theory that Takagi created is difficult and his proofs difficult to understand. Shokichi Iyanaga says, "Its proofs were when it appeared . . . often quite involved. Various attempts for the simplification of the proofs have been made."[36] Honda refers to "the long and difficult proofs of Takagi's theorems,"[37] as does Richard Brauer (1901–1977).[38]

With no one to talk to, Takagi was his own first obstacle :

> Since it was then quite an unexpected result, I thought it was surely incorrect. So I looked for where it was wrong. Perhaps, I was then a little neurotic. I often had dreams concerning the study. In a dream, I would believe that I had solved a problem, and then waking up, I would actually work it out at the desk, finding alas it was quite mistaken. I looked for counterexamples in vain. I had been searching for mistakes for a long while. I had hardly any confidence in my result even after I constructed the theory. If there is a hole made by ants in some place, the whole structure collapses due to that small hole. At least in mathematics, we cannot accept what is only roughly correct.[39]

Takagi published his results in Japanese and probably not a single person read and understood them. When the war ended, a new International Congress for Mathematicians was scheduled to begin September 22, 1920, in Strasbourg, which was now in France again. Takagi wrote a long paper in German, 133 pages, which he published in a Japanese journal set up to display Japanese results to the world in European languages. Even if the journal was not widely read in Europe at that time, reprints could be distributed. But Germans were not invited to the 1920 congress, and they were the only mathematicians who understood algebraic number theory of the type Takagi was presenting. He had to give his talk in French to an audience in no position to understand it. At the opening reception Takagi overheard a mathematician whisper: "I hear that Japanese fellow will talk on number theory. Perhaps, it must be Fermat's problem. We shall surely be amused!"[40] As we have seen, the Paris Academy itself had supplied entertainment of this sort. This cannot have lowered the level of "austerity" of the talk Takagi delivered three days later. He did not use the blackboard, which seems an odd omission in a mathematical presentation. Takagi's French must have been less than perfect, since he had never lived in the country, and most of his European contacts were German. The talk received no reaction.

After the congress Takagi returned to Paris, where he had already spent a month. Prompted partly by his love of Goethe's *Travels in Italy*, he traveled to Italy and particularly enjoyed his stay in Florence. He returned to Göttingen and visited Hilbert. While there, he heard of a brilliant young number theorist named Carl

Teiji Takagi (*courtesy of* Natascha Artin Brunswick).

Ludwig Siegel. Takagi does not appear to have met Siegel on his visit, but on returning to Japan he sent him a reprint of his long paper in German. Siegel was not working in this area. However, he saw the importance of the paper and in early 1922 loaned the reprint to Artin, who was exactly the right person to appreciate it. Where most mathematicians reacted to the theory's difficulty, Artin told Honda in 1962: "I felt strong admiration for it. It was not difficult to understand, since it was written very clearly."[41] In 1922 Takagi published a paper on reciprocity laws and proved some already known reciprocity theorems much more clearly and economically using his class field theory. The language he used to frame the reciprocity results helped suggest to Artin the formulation of what became Artin's general reciprocity law and the solution to the ninth Hilbert problem.[42] Artin formulated his general law in 1923 but was unable to prove it except in special cases.

Artin suggested to the young Hasse that he read Takagi's two papers. They inspired Hasse to present the theory in a simpler form to the world of mathematics at large. Takagi also found a partisan in Hecke.

Hasse's presentation of class field theory, starting with an address to the German Mathematical Society and including written reports from 1925–30, not only explained Takagi's work but framed it in very different terms. These papers also included accounts of Artin's contributions and solution of the ninth problem. Hasse based his treatment on what are called the p-adic numbers and introduced new ideas and notation. The p-adic numbers are alternatives to the real numbers that carry with them an alternative geometry. Physicists have found them interesting and work has been done to use them as the numbers that underlie quantum mechanics.

Class field theory excites strong emotions in its practitioners and just as it has a complex history, it has been redone many times. Hasse's recasting was only the first. A group of young French mathematicians from the École Normal began to make contributions, Weil, Herbrand, and Chevalley. All studied in Germany and Herbrand, Chevalley, Hasse, and Artin worked together to simplify the proofs. Chevalley, in his thesis of 1932, came up with an arithmetical approach to the theory. No complex analysis was used. However, this method wasn't entirely arithmetical. In 1940 Chevalley offered an entirely arithmetical theory using *ideles*, which he had invented in 1936. As we have seen, number theorists often want to see if they can accomplish something purely arithmetically. In the '30s Hasse found another approach, coming out of what is called the theory of simple algebras but again using p-adic numbers. Weil introduced the general concept of *adeles* (one notes the fanciful but utilitarian one-vowel variation of Kummer's original name selection—ideal, *idele, adele*). Artin and John Tate in the 1950s treated the subject using techniques of homological algebra, and Weil treated the subject again. This complex theory is visited lovingly again and again, as powerful new techniques and language come on the scene. Serge Lang, one of Artin's pupils, says:

> It seems that over the years, everything that has been done has proved useful, theoretically or as examples, for the further devel-

> opment of the theory. Old, and seemingly isolated special cases have continuously acquired renewed significance, often after half a century or more.[43]

Later in the book Lang writes:

> You should be warned that acquaintance with only one of the approaches will deprive you of techniques and understandings reflected by the other approaches. . . ."[44]

Takagi had now earned his high position in Japanese mathematics. He was the first Japanese mathematician to become well-known in Europe and America. By the 1930s world-class mathematicians were appearing regularly in Japan. Some of them were Takagi's students, like Iyanaga, whom we have quoted and who made his own contributions to class field theory in the '30s. Takagi's great achievement was class field theory, and the two papers in German in 1920 and 1922 account for half the total pages he published in European languages. Starting in 1930 he began to publish texts on many subjects that he had taught. His *Course in Analysis* is one of the all-time mathematical bestsellers in Japan. He also wrote *History of Mathematics in the 19th Century* and *Miscellaneous Notes on Mathematics*, which have had many editions.

In 1932 Takagi made a five-month trip to Europe. The occasion was the International Congress of Mathematicians held in Zurich in early September. Takagi was one of the vice-presidents of the conference. He stayed at the Hotel Eden, facing the Zurichsee, where he hosted a dinner in a private banquet room. The guest list is a *Who's Who* of class field theory, including Chevalley, Hasse, Iyanaga, Y. Mimura and his wife, M. Moriya, M. Nagano, Noether, Taussky, B. L. van der Waerden, and N. G. Chebotaryev. We encountered Taussky as a student with Gödel. She made significant contributions to class field theory herself. Taussky so admired Takagi's work that she began to learn Japanese in order to speak to him. At the dinner the Japanese were impressed by the general noisiness and exuberance of the Europeans. Hasse told war stories from the First World War and joked to the Russian Chebotaryev, "If the war had been going on further, I would have marched upon your city."[45] Noether enjoyed herself, talking the most, interrupting Chebotaryev with a stop gesture of her hands, and then, as the party ended, requesting that the Japanese teach her how to bow. Honda writes, "She asked, bending forward her fat, upper body: 'Like this, or more?' Her funny look remained long in the memories of the young Japanese mathematicians." It should be noted that Noether was the senior mathematician at this gathering. Honda writes:

> The dinner party seems to have been the best time in Takagi's long life. It is said that he wrote about this party in a letter to his wife in Japan, stating that he had specially selected the wines to be served.[46]

In Vienna, Taussky introduced Takagi to Furtwängler, who had also proved a series of important theorems about class fields. In Hamburg he met Artin. In Göttingen he visited a failing Hilbert with Noether. Takagi wrote, "Observing my old master grumbling as if speaking to himself, I wept in my heart."[47] Then he returned to Japan, soon to be cut off from the world outside for a second time. Thus ended what Honda calls "the unique gay period in the long quiet life of Takagi."[48] Takagi retired from the university in 1936. Many of his best books were written after his retirement. In 1951 his wife died of lung cancer. In his last decade Takagi spent time rereading the classics of Japanese literature that he had studied when he was young. In 1960 he died of a stroke.

The Ninth Problem

Emil Artin was born on March 3, 1898, in Vienna, the son of an art dealer and grandson of an Armenian merchant of handmade rugs. The city was infatuated with *Die Fledermaus* the year Artin was born. His mother sang the role of the soubrette. Later, when told that her son was a famous mathematician, Artin's mother expressed surprise and disbelief because he was clearly not recognized as he walked in the street. When Artin was four his father died, and he was sent to live with his grandmother. When she died, only two years later, he was sent to a school run by the Jesuits. The family is still in possession of some of the cutlery from that school—Emil was No. 43.[49] The word "lonely" has been used to describe Artin's childhood, but it would seem to understate the pain of these early experiences.

Artin was reunited with his mother when she remarried. Her new husband owned a woolen cloth factory in Reichenberg (now Liberec), Bohemia, in what is now the northwestern portion of the Czech Republic. At the time, it was the part of the Austro-Hungarian Empire called the Sudetenland that figured so prominently in the years leading up to the Second World War. Artin's mother continued to sing operetta in Reichenberg but was no longer so much in the limelight. She and her new husband had a son. The brothers were separated by quite a few years and so weren't close playmates, but the two branches of the family have remained in contact. In Reichenberg, Emil had a friend, Arthur Beer, who later became an astronomer and worked at the observatory at Cambridge, England. Emil made a telescope,[50] and the two boys rigged up a telegraph between their houses.[51]

Artin attended the local high school and was at first more interested in science, particularly chemistry, than in mathematics. Natascha Brunswick, Artin's wife for many years, dated his first strong interest in mathematics to a year he spent at the age of fourteen studying in Paris—certainly this year influenced him all his life.

Artin graduated in 1916 at the age of eighteen and spent one term at the University of Vienna before he was drafted into the Austrian army. Brunswick said that before he was drafted Artin found some mistakes in a Niels Nielsen book on function theory and wrote a letter to the author pointing out the errors. (There is a neat notebook in Artin's *Nachlass* with the solutions of all the exercises in this book.)

Gustav Herglotz (courtesy of Natascha Artin Brunswick).

When he returned from the war he found a letter of thanks and a package of reprints addressed to "Professor Artin." Artin went to study at the University of Leipzig under Herglotz, who lectured on many topics in physics, as well as number theory. As Charlotte John described Herglotz:

> There was something of the nineteenth century about him, which he cultivated, I think. He had the elegance of that century. In fact, his dress was actually like that of Goethe. And he had really luminous eyes. He looked like someone who enjoyed himself extremely—and who saw things. And as you heard him lecture, you had the feeling that you got a glimpse of what he saw.[52]

Though he also studied chemistry in Leipzig, Artin wasted no time in taking his degree in mathematics, receiving his Ph.D. at twenty-three, despite his military service. His thesis was a work of mathematical importance. In it Artin analyzed a field made of algebraic functions over a finite field and the structure of extension fields of it. The algebraic functions here can be added, subtracted, multiplied, and divided in a manner that satisfies all the field axioms. Any theorem that applies to all fields is true about this particular field. The strategy here is that many things about the particular field being examined can be understood without reference to the details—the essence of the axiomatic method used as a tool for investigation as practiced by Noether and her followers. In Artin, the push toward the recognition and study of formal structure, similar to the "high essentialism" of Takagi, found a new and powerful practitioner. However, Brauer observed that the thesis "discloses a very great skill with computations, a feature one might not have guessed."[53] Gian-Carlo Rota said that Artin loved calculation when he was at Princeton University many years later,[54] and Michael Artin also mentioned to me seeing a notebook containing graphs of various special functions his father had made as a youth. When MANIAC, one of the first modern computers, was built, Artin suggested to von Neumann and Herman Goldstine that they use it to try to calculate examples for a classification of primes flowing from a theorem of Gauss's, according to Joseph Silverman and Tate.[55] The style in much of Artin's published work is abstract, general, and formal, with the nuts and bolts of calculation not in evidence—no history, no motivation. But one of the sources of his mathematics was intimate calculational knowledge.

In his thesis Artin was particularly interested in the behavior of a zeta function that can be constructed for his field of functions that is analogous to the zeta function we saw in the eighth problem. Recall that the zeta function started with the Euler product formula, which equates a sum over all natural numbers with a product over all primes. The rules for division and for multiplying out the product result in all possible combinations of primes, which in turn are precisely all natural numbers. In algebra or number theory, any time one set of things is precisely all combinations of a second set of (prime) things, this can be expressed in a product formula, which can then be used to create an analytic zeta function. By including information about classes of functions in his zeta function, Artin raised the ante. This is common in mathematics. Find something that has worked before and see if you can make it work in the new context. Algebraic number theory has been worked on hard for a long period of time, so sometimes the layers of this are astonishing. In his thesis Artin looked at a limited group of cases but formulated his theory so that it could be opened up in the future.

After completing his thesis, Artin spent a year in Göttingen at the suggestion of Herglotz. Reid says that Courant recognized Artin's brilliance and tried to "draw" him into the inner circle of Göttingen (as he had tried with Siegel). He invited Artin to musical evenings at his house. This might have been a mistake. When Courant hit the keyboard it was with gusto, as would not be surprising in a wheeler-dealer analyst, who when the Göttingen Mathematical Institute was taken from him by the Nazis, moved to New York and built another. Reid writes, "Artin, whose music was as pure and rigorous as his mathematics, shuddered at Courant's untutored approach."[56] Artin stayed only a year at Göttingen before moving on to Hamburg as a privatdozent in 1923. He had his own strong vision of mathematics and its aesthetics and agreed with many of Noether's ideas but wasn't her "student." Hamburg was a new university where he could establish his own circle.

While Artin was in Göttingen, Siegel loaned him his reprint of Takagi's great paper on class field theory, and Artin has been just about the only mathematician to say that it was not difficult to understand. He set about generalizing Takagi's results. Class field theory characterizes and asserts the existence of abelian extension fields. This is a broad class. However, it does not include all extension fields. Artin started to investigate whether similar results could be found for any extension field. He constructed yet another zeta function and a whole new line of L-function, now called Artin's L-functions. These were different from Dirichlet's version and different from but related to L-functions that Takagi used. They are also different from L-functions for function fields he had used earlier, in his thesis. (L-functions proliferate like rabbits.) Artin also read Takagi's paper on reciprocity laws and, inspired by one of Takagi's observations, conjectured what is now called Artin's reciprocity law.[57] In 1923 Artin published this research in two papers. In the second he conjectured his general law of reciprocity and proved it in some special cases. Brauer says:

> In 1923, no possibility of proof for this new theorem was in sight. In reading [the paper on reciprocity], one has the feeling that though Artin then did not know how to approach his conjecture, he had full confidence that some day, sooner or later, he would have a proof. I wonder if this confidence in his own powers in a young man caused raised eyebrows among some of the older mathematicians of those days. If so, they knew better a few years later.[58]

Gauss's law of quadratic reciprocity concerns "quadratic residues": a prime q is a quadratic residue of a prime p if there exists an x such that $x^2 - q$ is divisible by p. Conversely, p is a quadratic residue of q if there exists an x such that $x^2 - p$ is divisible by q. These questions have a formal symmetry in the way they are stated but don't have an obvious connection. Gauss proved a deep connection that others had conjectured but been unable to prove. If you know whether q is a quadratic residue of p, you can make a simple calculation and find out whether p is a quadratic residue of q.

Using the Legendre symbol ( ) one can write Gauss's formula:

$$\left(\frac{p}{q}\right)\left(\frac{q}{p}\right) = (-1)^{(p-1)(q-1)/4}$$

The Legendre symbol,

$$\left(\frac{p}{q}\right)$$

is equal to 1 if p is a quadratic residue of q and -1 if it isn't. Gauss went on to prove a law of biquadratic reciprocity involving equations in x^4 rather than x^2. As mathematicians proceeded to higher powers of x the situation became increasingly complicated, so that the notation itself ultimately become a subject of study. It can be used as a powerful and elegant tool in proving results about whole numbers, which is part of its allure.* Already with Gauss at the turn of the nineteenth century, number theory had a more complex and modern feel than some other areas of mathematics.

With Takagi and Artin the reciprocity laws make another jump and take the form of statements about class fields. The old laws can be derived from the new formulation. As already mentioned, a central point of Takagi's class field theory was that the Galois group of an abelian extension is structurally identical (isomorphic) to a class group constructed using the ideals of the original field. The class group is a concrete embodiment of the way unique factorization breaks down. Artin supplied an explicit identification (or isomorphism) between the two groups and then observed that this could be used to calculate or generate any reciprocity law. It is generally believed that Artin's law captured everything that could be captured in this area (at least for abelian extensions). So far, it has not been generalized or reworked. In fact, Artin's reciprocity law is so general that once one proves it, many of the other results of class field theory follow.

Hilbert's ninth problem is titled "Proof of the Most General Law of Reciprocity in any Number Field." As was indicated in the quote from Brauer, Artin hadn't proved his reciprocity law as of 1923. We cut now to Russia. In 1922 N. G. Chebotaryev (1894–1947) proved what is called Chebotaryev's density theorem (on the density of prime numbers when divided into certain classes), using a highly original process that did not rely on class field theory, which had not yet reached him. (This is the Chebotaryev who attended Takagi's dinner ten years later.) *The Dictionary of Scientific Biography*, in an article by Yushkevich (in the 1970s, before truth could be uttered), says that Chebotaryev taught at the university in Odessa from 1921–1927. The narrative in "Chebotaryev and his Density Theorem" by P. Stevenhagen and H. W. Lenstra, Jr. in the *Mathematical Intelligencer* is more interesting. Chebotaryev's father was president of a district court in the Ukraine. With the Revolution, he was removed from office and the family plunged into poverty. Chebotaryev had been living in Kiev "like a student, earning money from private lessons and teaching in high schools."[59] He moved to Odessa in 1921 to help his

*Most treatments of Gauss's result fail to suggest why number theorists are so impressed with it, but Gerald Janusz supplies a good example on page 68 of his book *Algebraic Number Fields*: Suppose you want to know if 43 is a perfect square mod 3319. Checking each possibility would involve around 3000 individual calculations—not an attractive prospect. Quadratic reciprocity allows us to flip the numbers—that is, if we can determine whether 3319 is a perfect square mod 43, then we can find the answer to the first question. The question whether 3319 is a perfect square mod 43 is simply whether the remainder of 3319 divided by 43 (which is 8) is a perfect square. Is 8 a perfect square mod 43? This is a much simpler problem. Applying Gauss's law says the answer to the original question is yes. Any similar question, where one number is large compared to the other, is likewise clarified using the reciprocity result.

parents, but he wasn't doing the same kind of mathematics as the mathematicians in Odessa and couldn't get a job. His father died of cholera in 1922. His mother barely survived, selling cabbages. Chebotaryev wrote in a letter many years later: "I devised my best result while carrying water from the lower part of town (Peresypi in Odessa) to the higher part, or buckets of cabbages to the market, which my mother sold to feed the entire family."[60] In 1923 he married a physiologist and doctor, a former assistant to Pavlov, and since she was employed things looked up a bit.

Chebotaryev's density theorem was published in Russian in 1923. He was offered a job in Moscow, but when he got there found that everyone was mad at him. The position had been Egorov's, who had been forced to resign for political reasons. We met Egorov, with Luzin, the founder of the Moscow school of function theory, in the section on Gelfond. After seven months Chebotaryev resigned and returned to Odessa. He got "a badly paid and ill-defined position as secretary for scientific research at an Educational Institute."[61] He was able to put together a seminar and in 1925 traveled to Europe for the first time, where he met and impressed some of the German mathematicians. He eventually got an acceptable job at the university in Kazan, 450 miles east of Moscow. In 1931 Egorov was transported to a prison hospital there, where Chebotaryev's wife worked as a doctor. There is a story that she took him home to care for him and that Egorov died in the Chebotaryev home.

When Chebotaryev's paper was published in German in 1925, Artin realized he could use it to help prove his law of reciprocity. In 1927 Artin proved his result. In the fall of that year Chebotaryev came back to Kazan after spending the summer at his dacha with class field theory, which he had only now studied, excited with an idea of how he might prove Artin's conjectured reciprocity law. Immediately on return he saw Artin's article. Artin scrupulously recognized Chebotaryev's contribution, despite the fact that he had dramatically altered the way the idea was expressed. Chebotaryev, though disappointed, later said: "I was very touched by Artin's meticulousness in matters of attribution, as there is only an incomplete analogy between the ways in which the method . . . is used in the two papers."[62]

Though Hilbert's ninth problem helped make the search for a general reciprocity law a focus, the actual formulation and proof came out of the larger subject of class field theory, running by now on its own steam. Hilbert's original conjectures about class fields had given the subject a push, it is true, but until Takagi built the first full version of class field theory, the ninth problem had no real context or chance of solution.

Artin had been appointed extraordinarius in Hamburg in 1925 and full professor in 1926. He was only twenty-eight. His 1927 results on reciprocity gave evidence of the wisdom of this early promotion, but they were not the only important results Artin published that year. He also solved the seventeenth problem. No one else has cleanly solved two of Hilbert's problems, much less two in the same year. Brauer said that Artin's solution "was perhaps the first triumph of what is sometimes called 'abstract' algebra."[63] It involved a theory of "formally real fields" that

Emil Artin with the University of Hamburg in the background (photo by Natascha Artin *Brunswick).*

satisfied certain axioms: –1 in the field is not equal to a sum of squared numbers in the field. The real numbers are "formally real," while the complex numbers are not—because $i^2 = -1$. Artin used this abstract theory to gain information that could be applied to the specific example of the Hilbert problem. Again the move was away from detail, which can be confusing, to the abstract essence of the structures being investigated.

Artin continued to reconsider class field theory, as he would all his life, trying to discern its most simple and powerful formulation. He also published an impor-

Dinner at the Ratsweinkeller in the City Hall, Hamburg, 1927. From left to right: 1. Hans Petersson 3. Emil Artin 4. Gustav Herglotz 5. Hans Rademacher 11. Otto Schreier 12. Wilhelm Blaschke 15. B. L. van der Waerden (courtesy of Natascha Artin Brunswick).

Emil and Natascha Artin (courtesy of Natascha Artin Brunswick).

tant paper on the classification of braids in 1925 and significant papers on "hypercomplex numbers."

Artin supervised doctoral students with great warmth and success. His first student, Käthe Hey, finished in 1927 with a thesis on hypercomplex numbers and a zeta function that was significant enough to be mentioned in both Hans Zassenhaus's and Brauer's obituaries of Artin. In 1929 Artin married another student, Natascha Jasny, whom I have already mentioned as Natascha Brunswick. This young couple

Emil Artin in a housecoat reading (photo by Natascha Artin Brunswick).

Natascha Artin (photo by Emil Artin, courtesy of Natascha Artin Brunswick).

must have been one of the most attractive in mathematics. They soon had two children, Michael and Karin, to be followed by a third, Tom, after they emigrated to the U.S. Brunswick has a gift for understanding what starts a conversation and what keeps it going. She said that she had felt it would be improper to continue studying under her husband's colleagues, and so she never finished her degree. She continued to follow mathematics and was an editor for many years of *Communications on Pure and Applied Mathematics*, the journal Courant started in 1948 at NYU. In 2000 she was still doing translations from Russian for SIAM (Society for Industrial and Applied Mathematics).

Artin's mathematical production from 1921 to 1931 is startling, there is so much high-level work, with the year 1927 particularly remarkable. In the next decade Artin essentially published nothing. Brauer and Zassenhaus both give strong arguments that Artin continued to be mathematically productive. Brauer says, "Artin developed a strong aversion against the writing of papers for publication."[64] Both refer to unpublished lecture notes that circulated and contained important results. Both say that Artin gave away ideas to students (as Dehn did during the same period) or led them to discover ideas he was already aware of. During the 1930s Artin also attracted and influenced more mature mathematicians (some of them still young), including van der Waerden, Iyanaga, Herbrand, and Chevalley. Chevalley's important 1933 thesis, offering a quasi-arithmetical proof of class field theory, was heavily influenced by Artin's lectures. Van der Waerden acknowledged Artin's importance for his influential book *Moderne Algebra*. Other authors did likewise, including Zassenhaus and the noncorporeal Nicolas Bourbaki, which included Chevalley and Weil. (These French mathematicians choosing the name Bourbaki—from an obscure Napoleonic general—for their communal publications reminds me of young men naming a rock group.) Brauer gives a particularly effective appreciation of Artin's teaching, writing:

> There is the saying of G. B. Shaw about people who do things and people who teach things. If Shaw had ever met Artin, he might have grasped that there were men who were creative when they taught. Of course, Shaw could have discovered this by reading about Socrates.[65]

Artin acquired a nickname during this time in Hamburg that continued to be what he preferred. His students and friends took to calling him "Ma," short for mathematics.[66] He *was* mathematics. Zassenhaus devotes a number of paragraphs to describing his lecture style. Artin's lectures began with a discussion of motivation, followed by logical deduction done completely rigorously and in small, easily understood steps. At the end he returned to reexamine the structure and rigor of the process. Zassenhaus also emphasizes a kinesthetic component:

> The appeal to the sense of motion and progressing in time was powerfully supported by the external signs of motion. Artin allowed himself to use the whole platform as a substitute for peripatetic motion, he supported the sequence of sentences, carefully sculptured in time, by enormously expressive gestures of his hands, which in his later years occasionally seemed to invite a particular person in the audience to supply the answer and ultimately succeeded in

Emil Artin (*photo by Natascha Artin Brunswick*).

> generating true intellectual perception in almost every one of his hearers.[67]

Artin stated his own credo eloquently in 1953 in a review of a book by Bourbaki on algebra:

> We all believe that mathematics is an art. The author of a book, the lecturer in a classroom tries to convey the structural beauty of mathematics to his readers, to his listeners. In this attempt he must always fail. Mathematics is logical to be sure; each conclusion is drawn from previously derived statements. Yet the whole of it, the real piece of art, is not linear; worse than that its perception should be instantaneous. We all have experienced on some rare occasions the feeling of elation in realizing that we have enabled our listeners to see at a moment's glance the whole architecture and all its ramifications. How can this be achieved? Clinging stubbornly to the logical sequence inhibits the visualization of the whole, and yet this logical structure must predominate or chaos would result.[68]

The review ends:

> In concluding I wish again to emphasize the complete success of the work. The presentation is abstract, mercilessly abstract. But the reader who can overcome the initial difficulties will be richly rewarded for his efforts by deeper insights and fuller understanding.[69]

The emphasis on the unity of the whole and the goal of being able to see its structure "at a moment's glance" would tend to turn one away from narrow problems—and draw one to return to try to simplify the proofs in class field theory again and again. Artin's reciprocity theorem, in which he captures all past reciprocity laws as well as much of the structure of class field theory, is an example of a mathematical work that met these high goals, but such a moment will not come often. I think this is the context in which to understand Artin's later work and teaching, starting in 1931. Many mathematicians are like explorers opening up vast new territories. Poincaré and Kolmogorov are two wonderful examples of this wide-ranging modality. For Artin, the focus lay in seeing how the entire map of an area falls together. This has a mystical quality. Artin would have liked to open up a nonabelian class field theory, but that was too difficult.

Class field theory has remained difficult to see at a glance. P. Stevenhagen and H. W. Lenstra capture the situation as it has remained, referring to Chebotaryev's technique as used in proofs of Artin's reciprocity law: "It is widely felt that it works for no good reason, and that it is just as counterintuitive as most proofs in class field theory."[70] Modern class field theory dominated the mathematical careers of its creators and drew a major effort from some of their best students. Its importance has made generations of mathematicians devote significant effort to refining it.

Artin's emphasis on the whole led this most mathematical of personalities to be interested in many other aspects of culture. He was knowledgeable in astronomy, physics, biology, and chemistry. He had a lifelong love of music and musical history and played the flute, the harpsichord, and the clavichord. When he traveled to Japan he studied Buddhism. When he became a father he spent a lot of time with his children, particularly the first two. He was concerned about their proper education.

Germany in the 1930s became perilous for Artin's family. Natascha was half-Jewish; the children were quarter-Jewish. They were on the Nazis' list. The Artins left somewhat later than some others in equivalent positions—though sooner than Dehn. Artin was forthright in his denunciation of the Nazis, and Brunswick told me about him holding forth against them in his seminar as Blaschke, himself a Nazi, according to Brunswick, walked nervously in the corridor, closing the door to the room so Artin's opinions would not be broadcast to the world. She described to Reid, and repeated to me, her fear that the Nazis were coming every time she heard a car in the cul-de-sac where they lived. Natascha's father had been a Menshevik in Russia, and her youth was spent in flight from the Bolsheviks.

In 1936 and '37 the situation became more and more untenable, and Artin began to search for an appointment in the U.S. He received offers but couldn't get an exit visa—and he wanted the right offer. Finally the decision was made, but leaving was still not simple. By the time he could accept an offer, the only one available was from Notre Dame. According to Reid, Artin was afraid that the fact that he had been born Catholic would become known and the fathers would try to reclaim his children, though as it turned out, Artin had a good experience at Notre Dame.[71] Brunswick describes pressure from Hasse to stay, which she resented and felt was unethical. Hasse suggested he could get Artin's children declared Aryans, but that rather begged the question of Natascha.[72] At first, Artin was not allowed to leave because he was ruled indispensable. Then a few months later he was placed in *Ruhestand* (retirement). Then a visa from the American consulate was required. Artin, as a professor, was exempt from the quota, but when he went in to fill out the application the American official looked at how skinny he was and decided he might have tuberculosis. Artin had to go and get a doctor's certificate, but on the day he was to go in with it he had a dramatic-looking inflammation of the eye. He went anyway and was given a visa. The family finally sailed for America in the fall of 1937. When they arrived they were welcomed at the docks by Courant, Weyl, and Natascha's father, Naum Jasny—at this time working for the State Department as a consultant on Soviet agriculture and teaching.

Natascha was unambiguously glad to be out of Germany. For Emil, it was different. In Brauer's words, "The intellectual atmosphere of German universities of that period is remembered with nostalgia by all who knew it."[73] Culture for Artin existed in the medium of the German language. In the U.S. the family spoke German at home. Michael says that the German language still has warm associations for him dating to that time.

The Artins spent one school year at Notre Dame. Then Artin received a position at Indiana University, which was headed by a young Herman Wells (1902–2000). Wells saw that one of the quickest ways to improve the quality of his institution was to find positions for refugee academics. The resulting emigre community helped to make Artin more comfortable, and the whole Indiana University community was welcoming. Artin continued to commute back to Notre Dame and lectured there each year from 1938 to 1943. Brunswick described the time in Indiana as "wonderful." Karin Artin called it "nice, very nice," and Tom Artin used almost the same words. They had a large, open Arts and Crafts house with a big porch. When Thanksgiving came, Artin looked up how to carve a turkey in the *Encylopœdia Britannica*. Though they had not been allowed to take money out of Germany, the Artins had been allowed to ship their things in a huge box. The house in Bloomington was full of things from Germany—a harpsichord, a clavichord, and a *Tafelklavier* (a sort of table piano). Artin had repudiated the romantic composers of his youth and really did not like much music later than Bach—Tom says maybe Mozart, or occasionally even Beethoven. The *Tafelklavier* was sold and the proceeds went to buy a Hammond organ. The Hammond wasn't quite right. Artin spent months taking it apart, wiring and soldering to make it sound more like a baroque organ. He added five more foot pedals on the bass end and Karin remembers hearing all the Bach organ sonatas. Artin was generally horrified at American popular culture, and the children were not allowed to listen to the popular radio shows of the day unless they were sick. Instead he read to them every night, usually in German—Karin remembers *The Arabian Nights*, classics of German literature. Occasionally something in English would be allowed—Dickens' *A Christmas Carol,* something by Mark Twain, or *Alice in Wonderland.*

Artin began to get some good graduate students in Indiana. They gathered in his home after seminars. Papers began to appear. A series of collaborations with a young George Whaples in the mid-1940s was a kind of mathematical reawakening, and in fact, from that time on Artin published regularly again. In 1946, however, Artin received an offer from Princeton University too good to pass up. Wells was away overseeing postwar elections in Greece, and Karin says that this caused Indiana not to counter the offer as effectively as it might have. Indiana was left behind with reluctance.

Artin was an imposing and memorable figure at Princeton. Yet he was speaking a different language culturally. He wore an old black leather jacket or sometimes a long winter coat belted at the waist. Some students emulated him in dress and manner. Artin's dress was "a statement," but it was a statement from the Viennese *Kultur* of his youth. Clearly some people didn't quite know what to make of him.

The range of visitors to the home in Princeton was more consistently mathematical than in Indiana. Siegel was one, and the Artins had a dog they called *der Siegel Fresser*, "the Siegel devourer." The dog, a brindle mix, "very intelligent," with the head of a greyhound and body of a boxer, had been mistreated before the Artins got her, and, though harmless, barked at everyone. Frisky—the dog's other name—hated Siegel, and Siegel was terrified. He started bringing hot dogs when he visited to bribe the dog into silence, to no avail. Frisky often had to be locked in the basement. Artin built another telescope and took long walks with the children. He had trouble grinding the mirror so the parabolic shape was exactly right and checked the correctness of the shape again and again. Karin told me that when the Hubble space telescope had difficulties she immediately thought it was probably because they hadn't checked the parabolic shape carefully enough.

Tom, born in the U.S., also commented on Artin's dress—he remembers the coat, but was most impressed with his father's attire at home, if he had not gone to lecture that day. Artin would stay in his pajamas and a dressing coat cinched at the waist. He would either lie on a large sofa—almost like a day bed—in the main room, or pace back and forth the forty feet or so of the two front rooms, thinking about mathematics. Tom found this embarrassing when bringing home American friends after school. When he was about twelve he took up the trombone at school, because he thought it was the coolest-looking instrument. The trombone was not Emil's favorite instrument, as it was poorly suited to play the baroque and classical chamber music he favored and he associated it with Wagner. Tom says that he later found out that his father had been a "Wagnerite" as a younger man. Tom is the only member of the family never to have made a living from mathematics. (Michael Artin is a professor of mathematics at MIT and Karin Artin taught mathematics at high school and college levels.) Once, in Willy Feller's and von Neumann's pres-

Hermann Weyl on the Artins' front porch in Princeton (photo taken by Tom Artin).

ence, Tom told me he asked something like, "What is applied mathematics, actually? Is it the same as engineering?" He said, "There was a friendly rivalry between my father, the pure mathematician, for whom mathematics was an art not a science, and von Neumann, the father of the computer (among other things)."[74] Feller was famous for work in probability, including applications to genetics. Artin judged this a *bon mot* and gave Tom the habitual penny reward for such things. Tom studied comparative literature and got a Ph.D. from Princeton in medieval European literature. For fifteen years he was an academic, then gave it up and now makes his living as a jazz trombonist.

The list of Artin's students at Princeton is once again impressive—see Zassenhaus's obituary for a full list. If Artin had gone to Princeton partly looking for the best graduate students, the move was a success. His collaboration with Tate was especially fruitful. In 1951–52 Artin delivered lectures on class field theory from the perspective of cohomological algebra, which by then had reached a mature form. Artin published a revised version of these notes in 1961 in collaboration with Tate, who had himself made major advances in generalizing class field theory shortly after the lectures. Of all the versions of class field theory Artin had delivered over the years, which are all described as significant, this is the version he published. Karin married John Tate, so there came to be a family connection.

Artin's marriage had developed problems. In 1956 he took a sabbatical and returned to Germany for the first time. He taught for a term at Göttingen and a term at Hamburg. When he returned to the U.S., he had already decided to return to Germany permanently and in 1958 accepted an appointment at the University of Hamburg. This was difficult, because some of the mathematicians at Hamburg had been Nazis. Blaschke, already mentioned, had behaved well toward Artin. Brunswick says Ernst Witt (1911–1991) claimed that he read only *Mein Kampf* and the Bible. Hasse was in Hamburg. Michael said he thought that, in addition to the pull of the mother tongue, his father felt that he was "getting old" in the Princeton department and that he felt he was needed in Hamburg. Karin thought Princeton's mandatory retirement age of sixty-five was an issue for Artin. Tom told me:

> In my view, Emil never came to feel at home in the U.S. For him, I think, even at its best, the U.S. always seemed something of a cultural desert, in which he felt alien and ill at ease. Obviously, there had to have been great ambivalence attached to this feeling. . . . Nonetheless, Germany was still the land of Bach, Beethoven, Goethe, and Hilbert. Europe in general, and Germany in particular, remained for him the locus of Western culture, and when he returned there in 1958, I think he felt, for better or worse, he had come home again.[75]

Emil and Natascha were formally divorced in 1959. Hel Braun was in Hamburg and she and Artin had a relationship, both personal and mathematical. They collaborated on a book on algebraic topology. This last period in Hamburg is described as a time of intense mathematical activity. On December 20, 1962, Artin died of a heart attack suddenly, although he had suffered from angina for several years. Brauer wrote:

> I saw Artin for the last time in November 1958 in Hamburg. He spoke with satisfaction of his life and work in the United States. In Princeton, John Tate and Serge Lang had been his students. "This happens only once to a man. Not many mathematicians have been that lucky" were his words. He was content with his new life. There were vague plans of visits to America, but it was clear that Hamburg was to remain his home.
>
> We look a long walk one afternoon talking of old times. It was one of those misty, melancholy, and rather miserable days which all northern harbor cities know so well in late fall. We wandered endlessly through the streets searching, I did not know for what, until I realized, it was a Hamburg which no longer existed and times which were gone forever. Before Artin's eyes, I believe, there must have been the picture of the young Artin who had walked through the same streets thirty years before, full of life and strength.[76]

Michael remarked to me that his father's letters to him began to change; the structures of the German language began to overtake the English, though his father had spoken and written English easily in the U.S.

The German emigres from that time came to the United States well into adulthood, yet one can aspire to write English as well as some of them did. My quotations from Weyl, Brauer, and others are examples. The ability to write at this level in a language other than their first is a measure of their gifts. It was a boon to the United States to receive them. The French obituaries of Artin emphasize a side of his achievement that we have not yet fully captured. They describe his mathematical achievements, as all obituaries do, but also comment on his style from a distinctly French point of view. The word *algebriste* is used as a badge of honor, but qualified with the statement that Artin was so much more than just that. Chevalley writes that in Artin the play of the conscious with the unconscious, of methodical investigation coupled with flashes of intuition, is comparable to same traits in the great French poets Mallarmé and Valery. All of Artin's obituaries are very warm.

Tom said that his father would have hated having a funeral, but the university had one anyway. Since the ground was frozen when Artin died, his burial was postponed until spring. There was a procession through the twisting paths of the Hamburg cemetery with a graduate student carrying the ashes.

The Eleventh Problem

Helmut Hasse was born August 25, 1898, not far from Göttingen, in Kassel. His mother, Margaretha (Quentin), though born in Milwaukee, Wisconsin, was raised by an aunt in Kassel. His father, Paul Reinhard Hasse, a judge, was related on his maternal side to the composer Felix Mendelssohn. A connection to the Mendelssohn family is shared by other mathematicians who have figured in the history of class field theory—Dirichlet, Kummer, and Hensel. True to his ancestry, Hasse loved music. In 1913 Hasse's father received a judicial appointment that moved the family to Berlin, and Hasse attended the Fichte Gymnasium there. His mathematics teacher was Hermann Wolf, and Hasse probably decided on a career in mathematics at this time. In June 1915 he passed a special *Notabitur*, "emergency school leaving examination," and enlisted in the navy.[77] He saw action in the Baltic Sea and served aboard a number of ships. In 1917 he was stationed with the naval command in Kiel, where he remained until December 1918. Hasse's duties left time to study mathematics and he even matriculated at the University of Kiel and attended lectures by Otto Toeplitz (1881–1940). He also kept in touch with Wolf, who astutely recommended that he read the Dirichlet–Dedekind lectures on number theory.

When the war ended, Hasse went to Göttingen, where he attended lectures by most of the luminaries—Landau, Hilbert, Hecke, Courant, Carl Runge, and Peter Debye. Noether's lectures were too difficult for the young student and he attended only one. Hasse found Hecke's approach to analytic number theory most attractive, but Hecke left in 1919 to assume a position at the newly created University of Hamburg. In a used bookstore Hasse came upon *Zahlentheorie* (*Number Theory*), published in 1913 by his distant relative Kurt Hensel (1861–1941). Hensel had studied under Kronecker and was occupied editing Kronecker's collected papers after his death in 1891. He did relatively little significant mathematics of his own until, in 1899, he invented the p-adic numbers, which he thought would provide a different approach to certain questions and proofs in algebraic number theory. The p-adic numbers are extensions of the rational numbers, a different extension for each p, or prime. (For example, there are 2-adic systems, 3-adic systems, 5-adic, etc.) Unlike the extension fields we've been talking about, they are not finitely generated. Any given p-adic extension adds an infinite pile of fundamentally different extra numbers to the rational numbers. The p-adic numbers have an algebraic nature but a non-archimedean geometry—if we start taking equal steps to go somewhere, we won't necessarily arrive. Hensel spent the rest of his career elaborating and exploring the significance of his idea.

The p-adic numbers were regarded as a curiosity in Göttingen, but Hasse recognized their potential. Instead of following Hecke when he moved or remaining in Göttingen, Hasse went to study at Marburg University, where Hensel was a professor. By October 1920, Hasse had discovered his "local–global principle," which gave him more than enough material for his doctoral dissertation and his habilitation as a privatdozent, completed in 1921 and 1922. The local–global principle says that a general result about a rational number or even all rational numbers can (often) be established by verifying that the result holds true for each of the p-adic number systems. At first glance this would not seem a gain, because each p-adic number system contains the rational numbers as a subsystem with additional structure added. Checking all the p-adic systems would seem to involve inspecting all the prime numbers one by one. However, the additional structure actually serves to make many calculations more simple, and often one p-adic system behaves much like another, so that one can examine the behavior of all the p-adic systems using one general method. Hasse's local–global principle found a deep use for the p-adic numbers. The next dozen years were remarkably productive for the young mathematician, establishing him as a major force in number theory. Hasse solved Hilbert's eleventh problem in 1923 as one of the many problems he attacked with his new technique.

The eleventh problem reads in its entirety:

> Our present knowledge of the theory of quadratic number fields puts us in a position *to attack successfully the theory of quadratic forms with any number of variables and with any algebraic numerical coefficients.* This leads in particular to the interesting problem: to solve a given quadratic equation with algebraic numerical coefficients in any number of variables by integral or fractional numbers belonging to the algebraic realm of rationality determined by the coefficients

Three articles are cited in footnotes, two by Hilbert and, most importantly, one by Minkowski. The problem reads like one of Hilbert's programs, "*to attack successfully the theory of quadratic forms*," and in this guise some of Siegel's work in the 1930s would be part of the problem, as would much subsequent work. However, the footnoted Minkowski article has been taken to provide a definitive narrowing. Kaplansky writes, "The 11th problem is simply this: classify quadratic forms over algebraic number fields."[78] That is what Minkowski had done for quadratic forms with fractional coefficients.

A quadratic form (related to but not the same as a quadratic equation) is any polynomial in which each term has variables appearing exactly twice. All coefficients must be whole numbers; $ax^2 + bxy + cy^2$ is the general form for one with two variables (x^2 is x appearing twice). A given quadratic form is said to represent a natural number if substituting specific numbers for the variables gives the number. For example, $x^2 + y^2$ represents 5, because substituting 1 for x and 2 for y gives 5. It does not represent 6 because no whole number x and y substitution will give 6.

Gauss and those who followed found that if we change variables in certain ways (as we did in invariant theory), the new quadratic form represented the same natural numbers as the old, even though it looked different, and it might be more easily interpreted. He used this theory of equivalent quadratic forms to prove whole number theory results. For example, Lagrange had shown that any natural number can be expressed as the sum of four squares. Gauss proved this using his theory of equivalence relations by showing that the quadratic form $w^2 + x^2 + y^2 + z^2$ represents all natural numbers. Minkowski created and proved a similar theory for quadratic forms that had fractions as coefficients. As commonly interpreted, Hilbert's eleventh problem asks for a similar theory—essentially a mode of classification so we can tell if one form is equivalent to another—in the case where the coefficients can be algebraic numbers. Using his local–global principle and the fact that the theory is relatively simple for p-adic systems, Hasse accomplished this in a proof that is described as difficult "even today" (Kaplansky in 1977).[79]

In 1922 Hasse moved to Kiel as privatdozent. He married Clara Ohle—they would have a son and a daughter. Artin moved to Hamburg in 1923, not far from Kiel, and this resulted in the correspondence and collaboration between Artin and Hasse already discussed. In 1925 Hasse was appointed full professor at Halle. Hensel retired from his position in Marburg in 1930, and Hasse answered the call to fill his chair, which gave Hensel great pleasure. During this period, Hasse accomplished a great deal. In 1926, 1927, and 1930 he published a report in several parts on class field theory that simplified Takagi's methods and did much to disseminate class field theory in the mathematical community. In the early 1930s he gave a treatment of class field theory based on p-adic numbers and using "noncommutative algebras" (matrix algebras over a division ring). Edwards says Hasse's work on algebras and the role he found for them in his second version of class field theory have been called the "high point of the local–global principle."[80]

Much of the existing literature on noncommutative algebras was in English when Hasse turned to it. In 1930s Harold Davenport (who was served multiple courses of whole trout when he lunched with Siegel in 1966), a student of Hasse's English colleague Louis Joel Mordell (1888–1972), began spending his summers

From left: Clara Hasse, Kurt Hensel, the Hasses' daughter in a sailor suit, Frau Hensel, Natascha Artin, Helmut Hasse (photo by Emil Artin, courtesy of Natascha Artin Brunswick).

in Marburg teaching Hasse English and studying mathematics with him. They also read English literature. Mordell and Davenport were working on the properties of various congruence relations—polynomial equations of the form $y^2 = f(x) \bmod p$. Davenport challenged Hasse to look at the problem. Hasse realized that, though they at first didn't seem related, the congruence problem they were looking at was equivalent to the problem that Artin had framed in his doctoral thesis. Just as one can extend the idea of prime numbers in the natural numbers into similar ideas in terms of ideals in an algebraic number field, one can create an analog of the Riemann zeta function and the Riemann hypothesis for algebraic number fields. (We mentioned Hecke's 1917 result on this part of the eighth problem.) The equations for algebraic curves can be added, subtracted, multiplied, and divided and hence are a field themselves. Factorization can be done in them, and one can carry the analogy a step further by associating a zeta function with a field composed of algebraic curves. Artin's problem concerned a zeta function in this context.

Most mathematicians would probably call this "algebraic geometry," the study of the solutions to algebraic equations viewed as graphs. The solutions to an algebraic equation are thought of as corresponding to points in space—curves or, in higher dimensions with possibly more than one equation, varieties. This is a big tent and transforms almost any mathematics that might be of some use in investigating algebraic equations into algebraic geometry. In some cases the questions in "algebraic geometry" do focus on the geometry of these algebraic structures. Hilbert's fifteenth and the first half of the sixteenth problems are of this type. In other areas, the focus centers on the equations—solving them or using algebraic geometrical methods to answer questions about number theory. (In this guise it would have been better to call it "geometric algebra.") For example, draw a line through two points on a curve and see where and if it hits the curve again. This leads to algebraic equations that can be solved algebraically—that is, the picture makes a suggestion for strategies in algebraic calculation. However, even if you can't "see" the picture, pretending you can helps generate ideas. Artin's thesis had a zeta function for a field of functions defined over finite fields. Visualization here would seem impossible. Silverman and Tate put it well, though: "You may worry about your geometric intuition in situations like this. How can one visualize points and curves and directions [in these spaces]? . . . you can continue to think of the usual euclidean plane and most of your geometric intuitions concerning points and curves will still be true when you switch. . . ."[81] They followed this with the boxed saying:[82]

Think Geometrically, Prove Algebraically

Algebraic geometry is as a heuristic device of significant subtlety and complexity.

Using algebraic methods and language, Hasse was able to prove in 1933 a special case of Artin's conjecture on his new zeta function, which opened the door for Weil to prove the hypothesis in full generality in 1940. Hasse's theorem says that if we have an equation of the form $y^2 = ax^3 + bx^2 + cx + d \bmod p$, then the number of solutions to the equation is $p + 1$ with an error no more than $2\sqrt{p}$. (If p is small this error is big, but if p is large the error improves. For p around 100 the error can't be more than 20%. For 1,000,000 the error is at most .2%—for p around 100,000,000,000 the error is .0002%—and so on.) Much work in this area has followed. Hasse himself spent much of the rest of the '30s trying to generalize his result. However, politics, though not direct peril, also disturbed Hasse's work after Hitler rose to power.

Hasse was a conservative nationalist and naval veteran much aggrieved by the Treaty of Versailles and its aftermath, and he probably viewed Hitler's rise positively. The year 1933 was productive mathematically for him, still in Marburg, but the turmoil hit Göttingen immediately. On April 26, 1933, six Jewish professors were put on "leave," including Noether, Max Born (future Nobel laureate in physics), and Courant. James Franck, a Nobel laureate in physics, had already resigned in protest. Those put on leave found out about it in the local newspaper. There would be more leaves to follow. Many of the mathematicians placed on leave felt intense surprise, in addition to their pain. Courant wrote in a letter to Harald Bohr, "I have been harder hit by the turn of events and less prepared than I should have thought. I feel so close to my work here, to the surrounding countryside, to so many people and to Germany as a whole that this 'elimination' hits me with an almost unbearable force."[83] By the end of 1933 Göttingen was effectively destroyed, though this was not immediately obvious to many. Hilbert, after sending a bouquet of flowers to his Jewish doctor, said in April 1933, "It will not be long before the German people find Hitler out, and then they will put his head in the toilet."[84]

The mathematicians of Göttingen and their friends tried to work out strategies that would minimize the damage to Göttingen and German mathematics despite the storm sweeping over them. One of the first orders of business was to find a successor to Courant to direct the Mathematical Institute. Otto Neugebauer (1899–1990), who had been working closely with him, was appointed but had a very short tenure, perhaps as little as one day officially. Nazi students were against him and he was asked to sign an oath of loyalty to the new regime, which he refused to do. He was immediately suspended and branded *untragbar*—"unbearable."[85] Weyl, after Hilbert the next great Göttingen mathematician by near-consensus, was made acting head and tried to have the "leaves" for Courant and Noether revoked. He tried to gather testimonial letters for Courant, with some success. A petition for Courant was organized a little later, with mixed results. People were becoming frightened of signing such a petition. When completed, though, the impressive list of signers included Hasse.

Weyl would not be tolerated as head of the institute for long. His wife was Jewish and therefore his children were Jewish. He had been offered a position at the newly created—1930—Institute for Advanced Study in Princeton the year before, but after a period of time when both acceptance and rejection telegrams sat on his desk simultaneously waiting to be sent, he rejected the offer. He felt terrible about the risk to his family. He continued to salvage what he could in Göttingen, but he was not a natural administrator. Finally he was again offered a position at the Institute and he accepted. His son remembered his father walking around the house as the family packed for the move "intoning" lines of poetry, as he often did, this time the lines: "If the beast of tyranny takes over his country, he puts the torch to his own house and leaves."[86] Weyl contacted Hasse, urging him to take the directorship if asked. He felt that Hasse, though sympathetic to the Nazis, would try to keep up standards and be fair to mathematicians. By Christmas 1933, Weyl was in Princeton.

More faculty and staff were purged. Landau was still a professor in the fall of 1933 because of his long tenure (appointed under the Empire). During the tumultuous spring he had commissioned an assistant to deliver his lectures, but in the fall he resolved to begin teaching his calculus course again. He loved teaching and only during the forced hiatus had he realized how much. On November 2, 1933, when he went to teach, Landau found seventy students, some in Nazi dress, barring the way to the classroom. Students wanting to enter were turned away. Werner Weber, a privatdozent and once Landau's assistant, was the commander, and Oswald Teichmüller, called "notoriously crazy"[87] by Courant, but a brilliant mathematician, all of twenty years of age, was spokesman. The presence of good mathematicians in the mob must have made the experience all the more bitter. In 1934 the Nazi mathematician Ludwig Bieberbach would refer to this event in an infamous speech as a "manly rejection." The notion was that Landau's "alien" style of mind would pollute the natural "German" style of the impressionable students. By Christmas, practically no one of any standing was left at the Göttingen Mathematical Institute.

Another sinister character would soon be on the scene, Wilmar Tornier. Tornier had long been a secret Nazi. Prior to the events of 1933, which he saw coming, he had moved to the University of Kiel ostensibly to collaborate with Willy Feller (1906–1970). They had collaborated before. Feller was a Yugoslav whom Courant had discovered in his calculus course back in the 1920s and literally pursued to the students room to see what original mathematics he was working on. Feller eventually had a brilliant career as a probabilist in the United States. In order to get his transfer, Tornier had appealed to A.A. Fraenkel, who was ordinarius in Kiel. Tornier's sole purpose was to be present when the Jews were kicked out and assume Fraenkel's chair himself. Remarkably, he seems later to have admitted this in a letter to Fraenkel.[88] He also participated in establishing Feller's "non-Aryan" origin and so removed that obstacle to his appointment. He assumed Fraenkel's chair, despite his mediocre abilities as a mathematician. The period immediately following Hitler's rise to power was a time of insurgency by Nazi mathematicians, before what was left of the mathematical establishment found ways to protect the discipline. In early summer 1934 Tornier was appointed to fill Landau's chair, which had been Minkowski's, in Göttingen, perhaps the lowest point in this sad sequence of events. He was made acting director of the Mathematical Institute until the arrival of the new director, Hasse.

There were still enough real mathematicians in power so that Hasse had in fact received a call by Easter 1934, which Courant as well as Weyl had asked him to accept. His appointment was opposed by Nazi students led by Teichmüller. Reports of Bieberbach's speech on racial types in mathematics, made around the same time as Hasse's call, were filtering out, and the talk was turning into an international cause celebre, with Harald Bohr leading the charge against it. Bieberbach was trying to use the German Mathematical Society, the DMV, to rally support for himself and publish his responses. Hasse was on the editorial board of the *Jahresbericht*, the publication of the DMV, and was drawn into this conflict. Bohr wrote to Hasse soliciting his support against Bieberbach's outrageous speech in mid-May. Hasse postponed responding and then, in a letter of June 6, tried to paper over the disagreement. But Bohr responded that the situation was clear and he was outraged that Hasse wouldn't take a stand. Hasse wrote back admitting Bohr's point, but begged Bohr not to take his refusal too harshly and wrote that he couldn't really explain in a letter. He asked that Bohr not make what he did write public and suggested that they meet in the fall so he could explain his situation more fully. Bohr wrote to Veblen at Princeton:

> The situation in Göttingen is really extremely absurd, as the assistants in the Mathematical Institute declined to give the keys to Hasse when he came as the official new professor sent from the government.[89]

The narrative of this period is somewhat confusing. S. L. Segal—a mathematician and historian, not to be confused with C. L. Siegel—puts the incident of the keys at the end of May. Reid writes that in July, when Hasse came to Göttingen, he was greeted by "such an unpleasant demonstration by pro-Nazi mathematics students, led by Teichmüller, that he returned disgustedly to Marburg."[90] I don't know if the two incidents are the same. Tornier seems to have arrived in late June and that only made the situation worse. Segal publishes an excerpt from a letter from Bohr to Veblen written on August 11, 1934, that recounts an even more bizarre event:

> In the night of the famous 30, June, Tornier wrote an express-letter to Hasse, who was in Marburg, telling him that Tornier thought Hasse's life was in danger (since his so-called friendly attitude towards people like Courant and Emmy Noether made him suspectable) and advised him strongly immediately to take the train to Göttingen and come to Tornier's hotel so that he could protect him—what Hasse did. Tornier's purpose seemed to [F. K.] Schmidt to be to destroy Hasse's personality but to keep him as a mathematical sign for Göttingen.[91]

It is safe to say that Hasse encountered resistance when he tried to take his position in Göttingen. At times he probably felt he was in physical danger. However, by the end of summer he had assumed the directorship of the Mathematical Institute.

At the fall 1934 meeting of the DMV there was a showdown about the Bieberbach controversy and Bieberbach, supported by Tornier, tried to have himself installed as *Führer* of German mathematics. This effort proved to be the high-water line of the Nazis' explicit attempt to influence German mathematics and mathematical institutions. Hasse was part of the group that rejected Bieberbach. This broke Tornier's power in Göttingen. The ministry of education attempted to place him at another university, but no one wanted him. Finally, in 1936, Tornier was transferred to Berlin, which historian Herbert Mehrtens said "appears like a last resort for outspoken Nazis among mathematicians."[92] Reid writes, "There he embarrassed the mathematics faculty by being pictured in the newspaper walking on a fashionable boulevard with a notorious prostitute on his arm and a tame tortoise on a leash."[93] In 1935 Hasse made peace with Teichmüller, who participated in Hasse's seminar. An ardent Nazi, Teichmüller later volunteered for the army, wrote some of his best mathematics in a tent, and died at the Russian front.

From this point on, Hasse was able to, within certain parameters, run the Mathematical Institute. Political infighting continued. He was unable to get all his appointments accepted, in particular, Hlawka and van der Waerden. However, he succeeded in bringing good mathematicians to Göttingen, including Theodor Kaluza, Rolf Nevanlinna, and Max Deuring. Most remarkably, the outspoken Siegel came in January 1938 and stayed until he fled Germany. Hasse went on leave after the war started, to head a research institute that belonged to the navy in Berlin.

If these official actions had constituted the whole of Hasse's behavior during the Nazi era, he would have received little criticism after the war. He worked to uphold standards in mathematics and participated in the war efforts of his nation. However, he had a genuine belief in the Nazi cause. He applied for membership in the Nazi Party in 1937. He later said he applied to strengthen his political position in Göttingen and felt certain he would be rejected because of his distant Jewish relations. He was rejected, but there is some evidence that he later reapplied and was admitted.[94] Siegel wrote Courant on March 22, 1939:

> After the November pogrom [*Kristallnacht*], when I returned [from a trip to Frankfurt] . . ., full of nausea and anger at the bestialities in the name of the higher honor of Germany, I saw Hasse for the first time wearing Nazi-party insignia! It is incomprehensible to me how an intelligent and conscientious man can do such a thing. I then learned that the foreign policy occurrences of recent years had made Hasse into a convinced follower of Hitler. He really believes that these acts of violence will result in a blessing of the German people.[95]

In an interview with Reid in 1975 Hasse stated:

> My political feelings have never been National-Socialistic but rather "national" in the sense of the Deutschnationale Partei, which succeeded the Conservative Party of the Second Empire (under Wilhelm II). I had strong feelings for Germany as it was created by Bismarck in 1871. When this was heavily damaged by the Treaty of Versailles in 1919, I resented that very much. I approved with all my heart and soul Hitler's endeavors to remove the injustices done to Germany in that treaty. It was from this truly national standpoint that I reacted when the Faculty more or less suggested that such a view was not permissible in one of its members. It was also the background for my remarks to the Americans. They were talking about reeducating Germany, and I said some strong things against this. It irked me that everything against Hitler was desirable, and everything that he had done was wrong. I continued to be a national German, and I resented Germany being trampled under the feet of foreign nations.[96]

This statement, made with thirty years intervening for reflection, convicts Hasse of stubbornness at the least. He saw what Siegel calls the "pogroms" even if he did not see the death camps. Hasse displayed his dangerous innocence in his protest to the American occupying forces after the war. On Hasse's wartime activities Segal is insightful and charitable:

> Hasse never seemed to have realized the qualitative difference between bourgeois nationalism and the Nazi program. He was not the only such professor, nor, as Helmut Kuhn has remarked, was the confusion that surprising: the conservative nationalism of the educated bourgeoisie had acquired a touch of resentment-filled radicalism which narrowed the distance between it and the Nazi movement. This, coupled with the condescension of the educated classes for the "plebeian" Nazis, and the politics of the "Harzburger Front," made it easy for a highly educated conservative nationalist who was not greatly concerned with political distinctions to view the Nazis as just a little bit further to the right.[97]

The American forces would not allow Hasse to teach but allowed him to stay on as a researcher. Hasse refused this. In this period Hardy spoke up for him. Hasse had a difficult year and then in 1946 took a position as a research professor at the Berlin Academy. He gave private classes. In 1948 he was allowed to publicly lecture again and in 1949 was appointed professor in Hamburg, where he lived for the rest of his life. He received criticism after the war and fewer invitations to lecture abroad than he would have otherwise. His teaching continued to be excellent, and he supervised a number of good students. He wrote books and continued to do research that was significant, though not of the importance of his earlier work.

Helmut Hasse (courtesy of the Archives of the Mathematische Forschungsinstitut Oberwolfach).

Artin returned to Hamburg and their social relations were polite. Hasse retired in 1966, though he continued to be active, traveling as a guest professor to Hawaii, Pennsylvania, San Diego State College, and Colorado. He died in 1979, after a long illness.

Hasse was an excellent mathematician. But compare him to Hilbert in the same time and place, who saw clearly even from senility's downward slope. A political order that destroys mathematics and makes mathematicians flee for their lives is almost certain to be evil. To our knowledge Hasse never saw this.

The Twelfth Problem

Algebraic number theory and class field theory are now on a well-traveled interstate highway, while the twelfth problem sits by the side of the old road. Kronecker conjectured in 1877 that any finite extension of the rationals could be reached by substituting certain simple numbers into the exponential function e^z. Heinrich Weber proved this in 1896, though there was an error in the proof that was not uncovered until 1979. Hilbert also proved the result in 1896, thinking he was only offering an alternate method, but his proof was correct and so this was the first time the result was actually proven.[98] Kronecker's *Jugendtraum*, at the time a goal wilder than a conjecture, was, roughly, that any extension of an imaginary quadratic field could be generated and understood in a similar way, using elliptic functions. (The elliptic functions are not the equations of ellipses, but arise, in one approach, from a chain of reasoning that begins when one tries to calculate the length of a segment of an ellipse.) Hilbert took Kronecker's dream to a new level in the twelfth problem. He wanted to find analytic functions that would generate all finite abelian extensions starting from any algebraic number field. This problem is therefore a program, as each number field, barring some very good luck, would need a separate solution. This would involve understanding the automorphic functions, from which the answer might emerge, and the Riemannian surfaces they live on and would link complex analysis and algebraic geometry to the algebra and numbers involved.

Hilbert said he believed that Kronecker's *Jugendtraum* could be achieved "without very great difficulty," building on Weber's and his own recent work. He was too optimistic here. Hilbert's students pursued his program for algebraic number theory, but his stature was such that it wasn't discovered until around 1914 that his version of the *Jugendtraum* wasn't quite correct. (What does he mean by elliptic functions?—sometimes plural sometimes not. Does he also mean to include some modular functions?) For more on this see Norbert Schappacher's essay, "On the History of Hilbert's Twelfth Problem: A Comedy of Errors." Kronecker's youthful dream had to wait for Takagi's development of class field theory to be stated and proved properly.

Class field theory, as it was developed and elaborated, also handled a good part of the algebra and number theory of the twelfth problem and made connections to analysis, moving in the direction of a solution. Silverman and Tate, in their book published in 1992 and revised in 1994 say: "The class field theory alluded to above gives such a description [of abelian extensions], but it does so in a somewhat indirect manner."[99] More direct understanding of the functions did not follow, and over time the business of finding and understanding the functions that would actually do the job Hilbert asked for began to fall into the background—this knowledge no longer seemed essential. In the words of R. P. Langlands, writing in 1974: "Whether justly or not, the twelfth problem has received little attention."[100] (I think the reason for this is that the functions for the first layers of extensions—for example, the exponential function—were extremely powerful and easy to use. In the more complicated cases the theory became less elegant and the benefits of success were less clear.)

Rolf Holzapfel wrote a book directly considering Hilbert's problems from a contemporary perspective in 1995, *The Ball and Some Hilbert Problems*. He describes what amounts to a river of ideas that can carry modern solutions of the seventh, twelfth, twenty-first, and twenty-second Hilbert problems in the same context. These problems, originally separate, became tributaries to this river.* Holzapfel refers to a ball model solution of Hilbert's twelfth. This leads back to the somewhat

*The seventh problem is about proving that certain functions yield transcendental numbers with certain inputs. If we can understand this, perhaps the same or related methods will tell us when certain functions yield algebraic numbers. So there is a relatively straightforward connection between these problems.

ambiguous way that the problem is formulated, and I will let the new generation of workers in this area work out the status of Hilbert's twelfth—as he stated it—if they are interested.

Mathematics and physics have often had trouble capturing the direction of the flow of time—the equations usually work the same forward and backward. Where Holzapfel portrays different treatments of separate Hilbert problems as flowing into one unified treatment, Weyl uses the same metaphor in reverse. Writing more than forty years earlier, he said:

> Whereas physics in its development since the turn of the century resembles a mighty stream rushing on in one direction, mathematics is more like the Nile delta, its waters fanning out in all directions."[101]

If the flow of mathematics in time has turned away from Hilbert's twelfth problem as he stated it, and his statement was imprecise, Hilbert's vision for algebraic number theory was fundamental. When Fermat's last theorem was proven, class numbers, etc., were central in the proof. Algebraic geometry was used. Hilbert's basis theorem and *Nullstellensatz* were central in putting algebraic geometry on a rigorous footing. Simon Singh and Kenneth Ribet (Ribet made one of the major steps on the way to proving Fermat) wrote the following in a *Scientific American* article explaining the new result: "When the great German logician David Hilbert was asked why he never attempted a proof of Fermat's last theorem, he replied, 'before beginning I should have to put in three years of intensive study, and I haven't that much time to squander on a probable failure.'"[102]

Hilbert a "great logician"? That he was, but the currents of mathematics are strong enough that the waters of Lethe can flow in them also. This section began with a quote from Hardy, that Archimedes would be remembered when Aeschylus was forgotten. Archimedes' insights will be remembered and rediscovered countless times, but I suspect Archimedes himself might be forgotten—at least by mathematicians. Mathematicians often come to know Archimedes in general education classes, because Plutarch wrote about him, rather than in mathematics classes. But then that is how Sophie Germain got interested in mathematics, by reading about Archimedes.

Algebra and Geometry: A Miscellany

«14, 15, 16, 17, 18»

What Is Algebra?

The Fourteenth and Seventeenth Problems

Nagata

Each of the five problems in this section has some contact with algebra. The word "algebra" comes from the Arabic *al-jebr*, the reduction (as in setting a broken bone), and the equations and operations of algebra are an early example of a formal system. The reduction, or straightening out, of a possibly complicated equation uses well-defined rules to obtain a simple equation that says x equals the answer. Often a rote method can be discovered; sometimes not. Look at this simple example of the "basic black dress" of mathematics:

The problem in words:	The problem in algebraic symbols:	
A number is 1 more than twice another number.	$2x + 1$ ("a number")	x ("another number")
Together they add up to 73.	$2x + 1 + x = 73$	
What are the numbers?	The goal is to get x alone on one side of the equation.	
We are allowed to collect the x's.	$3x + 1 = 73$	
We are allowed to subtract 1 from both sides.	$3x + 1 - 1 = 73 - 1$	

We do the subtraction. $3x = 72$

We can divide each side by 3. $3x/3 = 72/3$

We do the division. $x = 24$

The equation is reduced (and we know that the two numbers are 24 and 49). When Leibniz conceived the goal of creating a *lingua characterica* and a *calculus ratiocinator* in order to clear up rampant confusion in philosophy, he was partly inspired by the success of algebra. Boole called his logical calculus an algebra.

Once one possesses the compact notation of algebra, one can write equations that would not have come to mind in a natural language. The introduction of algebraic symbols took place over a long period. After that polynomial equations became a focus. General methods of solution for quadratic, cubic, and quartic equations had been achieved by 1545. But there it stood. The quintic (a polynomial with x^5 as well as x^4, x^3, x^2, and x in it) did not yield.

In 1824 Abel used a complicated argument to proved that there was, in fact, no general method for solving quintic equations. Five years later, Galois introduced a more general argument that revolutionized algebra. He argued: Suppose we have the five general solutions to a general fifth-degree polynomial. (A fifth-degree equation has five solutions if none double. We will stick with the quintic, but the arguments hold for all higher degrees.) These solutions are five black boxes from our point of view. We can label them "first solution," "second solution," etc. This attaches labels to every member of the field extension they generate. If we swap the placards between the solutions, what happens? The equation is entirely general—it stands in for every possible fifth-degree polynomial. It has six coefficients, each of which can't be expressed in terms of any of the others—if they were related, the equation would not be general. If something about one of the general solutions (written with radicals, numbers, and coefficients) made it stand out from the others, we could use this fact to deduce a relationship between the coefficients, but we have assumed there is none. Therefore the general answers have to be formally interchangeable. Lagrange was aware of this in a paper published in 1770–72 but used it as a clue in trying to discover positive solutions—by the end of his search he thought a general solution was probably impossible.

Galois noticed that this swapping had an algebraic structure of its own, which he called a "group." We can do one swap, then another, and the result will be a third swap. Doing nothing is the identity swap. If we make a swap, we can swap back to the starting point (inverse). Swapping is associative. The Galois group of the general fifth-degree polynomial is the set of all possible swaps or permutations of its five "black box" solutions.

The next step is to look at all the possible expressions we can generate using the rational numbers; six letters (the six coefficients of a fifth-degree polynomial); and square roots, cube roots, fourth roots, etc. These expressions would have to include any general solutions to the fifth-degree polynomial if they existed. Can we find five expressions of the kind we are talking about that can be freely swapped? No. Therefore there are no general solutions to the fifth-degree polynomial formed in this manner, i.e., no "solutions by radicals." Galois had to investigate the structure of groups of these possible answers—this took detailed work and the concept of a "solvable group." (We *can* find four expressions of this type that can be swapped and there is, in fact, a general, if complicated, solution to a quartic polynomial.) Galois found a way to calculate the Galois group of any equation he was given, and he found a way to decide whether its group was solvable. This is a new kind of algebra: Groups are algebraic structures and yet don't follow the same rules as numbers.

Once they were understood, in the second half of the nineteenth century, Galois's ideas became fundamental. Klein, Lie, and others used the group concept to understand geometry and differential equations, and the notion of a group of transformations as a tool for classifying and understanding became insinuated throughout the body of mathematics. We saw it used in class field theory. At the same time, the structure of groups began to be studied in its own right in algebra.

In England, George Peacock (1791–1858), Duncan Gregory (1813–1844), De Morgan, and Boole were conceiving algebra as a calculus of symbols that did not have to be necessarily connected to the real or complex numbers. William Rowan Hamilton was also working in this milieu. He wanted to discover an analog of the complex numbers, which exist in two dimensions, that would exist in three dimensions and represent what we would now call vectors in space.[1] Such numbers would be immensely useful in physics, and Hamilton's greatest interest was understanding the mathematics of physics. In Germany, Grassmann came up with an even more complicated generalization of the complex numbers, called "the calculus of extension." His numbers were "hypernumbers." Grassmann actually had his ideas earlier but did not publish them until 1844, one year after Hamilton.

The third ingredient for the birth of modern algebra, after group theory and an expanded notion of algebraic objects, was set theory. The popularity of axiomatic systems became the fourth—to clarify what was important and what was not, to investigate questions of independence and consistency, and to generally provide a Good Housekeeping seal of approval.

All the components of modern abstract algebra were on the table in 1900, yet only two Hilbert problems are algebra proper—the fourteenth and the seventeenth. (This is partly an artifact of how things are labeled. The algebraic number theory problems are algebra in the obvious sense, but are seen as number theory because they focus on the character of numbers.) The fifteenth problem is about the geometry of curves and surfaces defined by algebraic equations, as is part of the sixteenth. The eighteenth, a geometry problem, has an algebraic solution. If we look at the Fields Medal winners for the years 1950–1994, we find a similar disparity in how problems are counted—only two of thirty-six prizes are for pure algebra—John Thompson (1970) and Efim Zelmanov (1994). However, seventeen are for work in algebraic geometry or algebraic topology, and there is a lot of algebra in what remains. Some have called algebra the "handmaiden" of mathematics, but that oversimplifies it.

It took a woman described as looking like "an energetic and very nearsighted washerwoman" with a "loud and disagreeable voice"[2] to assemble the new abstract algebra, as it emerged in Göttingen in the 1920s (swallowing various already existing topics from around the world). This was Emmy Noether (1882–1935). As the daughter of Max Noether (1844–1921), she was born into the unfolding story of invariant theory, wrote her thesis on invariant theory, and calculated specific invariants. She lectured to her father's classes when he was ill. When Noether's father retired and she needed to make her own way in the world, she appeared in Göttingen, in the depths of the First World War. Noether did not present an attractive social face. Weyl commented, "The graces did not preside at her cradle."[2] Her language skills were awkward and her lectures often impenetrable, attended by only a few loyal students. She hurried, skipped rafts of syllables, and erased the board almost before she was done writing.

One might think that Hilbert, with his interest in clarity and a habit of infatuation with pretty women, would have seen little in Noether, but he saw a great mathematician. Further, she knew things that he needed for his work on relativity theory. He fought to get her certified as a privatdozent. When that failed, he let her give lectures under his name. He continued arguing on her behalf, and she was eventually made a privatdozent. In the end she was also elevated to something with a long, unclear German title that amounted to not-quite-an-associate professor with no duties and no salary. Eventually her privatdozent position carried a small salary. Noether was a leading force from the 1920s until she was forced to flee Germany in 1934. She went to Bryn Mawr College, in Pennsylvania, but died soon after of a ruptured appendix. Noether's student, B. L. van der Waerden (1903–1996) of Holland, carried her work forward directly. Artin attended her lectures in 1923 and was influenced. Russian visitors to Göttingen, including P. S. Aleksandrov, particularly liked her. When the Russians began going around in shirtsleeves, probably to show up the coddled Germans, the outfit was dubbed the "Noether-guard uniform."[3]

Noether brought the axiomatic method to bear on algebra not as a method of sanitation but as a strategy of proof. As anyone who has ever done any school algebra knows, it is easy to create a swamp of complicating detail that not only is of no use in reducing or solving the problem, but actively makes it grow. In Noether's algebra this detail—like the Asian water lily that threatens to fill Lake Victoria—is the enemy of clarity. A set of axioms can capture the atomic truths of a class of algebras and, the choking detail now removed, can be reasoned with to establish general truths that apply to all the algebras that were described in the axioms. Set theory is the natural language of these axioms. An algebra is a set of objects. An algebraic operation is a rule—or more purely, a set of ordered triples. Noether built a theory of polynomial rings and fields and their ideals that was not tied to specific detail, but in which it could be proved that any ideal had a finite basis. This is a generalization of Hilbert's basis theorem, which arose in invariant theory. Hilbert's theorem becomes a special case of Noether's theorem. There are many other special cases that follow easily that no one had proven before.

As this method was clarified and applied, an explosion of research resulted. For example, traditionally a matrix was an array of real or complex numbers. What happens if we talk about arrays of members of a different field? We get a new area of research. This type of elaboration has been fruitful but controversial, partly because it moved algebra out of the handmaid role. Though Hilbert's algebraic and foundational work contributed to modern abstract algebra, his algebra problems were formulated in the older style.

The Seventeenth Problem

Hilbert was interested in the seventeenth problem, in part, because its solution would be useful in answering certain questions in geometric construction. Suppose we allow division or ratios of polynomials. For example:

$$\frac{x^2 + y^2}{x^2 + 2y^2 + 3}$$

These are called rational functions. We require that all the coefficients be real numbers. As always, they could be tremendously complicated—though our example is simple. Now, suppose we have a polynomial, and no matter what real numbers we substitute for the variables, when it is evaluated, it is never negative. Hilbert himself showed that there existed polynomials of this type that could not be expressed as sums of squares of other polynomials. Hilbert asked whether one of these polynomials can always be expressed as a sum of rational functions that are perfect squares. The algebra of a complicated example, much less a truly general polynomial form, quickly becomes difficult.

Though the problem is not inherently modern, Emil Artin's 1927 solution of it employed the new abstract algebra and should be counted as one of its first successes. Working with Otto Schreier (1901–1929), Artin undertook a treatment of "formally real fields"—the real numbers are a prominent example. In a formally real field, no sum of squared numbers can ever equal –1. Artin noticed that an element of a field is a sum of squares in that field if and only if it is positive in any ordering of the field. The real numbers themselves come to us with an order, but other formally real fields can be ordered in more than one way. A field of rational functions can be ordered by substituting real numbers for the variables. If these generate essentially all valuations, then a positive definite polynomial is expressible as a sum of squares of rational functions. Building on this abstract point of view, Artin proved the general result in a fifteen-page paper, thereby solving Hilbert's seventeenth problem.

Artin's proof is an existence proof. No method of taking an always nonnegative polynomial and expressing it as a sum of squares of rational functions was given. Thirty years later, in 1957, building on 1955 work of Abraham Robinson (1918–1974), Gödel's friend Georg Kreisel exhibited an explicit method of construction. He used machinery from symbolic logic and obtained other positive results in algebra. This was an achievement for foundational research. Artin commented that he preferred a clear existence proof to a construction with $2^{2^{100}}$ steps. (In keeping with our theme of useful defense spending, Kreisel's work was partially funded by the Office of Ordnance Research, U.S. Army, under Contract No. DA-36-034-ORD-1622.)

The Fourteenth Problem

A central aspect of Hilbert's work on invariant theory involved general polynomials with letters for coefficients and transformations or changes of variables that amounted to rotating the coordinate axes (not necessarily together). Certain combinations of the constants of the original polynomials didn't change under these transformations. Hilbert's basis theorem says that any combination of constants that doesn't change can be built from a finite number of basic combinations.

In the fourteenth problem Hilbert wants that work extended. It is titled "Proof of the Finiteness of Certain Complete Systems of Functions," and it begins: "In the theory of algebraic invariants questions as to the finiteness of complete systems of forms deserve, as it seems to me, particular interest." Here he talks about rational functions (ratios of polynomials—as in the seventeenth problem) instead of the more simple polynomials, and he allows transformations or variable changes generated by sets of rational functions. A rational function might be transformed into a polynomial, and Hilbert singles these out and calls them "relatively integral functions." Taken together these functions form a ring. Does this ring have a finite basis? David Mumford comments that after Hilbert's success with his own basis theorem, he "was overly optimistic about finite results in other algebraic contexts."[4] The general answer to the fourteenth problem and its more modern formulations is no. In 1959 Masayoshi Nagata produced an elegant counterexample that confirmed this, solving the problem.

Masayoshi Nagata was born February 9, 1927, in Obu, a town south of Nagoya. His father ran a small factory and for several years was a member of the town council. His schooling was ordinary: six years in primary school (1933–39); five years of middle school in Kariya (1939–1944), a nearby town; three years in "The Eighth High School" in Nagoya (1944–47). The high school was comparable to the final three years of the French lycée or to the senior year of high school and the first two years of college in the United States. Nagata said of high school, in correspondence to me: "Though I did not think that I was good in mathematics, I liked mathematics only."[5] He had several friends who he thought were better than he was. He said, "I do not think that I am a late-bloomer. I was a usual boy who liked

Masayoshi Nagata (photo provided by Masayoshi Nagata).

to think of problems in mathematics." The Second World War began while Nagata was still in middle school, and as able-bodied men were in the armed forces, the students were sent into the countryside to work on farms or to factories in the city. This shortened the school day. Nagata enrolled at Nagoya University in April 1947 and experienced the common post-war problems of scarcity of goods and inflation, but the difficulty did not impede his schooling. He graduated three years later. Graduate school at the time in Japan was ill-defined and did not lead to a Ph.D. degree. After six months Nagata was made an instructor. He taught in Nagoya for two-and-a-half years and then moved to Kyoto University in May of 1953 as a lecturer. He has been in Kyoto ever since, except for travel. When I corresponded with him he was at Okayama University of Science.

In 1953 Artin attended an international symposium on algebraic number theory held in Japan. Beforehand he visited Kyoto and, while there, asked Nagata what he was working on. On hearing Nagata's reply, Artin asked him if he knew of Hilbert's fourteenth problem. Nagata looked up the fourteenth problem and realized what he was working on was very similar. After that he thought of the fourteenth problem. Nagata received a "doctor of sciences" degree in 1957. He was a research fellow at Harvard when he found his counterexample in 1959. He has had a long career, and one of his students, Shigefumi Mori, won the 1990 Fields Medal.

The mathematics of this problem has a peaceful, calculational aspect. We encountered Paul Gordan walking through the German countryside calculating invariants. Hilbert walked along beside him on his visit, and the idea of how to prove a general basis theorem without calculation came to him, rendering Gordan's calculations largely unnecessary. With Nagata's work, though it is often more abstract, we are brought back to calculation. Hilbert's strategy would not always work. Nagata produced a specific counterexample that was ultimately based on the quiet details of calculation.

Schubert's Variety Show

The Fifteenth Problem

Hilbert's fifteenth problem is titled "Rigorous Foundation of Schubert's Enumerative Calculus." Its italicized portion (the first paragraph of two) reads:

> *To establish rigorously and with an exact determination of the limits of their validity those geometrical numbers which Schubert has determined on the basis of the so-called principle of special position, or conservation of number, by means of the enumerative calculus.*

Hermann Cäsar Hannibal Schubert (1848–1911) was a gymnasium teacher. We have made the point before, but this illustrates how elite the gymnasiums were. Schubert taught in Hildesheim from 1872 to 1876. Adolf Hurwitz, later Hilbert's close friend and colleague, was a student of Schubert's and a collaborator in this period (according to Hilbert in a memorial piece for Hurwitz). In 1876, Schubert was appointed *Oberlehrer* at the *Johanneum* in Hamburg, with the title of professor.[1] There was no university in Hamburg until 1919, and this was quite an institution. The baroque composer Georg Philipp Telemann was cantor and taught music at the *Johanneum* for more than forty years.

Schubert's calculus involved algebraic curves and surfaces and asked questions like: If we have this surface and that surface, how many times will they intersect, or more generally, "How many figures of a certain sort fulfill certain algebraic or geometric conditions?"[2] His impetus was thus "enumerative." Schubert's calculus was effective in solving a wide variety of problems that couldn't be solved before. However, by Schubert's own admission, its foundations were not completely secure, and therefore some mathematicians rightly felt uncomfortable about using it. Even at the fifteenth problem, Hilbert is still asking foundational questions.

The simplest result of the type Schubert was interested in would be that two lines in the plane meet exactly once. But what if they are parallel? The answer then is that they meet at infinity. The next simplest case would be that a line and a circle meet twice. But what if they don't?

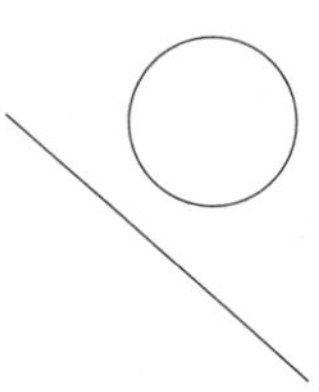

They meet in two imaginary points. (If we set up equations we find two imaginary solutions.)

What if the line is tangent to the circle?

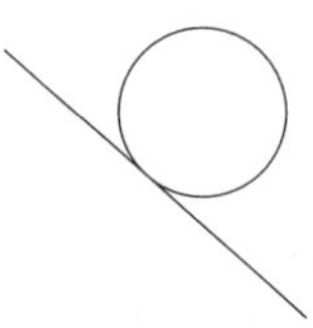

We must count the intersection point twice. (In factoring, the same factor appears twice.)

Two circles meet four times—two real and two imaginary. Tangent circles meet twice at the tangent point and twice at imaginary infinity, etc. The simple rules just given about lines and circles have coherent, consistent origins. The real difficulty comes with more complicated problems. These ideas had been developed and systematized around 1800 and later by Gaspard Monge (1746–1818), Lazare N. M. Carnot (1753–1823), and Jean Victor Poncelet (1788–1867), who wrote of imaginary points as well as points at infinity. Questions in this area can always be phrased as looking for solutions to a system of algebraic equations.

Thinking along these lines, Schubert calculated some spectacular numbers. In a letter to me, Steven Kleiman called Schubert's methods a "grand machine."[3] Schubert calculated 666,841,048 quadric surfaces tangent to nine given quadric surfaces and 5,819,539,783,680 "twisted cubic space curves tangent to 12 given quadric surfaces."[4]

Schubert had an almost radical faith in "the principle of special position, or conservation of number" as something that can be counted on in solving this sort of problem, and his calculus is based on it. An example shows how this principle is used. Suppose we start with four lines in three-dimensional space. How many lines can be drawn that intersect each of the four lines? Schubert himself solves this early in his book. The general problem is hard to visualize. He "specializes" the position of two of the lines so we have a somewhat more narrow problem than we started with. He says, suppose the first and second lines intersect, as do the third and fourth. The answer to this specialized problem is more obvious than the answer to the original—it is two. We can actually give a rule for drawing the two lines in this case. (One is the line through the two points of intersection, the other is the line that is the intersection of the plane determined by the first pair of lines and the plane of the second pair.) Schubert appeals to the "principle of conservation of number," which basically says that in these kinds of problems the general case answer is the same as the special, and so concludes that the answer to the general problem is also two.

One can always prove a particular case in mathematics by proving a more general case that includes it. Here one attempts to find the general by looking at a particular. Schubert never proved that this works all the time. A major purpose of adding imaginary points and points at infinity in algebraic geometry is to get a geometry and language in which this principle is true. This principle was, in fact, "furiously attacked by [Eduard] Study and [G.] Kohn,"[5] according to Kleiman, shortly after Hilbert posed this problem. Coolidge says, writing in 1940, "There has not been a more contentious topic in the history of geometry."[6]

This first approach to the problem that we just discussed is direct. This problem can also be approached more formally using Schubert's calculus. (An example can be seen in the appendix at the end of this chapter.) Schubert introduces symbols to stand in for statements like "a line meets two given lines" or "the line meets a given plane." He then introduces operations that combine two statements or symbols in a formal way, resulting in an equation. The plus sign means one or the other of the statements being added is true, while multiplication means that both statements are true. He now has a "calculus" (as in *calculus ratiocinator*) or an "algebra" of symbols. Problems in enumerative geometry can now be solved by formal calculation. Just as when algebraic notation became available there was effectively a great increase in what could be calculated, there was also an increase here—not because anything new had really been added, but because the new notation was so much less cumbersome than reasoning in the old way. Schubert's calculus is an ingenious system of generating, keeping track of, and reasoning about special cases. This idea of a formal system or calculus was in the air in the final quarter of the nineteenth century. Schubert was probably most directly influenced by the logician Ernst Schröder and mentions him by name. Using his calculus, Schubert calculated large numbers without directly constructing solutions for the original problems. This could theoretically have been accomplished without using the calculus, but as a practical matter never would have been. It would have been too much like wandering in a maze.

Kaplansky wrote in his 1977 summary of this problem: "Today's interpretation of Hilbert's 15th problem is this: One is to give a suitable definition of intersection-multiplicities [e.g., counting a circle meeting a tangent line as meeting it twice] for algebraic varieties [curves or "surfaces" defined by one or more algebraic equations], and prove the expected properties."[7] Taken broadly, this is tantamount to finding a rigorous foundation for a big hunk of algebraic geometry and proving that the law of conservation of number holds in it. Van der Waerden took a big step toward the rigorous grounding of Schubert's calculus with work published in 1930. Based on work done by Solomon Lefschetz (1884–1972), van der Waerden's point of view was topological. Francesco Severi (1879–1961), an algebraic geometer, and others had already laid the general groundwork for this development. Another point was to eliminate the non-elementary topological methods that van der Waerden had used. This was forwarded by a large number of mathematicians. By 1946, André Weil could publish *Foundations of Algebraic Geometry* (actually completed in 1944—that is the date mentioned in the introduction, which he has said the poet W. H. Auden helped write) and say: "Our results include all that is required for a rigorous treatment of so-called 'enumerative geometry,' thus providing a complete solution of Hilbert's fifteenth problem."[8] However, not all were satisfied, and Kleiman points out that Weil himself notes in the 1962 revision on page 331 that he hadn't closed all the gaps. Kleiman advanced a more balanced view in 1974:

> Schubert's ideas were slowly interpreted and made more precise, more formal, and more rigorous. Just some of the many, many mathematicians involved were Severi, van der Waerden, Ehresmann, B. Segre, Todd, Weil, Zariski, Chevalley, Samuel, Hodge, Pedoe, Chow, Serre, and Grothendieck. By 1960 an acceptably rigorous theory was available.[9]

A rigorous grounding of enumerative geometry did not necessarily bring with it an effective method for calculating numbers for actual complex situations. Kleiman wrote to me:

> In an actual situation calculating a number is only the first step; we must also investigate its significance. You see, even after the work of Severi, van der Waerden, and the others, it was still unknown whether Schubert's indirect determination of the two numbers via his calculus was valid. Did Schubert make any false assumptions or false calculations when he arrived at his value? Furthermore, justifying Schubert's indirect calculation is not enough: we must also show that there are precisely 666,841,048 quadric surfaces tangent to those nine given quadric surfaces, that none of them are cones and that each appears only once in the count.[10]

In the late 1970s and early 1980s a number of mathematicians, including Kleiman, applied these techniques to specific problems or classes of problems.

The communal aspect of the enterprise comes through in the following quote from Kleiman:

> The enumerative theory of conics and the greater theory of plane correlations were generalized to higher dimensions right from the start. They have continued to attract the attention of one generation of mathematicians after the other, more so than any other theories in enumerative geometry. Around 1980 a great new burst of activity began. Some of those involved to a greater or lesser extent are: Abeasis, Casas, De Concini, Demazure, Drechsler, Finat, Gianni, Goresky, Ihle, Kleiman, Kzawa, Laksov, Lascoux, MacPherson, Procesi, Springer, Sterz, Strickland, Thorup, Traverso, van der Waerden, Vainsencher, and Xambó.[11]

Investigation of cubics leads to a similar list: Abeasis, Coray, Ellingsrud, Harris, Kleiman, Piene, Sacchiero, Schlessinger, Speiser, Strømme, Vainsecher, Xambó, etc.

Schubert's numbers, 666,841,048 and 5,819,539,783,680, have now been justified. This allowed Kleiman to write in 1987:

> Tremendous progress has been made over the last decade, and now for the first time in over a century, a full understanding and rigorous treatment of all the material in Schubert's book appears close![12]

Summing up the meaning of all this he also writes:

> It is a tribute to Hilbert's vision that he could foresee that the rigorous development of Intersection Theory and Enumerative Geometry would be one of the great mathematical endeavors of the twentieth century.[13]

In the late 1980s and 1990s physicists began investigating superstring theory. In this theory there are ten dimensions; six curl up in Calabi–Yau spaces. In these little curled-up spaces, Edward Witten noticed patterns that came up and suggested that these structures from string theories be used to study generalized intersection numbers. In 1991, Candelas, de l'Ossa, Green, and Parkes published an important paper (it dealt with the mirror symmetry of the simplest Calabi–Yau space). In rushed a wave of mathematicians and mathematical physicists. Starting from physical principles and equations, they began computing some very large numbers similar to Schubert's. This started at least three cottage industries investigating the numbers that popped out of the physics and applying the new methods suggested by the physics to mathematical problems. One of these is quantum cohomology—a remarkable tool that has produced a lot of geometric numbers. One mathematician to be attracted, Maxim Kontsevich, won the 1998 Fields Medal. The fifteenth problem has opened up in a surprising way. Returning to the original problem, we can now say that Schubert's calculus itself is well-founded, and Hilbert's fifteenth problem is essentially solved. The biographies of the mathematicians who have worked on it, many of whom are still alive, will have to remain curled up and invisible for now, except to those who know them.

Appendix

(This is based on a letter from Steven Kleiman to me.)

The Problem: "Start with four lines in three-dimensional space. How many lines can be drawn that intersect each of the four lines?" This can be solved using Schubert's calculus as follows:

The translation from words to symbols is:

g	is	"a line meets a given line"
g_p	is	"a line goes through a given point"
g_e	is	"the line meets a given plane"
g_s	is	"a line meets a given pencil"
G	is	"the line is equal to a given line"

Next, Schubert proves the formal equation $g^2 = g_p + g_e$. (This equation means that the condition that a line meets two given lines is, for the purposes of enumeration, equivalent to the condition that either the line goes through a given point or it lies in a given plane. This equivalence follows from the principle of conservation of number because, if we specialize the two lines so that they intersect in a point, then the line meets the two if and only if, either, it passes through the point or it lies in the plane determined by the two lines.)

He derives also $gg_p = g_s$ and $gg_e = g_s$ and $gg_s = G$

He multiplies both sides of $g^2 = g_e + g_p$ by g getting $gg^2 = gg_e + gg_p$ which can be rewritten $g^3 = 2g_s$ (since $gg_p = g_s$ and $gg_e = g_s$)

He then multiplies $g^3 = 2g_s$ by g again on both sides getting $gg^3 = 2gg_s$ since $gg_s = G$ he gets $g^4 = 2G$.

This last says that there are two fixed lines. This is the result.

Graph That Curve

The Sixteenth Problem

A natural and commonly assigned type of problem in algebra is a request to graph this or that algebraic equation. The simplest type is of the form $2x + 3y = 5$, and its graph is a line. The next simplest equations have squared terms. These look like $2x^2 + 6y^2 = 3$ or $xy = 7$ (xy counts as a squared term). The general form is $ax^2 + bxy + cy^2 + dx + ey + f = 0$ and the graphs of these are the conic sections—ellipses, parabolas, and hyperbolas. The classical Greeks understood these equations well, and Apollonius' *Conic Sections* (c. 262–190 BC) stands as a high point. After the Greeks, little progress was made until 1665, when Newton (proving to me that he was as much a mathematician as a physicist) succeeded in classifying the graphs of third-degree equations (those with cubed terms) into nearly a hundred types. He included asymptotes, inflection points, mutual position of tangent lines at those points, etc., in his description, a *tour de force*. Direct application of Newton's strategy to more complicated equations, however, would be nearly impossible on a practical level and, even if successful, aesthetically unappealing.

The title of the sixteenth problem is "Problem of the Topology of Algebraic Curves and Surfaces." Knowing that detailed descriptions of every possible case would be tedious, Hilbert asks for a qualitative (yet rigorous) description of the shapes possible for the graphs of algebraic functions with only real numbers allowed as solutions. (At the end of the problem Hilbert asks a second question about limit cycles and differential equations that will make more sense after we have spent time with Poincaré, who originated this area of research. The chapter on Poincaré includes a discussion of this part of the sixteenth problem.)

Mathematicians discovered that it is easier to see similarities of form in the graphs of polynomial equations if we draw the graphs in the real projective plane.

Projective geometry arose, in part, out of artists' attempts to deal with perspective, though the mathematicians' version of this evolved several stages beyond what the artists needed. The projective plane is the euclidean plane with "points at infinity" added. Just as in a picture drawn with the aid of perspective, the vanishing point represents infinity and actually appears in the picture, infinity is represented in projective geometry as a point at the ends of each line and is in the "picture." A line is seen as meeting itself at infinity; parallel lines meet at infinity (at both ends). The resulting plane is itself geometrically complicated and cannot be embedded in three-dimensional space. If this sounds familiar, Schubert was also working in projective spaces. The big payoff for this exercise is that viewed in the projective plane, the graphs of even-order, nonsingular algebraic curves are made up solely of ovals.*

The approach Hilbert suggests in the sixteenth problem uses two strategies to recognize similarity of form in graphs—the projective plane and topology. Topology views a curve or any other geometrical object as stretchable: Roughly, two graphs are the same shape if they can be smoothly stretched, pulled, and/or shrunk into each other. The details of how this is done are technical. In this problem, the phrase "equivalence up to isotopy" (or "rigid isotopy") is often used. An "oval" divides the projective plane into two parts. The inside is just a disk. However, the outside is topologically a Möbius strip. Topologically, two ovals can be separate or one inside the other:

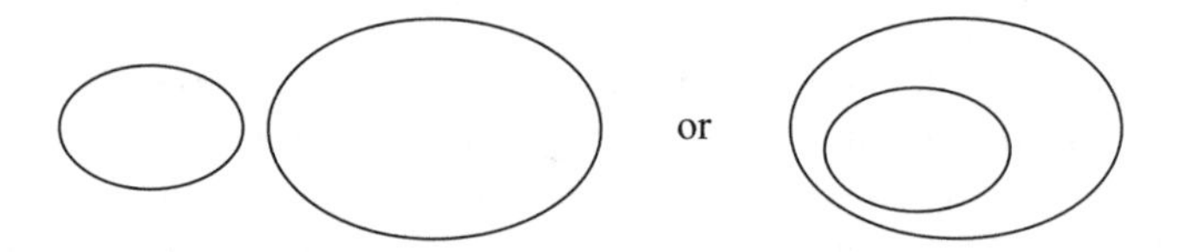

In 1876 Axel Harnack (1851–1888) showed that for a curve of degree n (referring to the largest sum of exponents in any one term of the equation), there were at most $(n - 1)(n - 2)/2 + 1$ ovals. For $n = 2$ this is 1. These are the conic sections; each one is a single oval in the projective plane. For biquadratic equations ($n = 4$) the graphs are at most four ovals; for $n = 6$ there are at most eleven ovals, etc. The highlighted part of the sixteenth problem is:

> *A thorough investigation of the relative position of the separate branches when their number is the maximum seems to me to be of very great interest, and not less so the corresponding investigation as to the number, form, and position of the sheets of an algebraic surface in space.*

*A parabola would seem to go off to infinity in two directions. As the two sides continue farther and farther out, each of their directions becomes straighter and straighter up and they become increasingly parallel. Therefore, they meet in this picture—the parabola is an oval, if we include the points at infinity. A hyperbola is also an oval, as is an ellipse (more obviously).

This problem is a rather large program, and yet it also has a beguilingly elementary character—the equations come from ordinary algebra and treatments have pictures like:

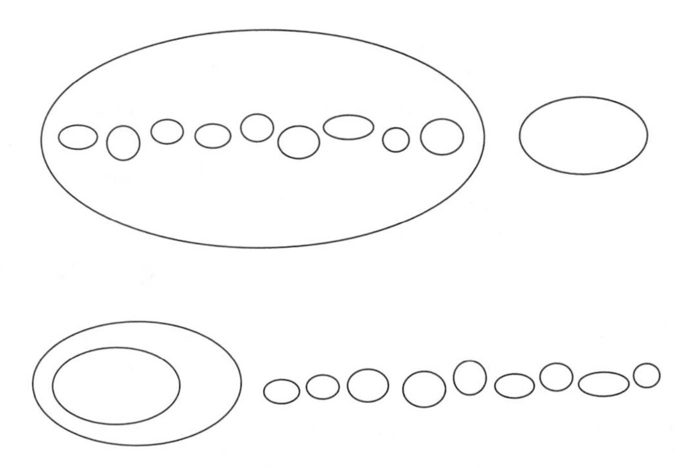

These are two possibilities for $n = 6$, which Hilbert himself had investigated. Hilbert thought these two arrangements exhausted the possibilities that could be realized for $n = 6$.

Now, a quick box score of what has happened:

Classical results of Hilbert, Harnack, and Klein flow into work by K. Rohn and Virginia Ragsdale, an American. Next came advances by L. Brusotti and A. Comessatti in Italy. The field, in large part, then moved to Russia, though Comessatti and Brusotti continued to be active. I. G. Petrovsky (1901–1973) made the major progress on this problem for $n = 6$ and proved some general facts about higher-order equations, in papers of 1933 and 1938. The 1938 paper contained a rigorous proof of Hilbert's assertion that there existed prohibitions for the case $n = 6$, though it is probable that Rohn had succeeded earlier with a less clearly complete proof. Petrovsky's papers made concrete connections between the algebra of this problem and the techniques of algebraic topology and provided a technical base for further developments. In 1949 Petrovsky and Olga Arsenevna Oleinik proved what are called the Petrovsky–Oleinik inequalities and brought the Euler characteristic into play—a concrete way for the algebra and topology to meet. (Ragsdale, who had traveled to Göttingen to study under Hilbert and Klein, had early—1906—understood that there should be a relationship between her conjectures and the Euler characteristic, but had not proved this.) Inequalities in this problem are often limits on how many ovals can be nested.

In 1969 D. A. Gudkov (1918–1992), pushing the approach of Rohn and Hilbert further, found a construction for a third possible configuration for $n = 6$ and proved there weren't any more. The third configuration is:

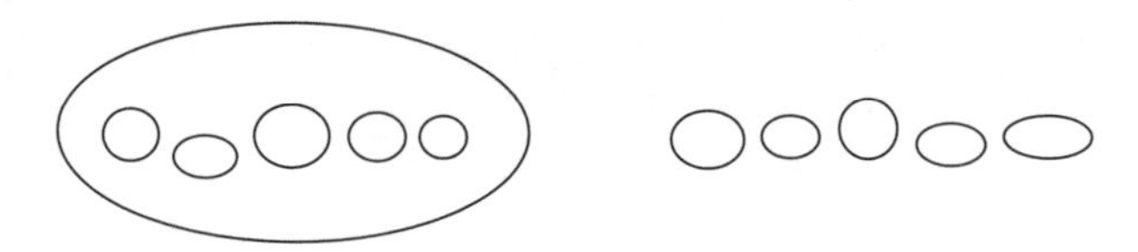

Gudkov also made a series of general conjectures that gave energy to the subject. (The flavor of these conjectures is conveyed by the example: The number of ovals lying inside an even number of ovals minus the number inside an odd number of ovals is equal to the remainder of half the degree of the equation squared when divided by 8. Or symbolically, $p - n = k^2 \bmod 8$.)

In 1971 V. I. Arnold proved a weakened version (mod 4) of Gudkov's conjecture. His methods were more important than the specific results. This problem is restricted to graphs of the real solutions of polynomial equations. Arnold looked at the "complexification" of the real curve. He found deep connections between knowledge of these complex surfaces and the real case. Though people like Klein and Hurwitz, for curves, and Comessatti, for surfaces, always looked at the real picture from a complex point of view, Arnold was the first to see how information visible in the complex picture could be carried over into the real picture in large quantity. A lot of work followed. We will treat Arnold shortly, in the chapter on the thirteenth problem (an analysis problem masquerading on Hilbert's list as algebra).

In 1972 Vladimir Abramovich Rokhlin (1919–1984) improved Arnold's methods, shed some of the topology, and proved that Gudkov's conjecture was correct without weakening. Viatcheslav Kharlamov found a classification of all nonsingular surfaces for $n = 4$. Oleg Viro found a classification of all nonsingular curves for $n = 7$. There has been a lot of progress 1975–2000, so this problem is still quite active and, like the fifteenth problem, has a communal aspect. Some names are: A. Degtyarev, T. Fiedler, S. Finashin, P. Gilmer, B. Gross, J. Harris, J. Itenberg, I. Kalinin, A. Khovansky, A. Korchagin, V. Krasnov, A. Marin, G. Mikhalkin, N. Mischacev, V. Nikulin, S. Orevkov, G. Polotovsky, E. Shustin, A. Slepian, G. Wilson, and V. Zvonilov. Before his death, Rokhlin proposed two detailed book-length surveys of the field. Kharlamov and Viro continued to work on one of these. The vigor of this problem is illustrated by the fate of this survey. Kharlamov says in a survey of Rokhlin's influence: "Pitifully, it seems that the book will never appear (partially due to the fact that the subject is developing faster than the text is being written)."[1] Those who want more detail can start by consulting Viro's 1986 survey and Degtyarev and Kharlamov's 2000 survey.

How Many Kinds of Crystals Are There, and Does the Grocer Know How to Stack Oranges?

The Eighteenth Problem

≪ Bieberbach ≫

The eighteenth problem, titled "Building Up of Space From Congruent Polyhedra," is generally treated as having three parts. The first two have been solved, while the third probably has been.

Hilbert emphasized the first part of the problem, which deals with crystallographic groups—it would not be unreasonable to identify this as the eighteenth problem. As the name suggests, crystallographic groups are of use in understanding the shapes of crystals observed in nature, and in the early papers in this area that connection is clear. E. S. Fedorov (1853–1919) in 1890, Arthur Schönflies (1853–1928) in 1891, and William Barlow (1845–1934) in 1894 independently and by different methods proved that the number of crystallographic groups in three dimensions (of euclidean space) is finite, 230 in a common way of counting. Their proofs flowed from a case-by-case examination of the possibilities, ending with the observation that they had checked them all—there were none left.

Hilbert asks whether the number of crystallographic groups in higher dimensions (of euclidean space) is also finite. Though I am aware of no four-dimensional crystals in nature, the question makes mathematical sense and naturally presents itself. As the number of dimensions goes up, the number of possibilities to be checked quickly becomes astronomical and geometrical intuition decreases at a similar rate. (In the 1970s a computer calculation yielded 4,783 or 4,895 kinds of crystallographic groups in four dimensions—depending on how you count.) Some kind of appeal to more general principles is in order.

We can look at the two-dimensional case. How many ways are there to "tile" the plane, as one would a very large bathroom, with identical tiles that are symmetric? (Each of the different ways would correspond to a crystallographic group.) Often in mathematics or physics the word "symmetry" has an abstract, formal meaning, but here the precise mathematical definition matches the usual sense. A crystal is a regularly repeating lattice of molecules. A fundamental region (or "tile" in two dimensions) is the area occupied by the smallest collection of molecules that repeated identically yields the whole crystal.

We may look at two examples of symmetrical tiles with which we can tile the plane:

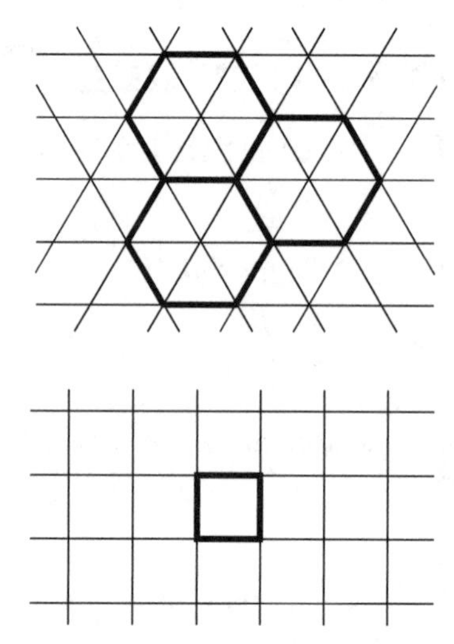

These are symmetric in that certain lines divide them into halves that are identical under reflection. They are also symmetric with regard to rotation. In the case of a square, rotations of a quarter-turn or multiples of a quarter-turn leave the figure unchanged, while with the hexagon sixth-turns have the same result. The equilateral triangles drawn in lightly in the first picture are themselves a tiling.

Pentagons have similar symmetries, but if we try to tile the plane with them the angles don't match up.

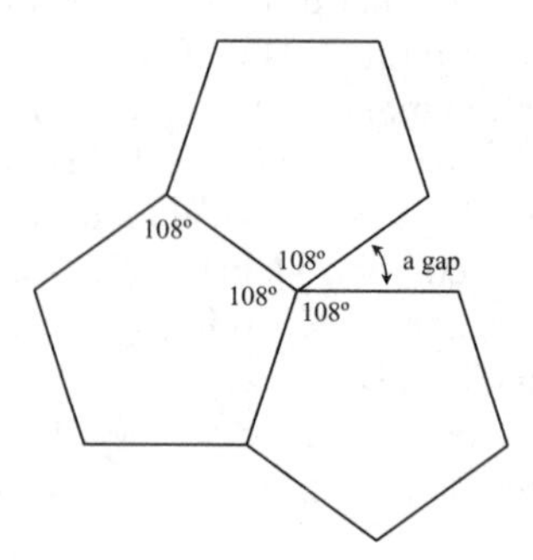

The language of crystallographic groups captures these geometrical requirements in terms of the theory of groups, specifically, groups of matrices with determinants plus or minus 1 and made up of whole numbers. The question becomes: How many different kinds of groups of these matrices are there that satisfy the restrictions of the problem? The restriction to whole numbers makes the problem discrete and combinatorial. The connection to matrices is natural because in analytic geometry matrices are the means by which rotation is accomplished. The key to the proof is to show that if you use whole numbers that are too big, you have no hope of finding matrices with determinant 1 that form a group. Ludwig Bieberbach solved this problem in 1910.

Bieberbach was born on December 4, 1886, into a comfortably situated family in a small town, Goddelau, near Frankfurt. His father was the director of the psychiatric hospital at Heppenheim; his grandfather was also a physician who had been employed at the same institution. After private tutoring, Bieberbach attended gymnasium in Bensheim. He showed an early interest in mathematics. In 1905–1906 he was stationed in Heidelberg for his compulsory year of military service. He attended mathematics courses at the university while there and became interested in function theory. After this year he went to Göttingen and studied under Klein and Paul Koebe (1882–1945). In 1907 Koebe solved Hilbert's twenty-second problem on uniformization by automorphic functions, and this made the subject a hot topic in Göttingen. Bieberbach wrote his doctoral dissertation in 1910 on automorphic functions.

Klein, however, made the deepest impression on Bieberbach. Constantin Carathéodory (1873–1950) wrote of Klein, "In his youth he used to look at the most difficult problems and *guess* their solutions."[1] Klein wrote his major papers in the nineteenth century, in a style more discursive than the more austere and rigorous twentieth century style then emerging. Bieberbach liked the older style, and his own writing was often looser and more intuitive than much mathematics in his own century.

Bieberbach first encountered the work of Schönflies on crystallographic groups in Klein's seminar. He quickly achieved his solution of the first part of Hilbert's eighteenth problem, proving that the number of crystallographic groups in each higher dimension was finite. He published the complete results in 1911 and 1912. These served as his habilitation. Schönflies became one of Bieberbach's active sponsors.

When Ernst Zermelo was called to Zurich from Göttingen in 1910 Bieberbach also went, as a privatdozent. He didn't stay the year, but instead moved to Königsberg to take a special professorship that Schönflies had arranged. Then in 1913 Bieberbach was appointed full professor in Basel. He married Johanna Stoermer in 1914 and accepted a professorship at the newly formed Frankfurt University in 1915. Schönflies had left his chair in Königsberg in 1911 to go to Frankfurt to take a position at the Academy for Social and Commercial Sciences, accepting apparent demotion for the sake of helping to found the new university. Opened in 1914 with donations largely from wealthy Jewish businessmen, Frankfurt was more open to

Ludwig Bieberbach

Jewish mathematicians than anywhere else in Germany, with the possible exception of Göttingen. Schönflies wasn't narrowly interested in promoting Jewish mathematicians, though. He was behind Bieberbach's call to Frankfurt.

In Frankfurt, Bieberbach did research in a wide area and wrote books on topics ranging from differential and integral calculus to conformal mapping and function theory. He is probably best known in this period for his work in function theory, specifically for work on *schlicht* (smooth, plain, homely) functions published in 1916.

Bieberbach had still been in Switzerland at the beginning of the First World War and so was not a professor in a German university, which would have exempted him from service. But he was not conscripted into the German army, because at some time after his early service in the military, he had become "unfit for military service."[2] His progress in the hierarchy of German mathematics was therefore not delayed, and he was left free to achieve mathematical results. This was good for his career but probably bad for his self-image. To the end of the war he passionately hoped for German victory. In the preface to his 1918 book on integral calculus Bieberbach refers to the date as "the '*siegerfüllten*' (filled with victory) spring of 1918."[3] The Armistice came as a shock, as Germany still occupied a substantial amount of territory and the war was still not being fought on German soil. This led to a widely shared belief that the war hadn't been lost militarily but by a betrayal of the leadership, which set the stage for the rise of Hitler. Patriotic or political statements increased in Bieberbach's mathematical publications.

His changing views on mathematics itself also foreshadowed Bieberbach's later actions. The early work on crystallographic groups of Schönflies, Fedorov, and Barlow is concerned with analysis of the specific properties of these groups in three dimensions. Joseph Wolf describes these works as "incredible displays of *ad hoc* synthetic geometry."[4] Early on, treatments of crystallography gave pages and pages of crystal types, with an impressive taxonomic vocabulary. Salt, which has cubic crystals, is a member of the holosymmetric class of the cubic system. On the other hand, we find that benitoite is the sole representative of the ditrigoanal bipyramidal class of the hexagonal system. There are large books of tables in the library containing the geometric details of the crystallographic groups, of particular use today in X-ray diffraction studies.

Bieberbach's 1911 paper about crystallographic groups, however, had little geometric information visible and was all group theory and matrices. Bieberbach wasn't slighting pictures—it is just that pictures and geometric intuition fail in 100 dimensions. Treating the problem as group theory works. In 1914 Bieberbach delivered an address on the philosophy of mathematics as a formalist following Hilbert. Pure mathematics was the investigation of systems of axioms. He was young and had just come out of Göttingen. During the war and after, there was often explicit hostility between the mathematical communities in France and Germany. Bieberbach's stance can be seen partly as a response to the French, who subscribed to a version of intuitionism that was opposed to the formal axiomatic position. (Brouwer was Dutch and decidedly pro-German, but there were French intuitionists and Bieberbach identified intuitionism with the French.)

But in 1919 Bieberbach wrote, "The tendency towards formalism must not cause us to forget the flesh and blood over the logical skeleton. . . . The formalism of university education in mathematics has forgotten about application and the cultural meaning of mathematics."[5] By the 1920s, he was increasingly sure he didn't like the new legalistic style in mathematics and wanted more pictures. He had passed out of his period of greatest productivity. His textbooks were criticized as mathematically sloppy but were popular and readable. Gödel checked out Bieberbach's book on functional analysis both in Brunn and in Vienna.[6] Bieberbach's books, translated into Russian, were important to the development of the Moscow school of analysis,[7] and they are still widely stocked in U.S. university libraries.

After the First World War, Germans were kept out of international societies and were not invited to conferences. Bieberbach became active in the German response to this at the same time that his unhappiness with the new style of mathematics steadily grew. In 1925 he published "On the Development of Non-Euclidean Geometry in the 19th Century," in which he talks about *Anschauung*. He was hardly the first German to do so. However, it gave him a hook to hang his discontent on. Herbert Mehrtens writes:

> This notion touches on visual perception, geometrical intuition or visualization, as well as 'intuition,' in the sense of Brouwer or Kant. It remained sufficiently vague to have different meanings in different contexts, and it sets a sharp contrast to formal operations and reasoning.[8]

Anschauung was good. Later Bieberbach specifically identified it with the German race. In 1926 he gave a lecture that identified two opposing styles, or ideals, of science and explicitly broke with Hilbert and sided with Brouwer in the foundational debate. Now he was influenced by a Frenchman, Pierre Boutroux (Poincaré's nephew), who wrote describing the wrong kind of mathematics as an "edifice of symbols which we pile up to infinity—like a clever juggler would do, who finds joy in multiplying the difficulties of his exercises."[9] The rhetoric was ratcheting up.

In 1921, Bieberbach was chosen to replace Carathéodory in Berlin. In the post only two years, Carathéodory had been called by the Greek government to help found a new university in Smyrna. (The Turks burned it down in 1922, so in 1924 Carathéodory was back, in Munich.) The post in Berlin was nominally one of the premier appointments in Germany, but Berlin was expensive after the war and the mathematical department didn't have the luster of earlier years. Brouwer, Herglotz, Weyl, and Hecke each refused the post before Bieberbach accepted it.

In Berlin, nevertheless, Bieberbach was comfortable, with a growing family (he eventually had four sons). He lived well and was at peace to work. Merhtens writes that "he was both physically and intellectually a very lively man, sometimes expressive to a comical degree. He was a somewhat muddled lecturer but very friendly and helpful to his students, a little conceited but not arrogant, all in all a slightly eccentric, but sympathetic character."[10] Helmut Grunsky, a student, said that Bieberbach sometimes had a rough or rude manner but attributed it to shyness or awkwardness. He said Bieberbach was helpful to his students if they had personal problems. Hans Freudenthal quotes a letter from Einstein in 1919 to Max Born's wife: "Herr Bieberbach's love and devotion for himself and for his muse is quite priceless. May God keep it with him, this is the best way to live."[11]

Meanwhile, Göttingen under Courant had consolidated its power within German mathematics, partly through connections with the publisher Ferdinand Springer, partly through connections with the Rockefeller Foundation, and partly by having attracted the best mathematicians. This shift occurred before the war under Klein but became more pronounced in the 1920s. There were battles over editorial policy—silly things. Should German journals publish French mathematicians? If so, which? International contact continued to be an issue. Hilbert pushed for open exchange. In 1928 the International Congress of Mathematicians was scheduled to be held in Bologna, and for the first time since the war the Germans were invited. However, the sponsoring organization had ties to an organization that had led the boycott against German science. When the invitations arrived, the Italians included an announcement of an "excursion to the 'liberated area' of Southern Tirol."[12] These were fighting words. A counterboycott was organized, led by Brouwer. This outraged Hilbert and a small war ensued. Brouwer was neither German nor had he fought in the trenches. Bieberbach, as secretary of the German Mathematical Society, the DMV, maintained official neutrality but privately urged mathematicians not to attend. Hilbert won and a sizable group of German mathematicians attended the congress (where Hilbert delivered his four problems on the foundations of mathematics that were solved by Gödel).

After the conference, Hilbert succeeded in having both Brouwer and Bieberbach removed from the editorial board of the *Annalen*. Bieberbach appeared a year later on the editorial board of the *Mathematische Zeitschrift*, another Springer publication, probably an attempt by Springer to keep order in the house. There were more battles over control of journals, which Bieberbach generally lost. By 1933 Bieberbach had been marginalized from the leading mathematicians of Germany. At Hilbert's seventieth birthday party, in 1932, hardly anyone talked to him. The unsettled nature of Bieberbach's personality comes through in the following from Freudenthal:

> At ours in Berlin Bieberbach was known as a moderate leftist, at least if compared with the rightist majority in university chairs. His "doctor father" had been the Jew Schönflies, to whom he was also indebted for his academic career. Once somebody in Berlin told me the only communists he ever knew as such, he had met in Bieberbach's home. Still as late as 1 April 1933, the famous boycott day, thus months after Hitler's takeover, Bieberbach voiced his sympathy for his dear colleague and friend Issai Schur, whose absence he complained [of] in public.[13]

(Schur was forced to retire in 1935. He emigrated to Palestine in 1939, where he died of heart disease in 1941.) Freudenthal also said of Bieberbach, "It was impossible to face him because if addressed he cast down his eyes."[14] Mehrtens reports that on April 18, 1933, a few months after Hitler took power, Bieberbach sent a birthday postcard to Richard von Mises (1883–1953), a Jewish colleague in Berlin. "The postcard showed a small German town over which a stylized sun inscribed with a swastika had risen."[15] Ironically, von Mises had supported the German boycott of the Bologna Conference. (Von Mises ended up at Brown University after a sojourn in Turkey.)

In his often-losing fights over mathematical politics in the 1920s Bieberbach had followed Brouwer and looked to him for succor. Now he would follow a more powerful leader. In 1933 Bieberbach marched along with his sons in a storm trooper march from Potsdam to Berlin. He officially joined the storm troopers, the SA, in November of 1933. His participation was so enthusiastic that he won the storm trooper sports badge. He took up lecturing and writing about styles in mathematics, but this time the good style was specifically identified as German, the bad style clearly Jewish or French. He delivered an internationally reported lecture to the *Förderverein* (Association of the Advancement of Education in Mathematics and Natural Sciences) on personality structures, blood, race, and mathematics. In one version he published he said:

> Thus, to have a current example, the manly rejection (*mannhafte Ablehnung*) of a great mathematician, Edmund Landau, by the Göttingen students was caused by the fact that the un-German style of this man in research and teaching proved intolerable to German sensibilities. A people who have realized how foreign desire for power has gnawed into its vitals and how enemies of the people try to force them into foreign ways, must reject those types of teachers who are foreign to them.[16]

Bieberbach agitated to remove Jews from German publications. He quit the editorial board of *Mathematische Zeitschrift* because Springer was too open to Jews. Brouwer had his own journal in Holland, *Compositio Mathematica*, and Bieberbach fought with him over the participation of Jewish mathematicians and led an exodus of Germans (some of whom would have been considered Jewish by the Nazis) from the editorial board. He also moved to change the policies of the *Jahrbuch*.

Harald Bohr published an article critical of Bieberbach's pronouncements in a Danish newspaper. Bieberbach delivered the coup de grace to his own standing within German mathematics in a sequence of actions beginning with his reply, an "Open Letter to Harald Bohr," in the DMV's *Jahresbericht*. The DMV was now embroiled in a nasty and personal international conflict with a mathematician of impeccable credentials. Bieberbach had created the impression that he represented the DMV, and he not only hadn't acquired permission but had published his intemperate letter over the objections of his co-editors. Freudenthal writes: "The majority of the German intelligentsia was conservative, monarchist, nationalist, or some mixture of these properties."[17] Bieberbach's behavior was becoming embarrassing and disruptive. The DMV decided to take up the matter at the annual meeting in September 1934. In Mehrtens's account, Bieberbach appeared with a group of his "student-followers" in storm trooper dress. The DMV found the nerve to remove them from the hall. Bieberbach's defense was that Bohr's letter was an attack on the German state and that as an officer of the DMV he had the duty to respond. The DMV compromised and "condemned" Bohr if he had, in fact, attacked the German state, but "regretted" Bieberbach's response. Bieberbach attempted to "have himself installed as '*Führer*' of the DMV," but he lost by a large margin.[18] After the meeting he continued his resistance, publishing the "condemnation" of Bohr, which he had been instructed not to do. The Nazi bureaucrats had what they wanted. Bieberbach's antics had forced the official governing body of German mathematics to go on record as supporting the Nazi state, and from that time on they were in official cooperation with the Nazis. In turn, mathematics in Germany continued under the Nazis in an attenuated condition, officially undisrupted.

A month after his loss in the DMV, Bieberbach applied to found a new journal, the infamous *Deutsche Mathematik*, which mingled race politics with "good German mathematics." Its first press run was 6,500, but by 1938 that had fallen to 700, many of which were never sold. He continued writing and lecturing about race-typology through 1940. According to Bieberbach there were two types of mathematicians, the S type, short for *Strahltypus* (Jewish, French, etc.—bad) and the J type (German—possessing *anschauung*—good). The J-type had three subdivisions J1, J2, J3. So the man who had helped clarify the classification of crystals into types, making his way in the world, extended classification into human life, an obscene and sloppy tiling.

People wonder why Bieberbach became a Nazi. Mehrtens's descriptions, "He was both physically and intellectually a very lively man, sometimes expressive to a comical degree" and "It was impossible to face him because if addressed he cast down his eyes," to me, have an almost neurological flavor, implying an element of mental illness. Alienated, aggrieved, reacting—Bieberbach could represent many Germans of that time.

After the war, like most Germans of any prominence, Bieberbach was interrogated by the occupying military forces as part of their program of denazification. He was very forthcoming to the American officer who interrogated him and explained his views on types and styles. He went so far as to complain that the Nazis had favored applied over pure mathematics. He had already lost his position, and his performance helped ensure he wouldn't get it back. He never again held a university position. Remarkably, though, he kept on publishing books and articles. His bibliography after the war is large. Springer, consistently inclusive, published or republished a number of his books. Bieberbach maintained contact with mathematicians he had known before, and some of them invited him to give talks. Alexander Ostrowski (1893–1986), one of the first to extend an invitation, to Basel, had been moved on hearing that Bieberbach "was earning his living by attending to the central heating of several houses."[19] Ostrowski, originally from the Ukraine and himself Jewish, had been studying in Marburg when the First World War broke out and had spent the war as a "civil prisoner." After the war he had made his way to Göttingen, where he excelled, and received a call to Basel in 1927, where he had ridden out the Second World War.[20] Other mathematicians were also surprisingly forgiving. Reid reports that Courant, near the end of his life, said:

> One cannot forget all the suffering, of course; but I was always in favor of being positive toward Germany, reestablishing contact. I felt very bad about Bieberbach once, but now it's been so long—so much time has elapsed—and I can always see that there were mitigating circumstances with all these people. I never had any contact

> with Bieberbach again, but if I would meet him here now I would be friendly with him.[21]

Mehrtens writes:

> When I visited Bieberbach shortly before his death, he lived in the house of one of his sons. I met a small, old and very sick man. His memory was very selective, but he still held that the discrimination between a "German" and an "alien" style had nothing to do with values.[22]

Bieberbach died on September 1, 1982. He was ninety-five years old.

In the second part of the eighteenth problem, Hilbert asks whether space can be filled with identical fundamental regions that are not associated with a crystallographic group. The answer is yes, and a three-dimensional example was described by K. Reinhardt in 1928. As it turned out, this was fairly easy to establish. It didn't open up new areas of research. John Milnor, in his article for the 1974 AMS conference on Hilbert's problems, describes it as "rather complicated" and then displays the following, more simple two-dimensional example supplied in 1935 by Heinrich Heesch (1906–1995).

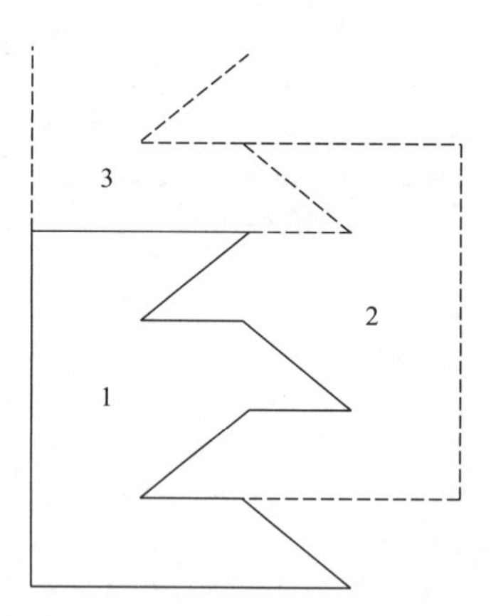

(This doesn't have enough symmetry to be a fundamental region for a crystallographic group.)

If Hilbert had loosened his focus here and looked at the question of repetition and symmetry in not perfectly regular contexts, he might have asked what kind of tilings are possible using two repeating shapes. In 1974 (Sir) Roger Penrose discovered a class of non-repeating tilings that nevertheless have symmetries.

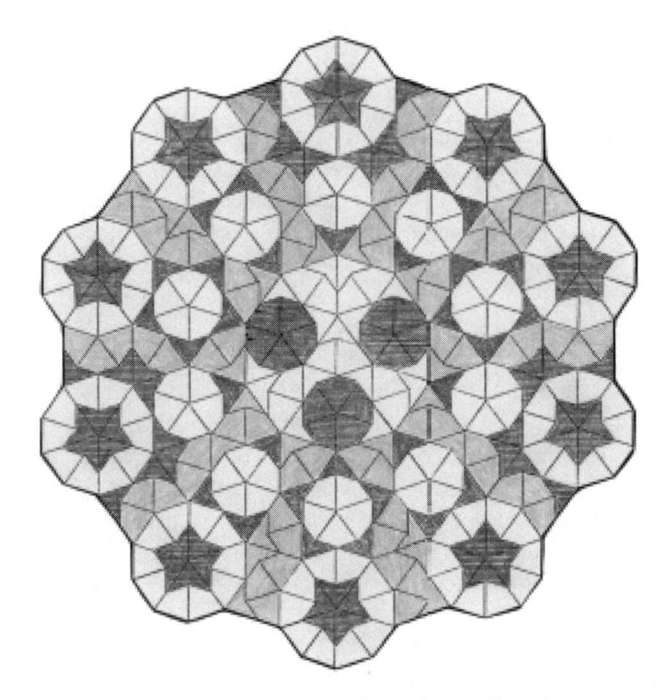

Penrose tiling (courtesy of David E. Kullman).

The discovery of "quasi-crystals" in the real world following this kind of pattern has added to their interest. This kind of fit with outside reality is closer to the original motivation of crystallographic groups. This is probably the only work with a connection to Hilbert's problems that has resulted in a patent suit about toilet paper. In 1997 Penrose filed suit against a British toilet paper manufacturer alleging the wrongful appropriation of his patented designs for use in a quilted toilet paper.[23]

At the end of the eighteenth problem, Hilbert asked how space can be most efficiently packed with various regular solids. In addition to being beguiling theoretically, this is of great practical interest: The modem on the reader's computer may have been designed with reference to this theory. The most historically important and resonant of these questions is how to most efficiently to fill space with spheres, in a packing with the least air or empty space.

The probable answer is obvious and, in fact, is the way the grocer stacks oranges. The difficulty is in *proving* what seems intuitively obvious. Milnor stated that lack of proof here was a "scandalous situation."[24] (That a relatively obvious mathematical fact cannot be proven outrages mathematicians.)

Why should a proof be difficult? In *Sphere Packings, Lattices and Groups*, J. H. Conway and N. J. A. Sloane write:

> As Rogers remarks, "many mathematicians believe, and all physicists know" that the correct answer is .7405 . . . [i.e., the way the grocer does it]. The situation is complicated, however, by the fact that there are partial packings that are denser than the face-centered cubic lattice over larger regions of space than . . . [the way the grocer does it] . . . .[25]

Strategies that contain irregularity cannot be easily ruled out. However, even when limited to regular lattice packings, the problem is difficult. As one pursues it into higher dimensions as Conway and Sloane do, a great deal of surprising detail appears. Just as counting 1,2,3,4,5, . . . gives us all the complexity of number theory, so does adding dimensions give us a lot of complex structure. Conway and Sloane write in their preface, "At one point while working on this book we even considered adopting a special abbreviation for 'It is a remarkable fact that,' since this phrase seemed to occur so often."[26]

In 1990 Wu-Yi Hsiang of UC Berkeley announced a proof that the grocer's packing is the best. Published in 1993, his proof is long and complex and, as most papers do, skipped details on cases he thought obvious. It has not been widely accepted. I spoke with Conway in October of 1998, and he said he was not sure whether Hsiang was still standing by his assertion. Conway said Thomas C. Hales was circulating a 250-page manuscript that recounts a different, and very large, computer-assisted proof. Hales told me that it involved 100,000 linear programming problems each with 100–200 variables. Nonlinear inequalities were proved by computer from geometric considerations in 5,000 separate cases. This was not done on a supercomputer but on "a lot of" Sun computers.[27] This level of industry may be laudable, but in the prior conversation Conway had said that no one was eager to volunteer to referee Hales's paper because it would involve so much time. From what Conway had seen, he thought that Hales had probably succeeded. There is growing agreement that Hales's proof is sound. In 2000 an article by Hales outlining his work appeared in *Notices of the AMS*. In late 1998, when I talked to Hsiang, though, he had been critical of the nature of Hales's proof. It provided no direct geometrical insight. He said, "We try very hard to understand how nature operates,"[28] emphasizing the world understand. Further, how could Hales's giant proof be really checked? This will be decided. Hales's computer-assisted proof is either right or wrong. Time and work will get the situation sorted out, and I suspect that Hales's proof will stand.

When Hilbert presented his list of problems, he was concerned with the proper resolution of the contradictions that had been found in the foundations of mathematics. He assumed that finitist proofs that would have satisfied Kronecker were solid and instead was concerned with fundamental problems raised by Cantor's expansion of mathematical argument and exploration into the realm of the infinite. Hilbert never got his proof that the axioms for the new systems or even arithmetic were consistent, yet the problems in the foundations were resolved and mathematics continued, using the new style of argument in the language of infinite sets. What the controversy about sphere packing illustrates is not some fundamental problem involving consistency, but the problem of finite proofs that are so long that they are difficult or impossible to grasp or check. The details of Hales's proof reside inside those Sun computers and can't be explained in full detail to anyone. The steps will never be printed out on paper; no one will ever even look at them. The proof will be accepted or rejected on the basis of the correctness of its architecture and by reduplicating its details in other actual programs run on other computers. This then hinges on the fundamental reliability of the computers and their software. These calculations involve so many individual calculations that even a perfectly programmed computer could be expected to generate a few small errors because of hardware error and, possibly, even for quantum mechanical reasons.

On the side of the divide that includes proofs that are actually written down and published, we can look at the series of papers constituting a proof that we know the structure of all finite simple groups. It has been estimated that this proof, one of the great achievements in abstract algebra in the second half of the twentieth century, would run 15,000 pages if gathered together all in one place. How can we be absolutely certain that there is no small hole that will let Takagi's ants in, reducing the whole to rubble? The fact that mathematics builds on its already proven results to obtain new results makes it likely that no mistake about anything important will remain confined in a small error. And all of these concerns sidestep the issue of "understanding" that Hsiang brought up.

Gleason said (as already quoted in the chapter on the fifth problem), "Proofs really aren't there to convince you that something is true—they're there to show you why it is true."[29] Hales's proof certainly doesn't do this, even if we believe it. Kenneth Appel and Wolfgang Haken's proof of the four color theorem doesn't give us this direct kind of insight. Are these long proofs merely inefficient? Perhaps an elegant and comprehensible proof of the four color theorem is possible, and we just have not thought of it. However, if we think of an axiom system that would encompass the sphere packing problem or the four color theorem, we can in principle make a list all possible proofs by the length of the proof. It is possible that we will only find proofs of these simply stated facts that are very, very long. The sum of human culture grows exponentially more complex, yet possibly not more profound. Homer composed a pretty good poem. Despite the success of modern science, which as a basic assumption posits there are "simple" explanations, there is no guarantee that simply stated truths can be proved by simple arguments.

The Analysis Problems

13, 19, 20, 21, 22, 23

Analysis Takes at Least Seven Years

Hilbert's list ends with five problems in analysis, the nineteenth through twenty-third. In addition, the thirteenth problem is really a problem in analysis, though its title presents it as an algebra problem. Most of the mathematics involved in axiomatizing physics, the sixth problem, would have to be classed as analysis at root. The second half of the sixteenth problem is about differential equations. That makes seven-and-a-half of the Hilbert problems. However, a census of mathematicians, including applied mathematicians, particularly if we include probability as a part of measure theory, would show that this proportion understates the amount of energy spent on analysis in the twentieth century. Only two of Hilbert's analysis problems, the thirteenth and the twenty-second, have been cleanly solved by specific mathematicians. One more, the twenty-first, was solved in 1908 and then re-solved in the negative in 1989. Its solution is clean, but the path to it was not. Why this relative lack of clean successes?

First, the structure of analysis is less clean, certainly historically, than other areas we have dealt with. Analysis freewheels with the process of calculation more than most other branches of mathematics. If a calculator has gone wrong in this area, nature will let us know—analysts work with a net. And if some relatively outrageous calculational gambit is consistently working in explaining nature, it becomes the mathematician's job to figure out why. That "why" is a clear statement of what is being done and a proof that it works mathematically. Analysis can take on some of the character of an experimental science, especially when done on a vast scale using computers.

Second, as of 1900 Hilbert had not been an analyst for long—he wasn't really ready to write these problems. His analysis problems are less sharply drawn than the problems in the other areas. The nineteenth, twentieth, and twenty-third problems call for programs of research. With Poincaré and Kolmogorov in the chapters on the twenty-second and thirteenth problems, respectively, we will touch on some of the major themes of analysis.

How Famous Can a Function Theorist Be?

The Twenty-Second Problem

Koebe and Poincaré

When Paul Koebe traveled he registered in hotels under assumed names, lest the waiters and chambermaids besiege him with questions about whether he were related to the famous function theorist, Koebe. If they were to find out that he was the great man himself, he would have no peace. It is safe to say that few mathematicians have had to be so careful. Hans Freudenthal, whose article in *The Mathematical Intelligencer* is the source of much of the biographical information in this short section, wrote:

> When on the first day of my study I was inscribed at the Mathematical Institute of Berlin University, it happened that Bieberbach was around and heard that I came from Luckenwalde. He turned and asked me: "So you are one of the Luckenwalde streetboys who run after Koebe to call 'there goes the famous function theoretician,' are you?"[1]

Koebe did hail from Luckenwalde, where his family, in Freudenthal's words, "ran a renowned fire engine plant."[2] One wonders if something in the water of Luckenwalde encouraged native sons to talk of "renowned" fire engine plants (or function theorists). Luckenwalde provided Koebe many advantages. Born in 1882, he enjoyed the prosperity of his family and received an excellent education, attending gymnasium in Berlin and then continuing at the University of Kiel in 1900. He finished at Berlin University in 1905. He also attended Charlottenburg Technische Hochschule, a more practically oriented school, perhaps with an idea of picking up useful skills for running a fire engine plant. He habilitated at Göttingen in 1907, which placed him in a setting where many mathematicians were thinking about uniformization theory and Hilbert's twenty-second problem. In 1907 Koebe, independently of Poincaré, solved Hilbert's twenty-second problem and achieved a measure of genuine fame—at least within the mathematical community. He then continued his work as a function theorist with single-minded determination and never strayed, doing other important work in this area. Koebe remained at Göttingen until 1910, when he moved on to Leipzig as extraordinarius. Thence to Jena in 1914 as full professor and back to Leipzig in 1926.

Koebe was respected, due to his talents, but considered "conceited and disagreeable" at Göttingen.[3] He had a reputation for stealing the ideas of younger mathematicians. Reid quotes Courant to the effect that Koebe stole the idea of his thesis, not yet published, of which the two had spoken. Koebe rushed off a paper on the same subject and thus claimed independent credit for the result. Then Koebe showed up at a seminar where Courant was scheduled to speak on his thesis and by virtue of seniority took the position of speaking first. This was not well received by Courant's friends, according to Reid. They rigged up an elaborate apparatus and hid it in a chamber pot under the lectern of a class Koebe was teaching. An alarm sounded at erratic intervals as Koebe lectured. When he finally pulled the mechanism out of the chamber pot, it was greeted with much laughter. One of the mischief makers, Kurt Hahn, saw to it that an article describing the incident appeared in the local paper.

Paul Koebe.

Freudenthal says that Koebe's eccentricities were so widely known that "younger people, when introduced to Koebe, almost automatically reacted: 'Ah, the great function theoretician.' "[4] Such a greeting, instead of being recognized for what it was, succeeded in getting an assistantship for a friend of Freudenthal's.

Freudenthal also says an "unidentified" man wearing a big hat, a turned-up collar, and dark blue glasses gained access to a Göttingen printer and inserted a footnote into a paper of Brouwer's on uniformization, acknowledging priority to Koebe. This incident supposedly happened in 1912, when Koebe had already moved on to Leipzig, so it contains the additional oddity of a possible special trip. Freudenthal does not vouch for the truth of this story and adds that later Koebe said it was a practical joke played on him. At the least, it captures something about how Koebe was perceived in Göttingen.

Freudenthal also wrote the entry on Koebe in the *Dictionary of Scientific Biography*, which ends:

> Koebe's mathematical style is prolix, pompous, and chaotic. He tended to deal broadly with special cases of a general theory by a variety of methods so that it is difficult to give a representative selective bibliography. Koebe's life-style was the same: Koebe anecdotes were widespread in interbellum Germany. He never married.[5]

Koebe died August 6, 1945.

A national hero of France and a best-selling author of explanations of mathematics and science to the layman, Jules Henri Poincaré really was famous. By some, he was considered an important philosopher. He was admitted to the French Academy, a body of forty writers that passes on the correctness of words and usage in the French language, despite the claim that he never revised a sentence. Poincaré was the greatest mathematician of his time, his only competition being Hilbert. Mathematicians compare him to Gauss, which is the highest honor. Blink at his solution of the twenty-second problem in the context of his whole mathematical career and you will miss it, and it was he who first asked the question.

Poincaré published around 500 papers in his lifetime, counting only papers reporting original research. His *Oeuvres* are eleven large volumes, each labeled a *tome*. While the first appeared in 1916, four years after Poincaré's death, the last did not come out until 1954, so monumental was the labor involved in selecting what was to be published. A separate three-volume set on celestial mechanics is important. Poincaré lectured on a different topic in mathematical physics every year at the Sorbonne and other institutions, like the École Polytechnique, and auditors turned many of these lectures into books.

Though prolific, Poincaré was a deep mathematician. Just as it took decades for the mathematical community to begin to understand and use the ideas contained in Riemann's eight-page paper on the zeta function, mathematicians find the same kind of depth in Poincaré again and again. Insights appear that are clearly expressed for their time, yet for which a precise mathematical vocabulary and technique would not be available for decades.

Poincaré came from a distinguished French family full of professionals, often of a technical type, and had a highly intelligent, devoted mother. Nearly all of my account of his early years comes from Gaston Darboux's "*Éloge Historique D'Henri Poincaré*," which appeared as a long obituary in *Mémoires de L'Académie des Sciences de L'Institut de France* in 1914. It also appears at the beginning of the second volume of Poincaré's *Oeuvres*. Darboux was the French mathematician whom Hilbert had the temerity to eulogize during the First World War. He had known Poincaré a long time, having been one of the examiners for his doctoral degree and for the École Normale. His account has a tone of Olympian praise, but I believe it was written with care and is accurate.

Poincaré is often mentioned and there are many books about aspects of his work, but astoundingly, nearly ninety years after his death, there is no comprehensive, modern scientific biography of him in English or any Western European language that I have been able to find. E. T. Bell wrote in 1937: "To give an adequate idea of this immense labor one would have to be a second Poincaré, so we shall presently select two or three of his most celebrated works for a brief description."[6] This echoes H. F. Baker in his 1915 obituary for the Royal Society of London: "To give any complete account of his work is a task well nigh impossible on account of its vast range."[7]

Poincaré's parents were "*lorrains tous les deux*,"[8] from the Alsace-Lorraine region in the northeast corner of France. Poincaré's father's family was originally from Neufchâteau. An ancestor who died in 1750 was an official of the court, *conseiller au bailliage de Neufchâteau*. Another member of the family was a mathematician at the *collège de Bourmont* near Neufchâteau. Poincaré's grandfather, Jules Nicholas, a pharmacist, was born in Neufchâteau in 1794. He was stationed at the San Quentin military hospital in 1814 during the Napoleonic Wars. In 1817, at twenty-three, he moved to Nancy, taking along his parents and sisters. The family acquired a distinguished house, which sat between the ducal palace and the *porte de la Craffe*. When Poincaré was accepted into the French Academy, a description of this house was included in the welcoming speech by Frederic Masson: "solid, massive and without ornament."[9]

Into this house was born Leon Poincaré, Henri's father, in 1828. Leon became a prominent physician and professor of medicine in Nancy. Darboux writes that the Academy of Sciences received his reports with high appreciation and describes him as having a spirit "*très original*."[10] Leon's younger brother Antoni, born in 1829, achieved a brilliant rank at the École Polytechnique and sent many communications to the French Academy about meteorology. He became inspector general of bridges and embankments. Henri's first cousin Raymond Poincaré, one of Antoni's sons, was president of the French Republic during the First World War and four times premier. Another son of Antoni's, Lucien, became "*directeur de l'En-seignement secondaire au Ministère de l'Instruction publique et des Beaux-Arts*." Darboux lists other distinguished Poincarés and in a footnote even traces the name (possibly) back to 1403 and a Petrus Pugniquadrati, a student at the University of Paris.

Darboux does not list an equivalently rich family tree for Poincaré's mother. In fact, as Bell points out, he neglects even to mention the family name (Launois), at least when she is first introduced, but he does mention her family's country estate in Arrancy, where Henri spent some of the best times of his childhood. Poincaré's mother is described as "*très bonne, très active et très intelligente*."[11] Gaston Julia writes in *Le Livre du Centenaire de la Naissance de Henri Poincaré* 1854–1954 that Poincaré's maternal grandmother had a gift for mathematics.[12]

Henri was born on April 29, 1854, into this lively and intellectual family. He had one younger sister. Her son Pierre Boutroux (1880–1922) became a mathematician.

Darboux says that Poincaré had an exceptionally happy, even blissful, childhood. "His mother watched over him with a solicitude full of intelligence."[13] He spoke very young and had trouble at first expressing the rush of his thoughts. When he was five he caught diphtheria. Today we have largely forgotten what this illness entails. It starts as a mild sore throat but progresses rapidly as a false membrane forms on the surface of the throat and toxins poison the body. Death often came rapidly. Poincaré suffered from a paralysis of the larynx that lasted for nine months. The illness left him weak and timid for a long time. He wouldn't even dare to go down stairs by himself and couldn't stand the rough play of children his own age. During this period he became an avid reader. He never read the same book twice, and at this point the power of his memory became evident. Darboux says that he could always remember where in a book, on which page and on which lines, he had read something. Though perhaps exaggerated, the spatial aspect rings true. We hear that Poincaré was particularly taken by a book of natural history, *La Terre avant le déluge*, by Louis Figuier. This early bestseller of science is a provocatively written and gorgeous book. Beautifully colored maps of the continents in surprising arrangements and shapes during different paleological periods appear throughout.

Poincaré's schooling began with a tutor named Hinzelin who had published well-respected educational books and who opened up many subjects to young Henri without requiring much written work. Poincaré's mother was nervous about this when Henri entered the lycée at the age of eight, but he received the highest grade on his first essay. With a few notable exceptions he would always be first in school. A notebook survives from this first year of school that Darboux says makes clear Poincaré's exceptional talents. His work on history and geography stands out, and a French composition is particularly noted as a "little masterpiece."[14] The exceptionally poor quality of his handwriting is also remarked upon.

School did not seriously inconvenience Poincaré. Homework got done in a haphazard though effective way. From a General (Paul) Xardel, who grew up with Henri (Darboux refers to him only as General Xardel throughout), we have the image of books and homework on a table, with Poincaré floating around the room, in and out, talking when there was conversation available, answering questions about the homework when asked. When he went to write he would pick up the pen with whatever hand was available.

Instead, the future member of the Academy liked to play. Since he was neither strong nor coordinated, he preferred games for which imagination and spirit were more important. Sometimes he played with the sisters of his friends. During summer vacations with his maternal grandparents at Arrancy, Henri was allowed freedom to wander in the large garden or the countryside. He liked to walk rapidly through the garden with a walking stick at his side. Sometimes he would stop and draw with the stick in the sand. He particularly loved animals and had a chance to observe them in the garden. Once he accidentally shot a bird in a tree and was greatly saddened. After that he would never shoot a gun.

Darboux reports that Poincaré was a loving son and brother. He was modest and didn't try to prove his superiority over others. However, when he had good reasons he could be immovable in his "passive resistance."[15] Darboux quotes General Xardel, narrating a vacation the two families spent together in 1865: "Henri wanted to see everything, understand everything and explain to us everything."[16] At the famous Echo of Ramberchamp Henri delivered a complete theory of the echo. Xardel said of Poincaré, "He had the gaiety and the expansiveness of a child, the reason of a man."[17]

When he was thirteen, Poincaré went to the Exposition in Paris and emerged with ideas about politics. This led to the founding of a republic called the "Trinasie" in the gardens of Arrancy, with his sister and a cousin. It was a federation of three parts, one for each child, and each had their own language as well as a common language, which they used to conduct the government. Poincaré found that without seeming to he could effectively take all the power, but after discovering this he never did it again. The Trinasie lasted a number of years and was the trellis upon which many games grew.

Much of the summer would pass with this free play in the garden with just one or two other children. But then there would be a week of visiting and parties with relatives in the region. Poincaré participated actively in the amateur theatricals and charades. When he was thirteen or fourteen he wrote a verse drama on Joan of Arc—also a *lorraine*. Poincaré liked to dance; Darboux says he never tired. This active social life continued into the school year and his studies never suffered.

During this period of play Poincaré was demonstrating the ability to concentrate when something grabbed his attention—a game, a book, an idea. When interested he would not let go, even if the time set aside for the subject was over. This focus caused him to sometimes forget to pay attention to the world outside. He never could remember if he had eaten. Later in life Poincaré described himself as having a poor memory, and this is a way to understand that evaluation from a man who remembered everything he read. Darboux tells of a time when Poincaré was seven or eight and was walking with his mother and sister, lost in thought, when he discovered he was on the other side of a small river from them. He remedied the situation by wading in a straight line through the river up to his waist to get back on the correct path.

When he was fifteen Poincaré discovered his passion for mathematics. Darboux writes, "From this moment this vocation did nothing but grow and become more and more dominating and absorbing."[18] Unlike the extreme case of Ramanujan, though, it did not keep him from pursuing his other studies. David Ruelle says that Kolmogorov had a theory, offered as an explanation for the childish behavior of a colleague, that psychological maturation stopped the moment mathematical talent emerged.[19] By this account Poincaré was allowed to mature to a relatively advanced mental age.

Poincaré passed his first set of exams with distinction, acquiring a *baccalauréat* in letters first and following it three months later with a degree in science. His Latin exam was even better than his French exam. Oddly, he did poorly on the written exam in mathematics. He arrived late and appeared to not understand the question, a simple one on a convergent geometric series. Poincaré did very well on the oral examination and the examiners knew of his reputation. One of the examiners said that if any other candidate had presented such an exam he would have failed. The *baccalauréat* degrees, earned at the age of seventeen, marked the end of what Americans call high school. However, for students headed for the *Grandes Écoles*, the most elite schools in France, two more years were spent at the lycée. Poincaré passed on to a course called "elementary mathematics."

The Franco-Prussian War, fought in great part in the northeast corner of France where Nancy lies, disturbed Poincaré's years in lycée. The French army attacked the Prussian army. The Prussians had been studying the lessons of the American Civil War and understood that with the new generation of firearms the defending army could sit back and slaughter the advancing forces. The French were slaughtered and surrounded. The French regime fell and the Commune of Paris (1871) had its day. Then the Germans took Paris and the weak and vacillating Third Republic was installed. The Prussians annexed the Alsace and part of the Lorraine, but luckily for Poincaré, Nancy remained as part of France, though closer to the border than it had been. The occupation was not polite. Poincaré had been too young and delicate to serve in the army but had accompanied his father in an ambulance to tend the wounded. He saw the brutality of war close up, but saw more personally compelling destruction when he accompanied his mother and sister to Arrancy. The battle of Saint-Privat had been fought nearby. On the journey they were forced to pass "in glacial cold, through burnt villages, empty of inhabitants."[20] They found the family home where they had spent so many summers wrecked and looted. Poincaré's grandparents were fed on the day of the looting by a poor villager. In an interesting side note, the Japanese, who had been studying military strategy under the French, switched to Germany, which is partly why Takagi ended up in Göttingen rather than Paris.

It is said that these events made Poincaré a lifelong patriot. They certainly had an effect on the attitude of his more political cousin, Raymond. During the period of occupation Henri learned German so he could read the only newspapers available. Despite the poor behavior of the German soldiers, Poincaré went out of his way throughout his career to apportion credit fairly to German colleagues. In some of his first major work he called what are now more commonly referred to as automorphic functions, at least in English, Fuchsian functions, after the German mathematician Lazarus Fuchs, who had some early involvement with them. In the same context he referred to Kleinian groups, since Felix Klein was working on the subject at the same time as Poincaré. When Poincaré delivered a series of six lectures in Göttingen in 1909 at Hilbert's invitation, he delivered five in German (although some of the Göttingers were a bit miffed that he had the nerve to talk about integral equations, about which Hilbert was in the process of making substantial advances).

France and Poincaré continued with remarkable continuity after the interruption of 1870–71. In 1872, Poincaré competed in a *concours général* given to all lycée students in France and won first prize. The next year he took the "special mathematics" course and again took first prize. During this last year at the lycée, Poincaré met two young mathematicians, Paul Appell (1855–1930) and Colson (probably Clement, 1853–1939), who also became prominent. Appell had moved to Nancy to escape the Germans, as his home town of Strasbourg had become part of the German Empire. On the first day of class Appell and Colson sized up Poincaré. They were perched above him in the tiered classroom and could observe the notes Poincaré was taking, or not taking. He was using a funeral announcement for paper. The notice appeared day after day and occasionally Poincaré would scrawl (*griffonner*) a line or two on it. They were astounded. They thought perhaps he wasn't serious. However, he had won first prize in all of France the previous year in the *concours*. They approached a more advanced student (the lycée in these final two years flowed into or blended with École Forestiere in Nancy) and persuaded him to ask Poincaré a difficult and "particularly obscure" question. Poincaré gave the answer "without thinking a minute."[21]

Appell described Poincaré as a student:

> From the beginning of term his intelligence revealed itself immediately to our professor Elliot as it did to us schoolfellows; he had the genial gift of immediate comprehension of the particular detail of each question, the general idea from which it came, and the place it occupied in the whole. He also had this simplicity, this horror of effect, this good sense of someone from Lorraine, this sureness in friendship that he retained all his life.[22]

Poincaré took the exams for the École Polytechnique and the École Normale, the two most elite schools in France—doorways to the highest positions in French society, including the civil service. There is a bewildering array of institutions of higher learning in Paris, and many scholars are on the faculty of more than one.

Ph.D.s are granted by the universities. Poincaré chose the École Polytechnique, partly as a family tradition, entering first in his class. He did more poorly in the exam for the École Normale, finishing only fifth (in France). Mathematics occupies a particularly high place in these exams. Darboux regarded the fifth-place finish as a small failure. Appell finished second and chose the École Normale.

Darboux, writing forty years later, was able to find many people who remembered specific details of Poincaré's performance in his school exams. In one story, Poincaré spoke slowly, then closed his eyes. He asked if he could stop his demonstration. After examining the corner of the table he announced he wanted to deliver a different proof that was shorter and more elegant. Another story concerns an examiner, Tissot, who felt that Poincaré was going to do too well, so he suspended the examination for three-quarters of an hour in order to come up with a harder problem, which he found in an advanced geometry book. Poincaré listened and then drew the figure. He stood and stared at the ground for quite a long time and then started with a solution using advanced methods. Tissot announced that he wanted a more elementary solution. Poincaré then left the blackboard and delivered a trigonometric solution right at Monsieur Tissot's desk, in his face as it were, which prompted Tissot to say, "I desire that you not leave elementary geometry."[23] Poincaré complied and the examiner was happy (and by this time undoubtedly happy to be done). He gave Poincaré the highest grade.

At the Polytechnique, Poincaré continued to sit in lectures without taking notes. One preferred method of working was to march up and down the halls of the school deep in thought. Sometimes in the off hours he would join friends who had also come from Nancy. They would all join arms and walk and talk, but Poincaré would also often sit without speaking or moving, lost in his own thoughts as his friends heatedly discussed the events of that "epoch of crisis."

The Polytechnique's requirements included physical exercise, gymnastics, drill with arms, and drawing. Poincaré did awfully at all of these. His friends were particularly amused by his attempts at drawing and put together a public exhibit of his drawings, with labels in Greek such as "This is a horse." Sometimes they were wrong. "Young men are without pity like children," Darboux says.[24] In a much-repeated story, Darboux says Poincaré's drawing (*dessin*) was so bad that he scored a zero in his examination for the École Polytechnique. Normally this would stop the entry of a candidate. However, the zero was changed to a 1, which is what appears on the official paper that survives. This story is not quite correct. He received a 1 in *lavis*—wash or tint. Under the category of *dessin*, or drawing, Poincaré received a 12 out of 20. Despite his difficulties in these areas his margin of victory on the total entrance exam was about 7% over his nearest competition. His clumsiness in drawing figures later did cause him fall out of first place when he graduated, because he fouled up the drawing of a figure during the geometry exam and annoyed the examiner. We don't hear of Poincaré being overly concerned with the exams. He was governed by an internal standard and internally regulated interests.

Upon finishing at the École Polytechnique, Poincaré enrolled at the school of mines and gave every appearance of becoming a mining engineer. Many mathematicians and theoretical physicists head off to college with the intention of studying engineering, which seems solid and practical to parents. We have seen Cantor and Koebe, for example, at least momentarily on an engineering track, and the physicists Dirac and Feynman come to mind. However, most are on a career path in mathematics or physics when they leave college. Here, however, we have Poincaré completing school, passing a postgraduate engineering course, and then working as a mining engineer. He was happy and self-confident, and he was expected to be productive. He performed the duties of his job well. From 1875 to early 1879 Poincaré was mostly in Paris. He made trips, including one to Hungary, to investigate mining techniques. In early 1879 he was sent to Vesoul, seventy-five miles south of Nancy, as a practicing mining engineer. When there was an explosion in a coal mine that killed sixteen people, Poincaré promptly descended the mine shaft to inspect. Darboux says that he showed considerable *"sang-froid"* in his love of duty.[25]

During this period as an engineering student and engineer Poincaré's interest and capabilities in mathematics grew, and he followed a parallel track at the University of Paris, preparing for a Ph.D. degree. In 1878 he published his first original paper, which was on differential equations, and submitted a more substantial thesis, also on differential equations, for the degree. (Poincaré did publish a paper in 1874 while a student at the École Polytechnique, on geometry, but it was confined to giving new proofs of already established results.) Darboux was assigned to read the thesis and was impressed—there were enough original ideas for many theses. He found many points, however, that needed clarification or revision. Poincaré would never be very concerned with details unless he felt they were important. Darboux reports that Poincaré was accommodating about making corrections that were asked of him. He later told Darboux that his mind had already moved on to other things. The thesis was published in 1879, and Poincaré received his doctorate in August of that year, while he was working in Vesoul.

Poincaré received an appointment to teach at the University of Caen in December 1879, though some duties as an engineer lingered on. He was a promising young mathematician who had done some original work on differential equations. But this first work was like the odd rumblings that might precede the eruption of a volcano, because once started, only a word like "eruption" can capture the work that came in the next three or four years.

At the beginning of May 1880, Poincaré read a paper by Fuchs about a specific kind of differential equation. On May 28 the Academy of Sciences in Paris received from Poincaré a paper that contains many of the key ideas of the theory of automorphic functions, as an entry for its 1880 prize competition. John Stillwell writes, "Poincaré begins in the wilds of differential equations. After a long hack through this jungle, he finally emerges into the light of geometry."[26] (In this paper he drew a figure that was incorrect, noticed it, and told the reader to ignore the part of it that was wrong. He did not recopy the page and draw the figure correctly.) Poincaré himself supplied a piece

Jules Henri Poincaré (courtesy of the Archives of the Mathematische Forschungsinstitut Oberwolfach).

of the picture of what his life was like at this time and what his path was to full production as a mathematician. He wrote in *Science and Method*, published in 1908:

> For fifteen days I strove to prove that there could not be any functions like those I have since called Fuchsian functions. I was then very ignorant; every day I seated myself at my work table, stayed an hour or two, tried a great number of combinations and reached no results. One evening, contrary to my custom, I drank black coffee and could not sleep. Ideas rose in crowds; I felt them collide until pairs interlocked, so to speak, making a stable combination. By the next morning I had established the existence of a class of Fuchsian functions. . . . I had only to write out the results, which took but a few hours.
>
> Then I wanted to represent these functions. . . . I asked myself what properties these series must have if they existed, and I succeeded without difficulty. . . .
>
> Just at this time I left Caen, where I was then living, to go on a geologic excursion under the auspices of the school of mines. The changes of travel made me forget my mathematical work. Having reached Coutances, we entered an omnibus to go some place or other. At the moment when I put my foot on the step the idea came to me, without anything in my former thoughts seeming to have

> paved the way for it, that the transformations I had used to define the Fuchsian functions were identical with those of non-euclidean geometry. I did not verify the idea; I should not have had time, as, upon taking my seat in the omnibus, I went on with a conversation already commenced, but I felt a perfect certainty. On my return to Caen, for conscience' sake I verified the result at my leisure.
>
> Then I turned my attention to the study of some arithmetical questions apparently without much success and without a suspicion of any connection with my preceding researches. Disgusted with my failure, I went to spend a few days at the seaside, and thought of something else. One morning, walking on the bluff, the idea came to me, with just the same characteristics of brevity, suddenness and immediate certainty, that the arithmetic transformation of indeterminate ternary quadratic forms were identical with those of non-euclidean geometry.
>
> Returned to Caen, I meditated on this result and deduced the consequences. . . [but ran into a problem].
>
> Thereupon I left for Mont-Valérien, where I was to go through my military service; so I was very differently occupied. One day, going along the street, the solution of the difficulty which had stopped me suddenly appeared to me. I did not try to go deep into it immediately, and only after my service did I again take up the question. I had all the elements and had only to arrange them and put them together. So I wrote out my final memoir at a single stroke and without difficulty.[27]

This is a striking passage. Poincaré seems to have an official workday for mathematics of one or two hours during this period. There is also remarkable evidence of self-confidence. When he sees the answer to major difficulties he has other things to do, so he does not verify his insights immediately. He has seen the answer.

Poincaré's creation of the theory of automorphic functions secured his reputation as a mathematician (though it did not win the 1880 prize competition). He published two brief accounts of his results in *Comptes Rendus* in 1881 and was called to a position at the University of Paris in the same year. He published a full account of his results in a series of four papers in *Acta Mathematica* in 1882-84. Poincaré's theory of automorphic functions and the discrete groups that accompany and generate them are the ground of Hilbert's twenty-second problem.

At the beginning of the nineteenth century there was a relatively small roster of functions in the world of mathematics. Algebra supplied the bulk of them, but there were also trigonometric functions like sine and cosine, the exponential function, and special functions created by calculus and/or physics—all functions of real numbers. As the nineteenth century progressed and the real numbers were understood and the imaginary numbers accepted, there was a population explosion in functions. Function theory is an attempt to order, understand, and contribute to this explosion.

What is a function? At first the idea was intuitive. A function took one number as an input and gave back another number as an output. One could draw the graph of a function. Usually a function was identified with a specific rule by which it could be calculated. A simple example is $f(x) = 2x + 3$. As x changes, the result of multiplying x by 2 and adding 3 changes. Sine(x) has a more complicated rule. In its modern, set-theoretical formulation a function is a map (it could be totally arbitrary) from a set that is the domain to a set that is the range. It doesn't have to have any smoothness or even have a method of calculation. The only constraint is that it take each member of the domain to a unique element of the range.

We can illustrate a great deal by examining what happened to the sine function.

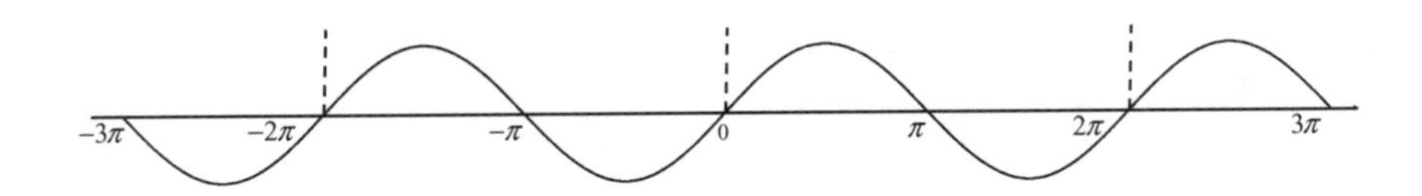

Its graph repeats itself. We could build the whole graph by repeating the section from 0 to 2π. One can come up with the sine function in many ways. However, the historical source and the most natural way to get the sine function is to look at a circle with radius 1:

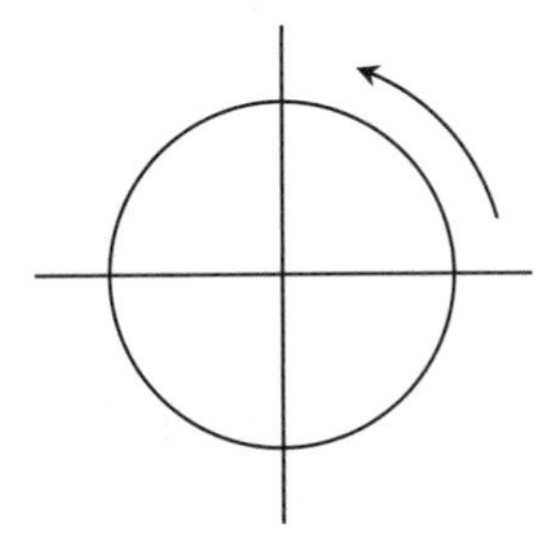

We place the circle in a coordinate system and then walk around the circle counterclockwise. If we go 2π around the circle we are back to where we started. (We could use degrees, but mathematicians usually use the distance-walked setup.)

The sine function is naturally understood by the picture of the circle and from its first appearance was understood in this way. If we had historically come up with the sine function in some other way, it would have been a surprising moment of clarity when someone drew this picture and pointed out the connection. In the first half of the nineteenth century something like this happened with what are called elliptic functions—not when they acquired the name "elliptic," but when they were later connected to the torus. They were discovered when mathematicians tried to calculate the distance around an ellipse, or a portion of the curve of an ellipse. Planets orbit the sun in elliptical paths, so this was a natural thing to try to do. These calculations led to an integral* that could not be solved using any function that was available off the shelf. The name "elliptic" stuck for the function they discovered, but eventually mathematicians were investigating elliptic functions that had very little to do with an ellipse.

Elliptic functions are analytic functions on the complex plane that repeat, like the sine function, but in two separate directions, which we can call w' and w''. Just as Sine(x) = Sine($x + 2n\pi$) for any integer n, an elliptic function $f(z)$ follows the law: $f(z) = f(z + mw' + nw'')$ for any integers m and n. We can't draw graphs of such a function (domain together with range) because the domain has two dimensions, as does the range, making a total of four. We can draw the domain, though, and mark the pattern of repetition:

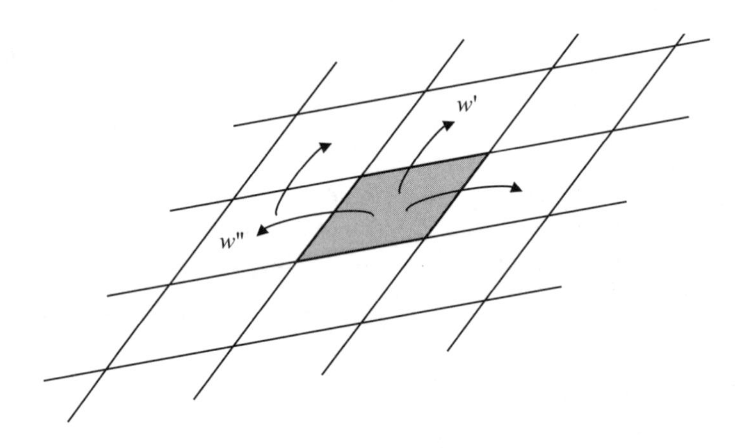

The shaded area is repeated all over the complex plane. A region that can be copied and used to tile the whole plane is called a "fundamental region." Just as the sine function "lives" on a circle and we walk around the circle to get the whole function, a similar thing can be done here. We can cut out one fundamental region—a parallelogram—and glue the edges to each other where the function starts to repeat, first getting a cylinder for w'' glued to itself and then a torus (donut) when w' is glued:

*This integral was of the form: $\int \frac{dx}{\sqrt{(1-x^2)(1-k^2x^2)}}$

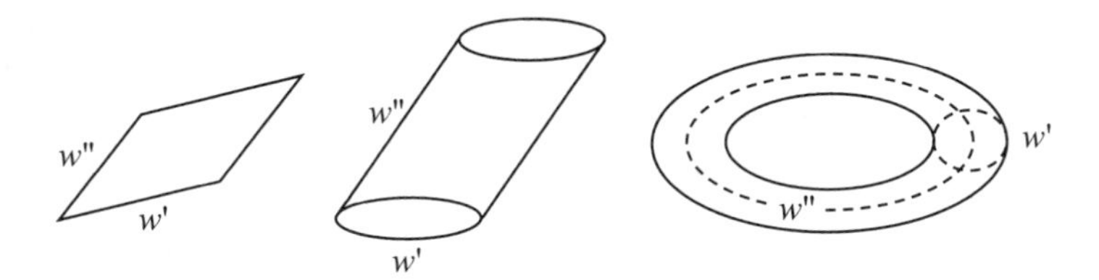

We go around one dotted circle (through the hole of the donut) and the function repeats with each revolution. We go around the other dotted circle (walking the top surface of the donut) and the function repeats. Riemann noticed this fact about elliptic functions in his epochal thesis of 1851, which opened up the full range of complex function theory for the first time.

One of the central, remarkable facts about complex analysis is that if we have a little piece of an analytic function we can continue it, building a function uniquely determined in all its particulars by the scrap we start with. As we continue, the function (and its domain of definition) grows to its own pattern, as its own thing, and we are inclined to think at first that we are wandering in the complex plane, because that is where analytic functions originally lived. We wander on our creation in what we, by local landmarks, think to be a circle and assume we are back to where we started. However, the function has a new value. This is curious and difficult, because functions must give only one value for a given input. Riemann's solution was to say we don't really return to where we would calculate we are, but are instead wandering on a more complex Riemannian surface that is the natural home of the function we are investigating. Sometimes it has interconnected leaves, a bit like a screw that we are winding up or down. One of the main tasks of the theory of functions from this time on was to determine the shape and nature of the surface on which a given analytic function of a complex variable lives. The elliptic functions, which were the clue to all this, repeat so reliably that with them all we need is a torus.

Poincaré was blissfully unaware of Riemann's ideas about surfaces when he began to work on automorphic functions almost thirty years later. In the words of Jean Dieudonné:

> Poincaré's ignorance of the mathematical literature, when he started his researches, is almost unbelievable. He hardly knew anything on the subject beyond Hermite's work on the modular functions: he certainly had never read Riemann, and by his own account had not even heard of the "Dirichlet principle," which he was to use in such imaginative fashion a few years later.[28]

Freudenthal gives a more detailed account of Poincaré's "*ignorance merveilleuse*" in his piece in the volume celebrating Poincaré's centennial. (This volume also contains praises of Poincaré's patriotism, a photograph of his family home, a reproduction of early exam results, and a facsimile of his handwriting.) Poincaré himself says laconically, "I was then very ignorant."

The modular functions Poincaré had encountered in Hermite's work "repeat" in a more complex way than elliptic functions under a group of nonlinear transformations of the form $z \rightarrow az + b/cz + d$ where a, b, c, and d are integers and $ad - bc = 1$. Fuchs's 1880 paper precipitated Poincaré's work on automorphic functions by leading him to investigate questions about a type of differential equation and the functions that come from it. Assuming they existed, these functions would remain invariant under discrete groups of transformations of the form $az + b/cz + d$ with a, b, c, and d allowed to be real numbers. In a discrete group no transformation (be it a member of the group or not) can be arbitrarily well approximated by other members of the group. Just as in the case of elliptic functions, where there is a fundamental region that, repeated, covers a domain, the groups that Poincaré was thinking about in his work on automorphic functions take a more complicated polygonal region (a fundamental region) and cover a domain, which for automorphic functions is the upper half of the complex plane.

Poincaré came at this from the direction of equations and algebraic transformations, so he did not have the clear drawings we have today, in a way analogous to (but far more complicated than) not knowing how the sine function relates to a circle. This was early 1880s work. One may be led to think of crystallographic groups, but Hilbert referred to Poincaré's work in setting up the eighteenth problem.

As we saw in the passage quoted, after drinking too much coffee Poincaré saw how to construct functions that remain invariant under a group of these transformations. It was in the air that a discrete group of this type has a fundamental region that is reproduced in a distorted way across the upper half-plane. As he put his foot on the step to the omnibus in Coutances, it came to him that these algebraic transformations were of the same form as transformations that appear in non-euclidean geometry. Here is a drawing of this for the modular group:

Tessellation of the modular group. A dark and a light region together make up a fundamental region.

He realized that this picture (and any more complicated picture of a Fuchsian group's division of the upper half-plane) was a distortion of a whole non-euclidean plane and the fundamental regions were exact copies of each other. It is particularly amazing that Poincaré was able to "see" this coming from the direction of the equations.

With the insight that non-euclidean geometry provided, Poincaré proved a wide range of results. He showed that if one starts with a polygon that has appropriate geometrical properties, one can then find a Fuchsian group that generates it. He discovered a classification system and explored some examples in detail. He observed that the edges of a fundamental region are naturally connected to each other by the action of the generating group. Therefore he imagined cutting out the fundamental region and gluing the edges together, just as the edges of the parallelograms are glued together in the case of elliptic functions to get a torus. He quickly observed that the resulting two-dimensional surfaces were topologically complex and could not exist embedded (without distortion) in normal three-dimensional space. These more complicated surfaces are the natural home for automorphic functions, just as elliptic functions live on a torus. Poincaré proved that if one starts with a compact fundamental region one obtains a surface that is characterized topologically by its genus—the number of holes it has. He had rediscovered Riemannian surfaces. By 1881 Poincaré was corresponding with Klein, who had been working on this problem also, and Klein informed him of the existing literature, so he didn't have to reproduce all of Riemann's work. There was a rivalry between Poincaré and Klein in this period, felt more strongly by Klein.

Poincaré published his work on discrete groups in 1882 in *Acta Mathematica* in a substantial setting of the stage. In his second *Acta* paper, he showed how to construct an automorphic function for any discrete group. This paper is a virtuoso performance of classical analysis. Then came a third *Acta* paper exploring a generalization of Fuchsian groups that Poincaré named in honor of Klein. (When Poincaré honored Fuchs by calling automorphic functions Fuchsian, Fuchs hadn't worked on them. Now, Klein had not yet worked on Kleinian groups.) A Kleinian group is like a Fuchsian group except that a, b, c, and d are now allowed to be complex numbers instead of real numbers. This has large consequences, and Stillwell put it, writing in 1985:

> A Kleinian group action on the plane turns out to be just the shadow of a discontinuous group of three-dimensional non-euclidean motions. . . . the difference between Fuchsian and Kleinian can be understood as the difference between two and three dimensions. This is of course a vast difference, and the theory of Kleinian groups after Poincaré had to wait for further progress in three-dimensional geometry and topology, much of it coming only recently.[29]

Here is part of a picture of the limit set of a Kleinian group drawn with the aid of a computer. (The limit set is the set of points that are approached by repeated actions of the group on the points in the domain.)

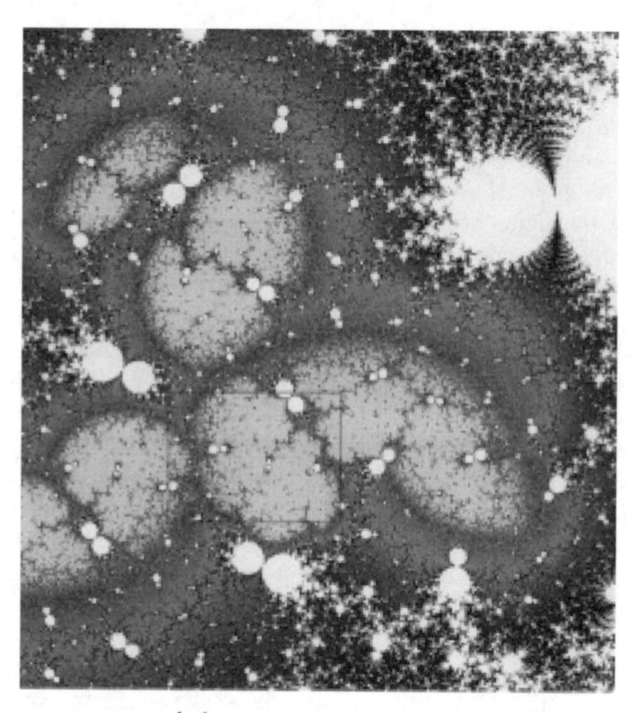

Limit Set of Kleinian Group (courtesy of Peter Liepa).

Poincaré's fourth *Acta* paper on automorphic functions is the most open-ended, suggestive, and least rigorous. In a paper like this one, and there were many others, one can get an inkling of how deeply Poincaré was looking into the future of mathematics, as if he were building a ship in a bottle through a wormhole to the future. Part of this paper is about uniformization by automorphic functions, the subject of Hilbert's twenty-second problem.

During the same time period as the papers on Fuchsian functions, a long paper (also in four installments), "Memoir on the curves defined by a differential equation," was conceived, written, and guided to publication. Many of the ideas were already present in Poincaré's thesis. Darboux informs us that one of the reasons Poincaré was able to write so fast was that he never revised. (Hilbert, on his visit to Paris as a young man, expressed reservations about Poincaré's work partly for this reason.) Dieudonné says:

> The most extraordinary production of Poincaré, also dating from his prodigious period of creativity (1880–1883) (reminding us of Gauss's *Tagebuch* of 1797–1801), is the qualitative theory of differential equations. It is one of the few examples of a mathematical theory that sprang apparently from nowhere and that almost immediately reached perfection in the hands of its creator. Everything was new in the first two of the four big papers that Poincaré published on the subject between 1880 and 1886.[30]

A differential equation relates the rate of change of something (e.g., physical position) to something else (e.g., force applied). The most famous of all differential equations is Newton's $F = ma$, force equals mass times acceleration. The force in this equation is specified in some way, the mass is the mass of the body being pushed, and acceleration is the rate at which the velocity of the body being pushed is changing. In turn, velocity is the rate at which the position of the particle is changing. The goal is to move from this equation, which involves acceleration and force in the moment, to a description of the movement of the body as time unfolds. We can write differential equations for many things—fluid flow, the orbits of the planets, the propagation of electromagnetism, the flow of heat, or even the growth of a population of fleas. The equations for the sun, earth, and moon cannot be solved exactly, and those for more complex situations are even more difficult. The central mathematical tool of science, differential equations have been studied and studied hard. They were clearly the center of Poincaré's interest in mathematics. Even when he does epoch-making work in geometry or topology, we find he was usually led there by a differential equation. Poincaré's first paper was on differential equations, his last paper was on differential equations, and he wrote at least one paper on differential equations just about every year of his mathematical life (as Dieudonné observed). His work on celestial mechanics is about differential equations. Poincaré's universality as a mathematician did not lead him to differential equations. Rather, it could be argued, his interest in differential equations flowered into universality.

The problem with differential equations is that most of them are too difficult to solve. When one attempts to calculate the length of a curve, or the area bounded by a curve, at first everything works wonderfully (as it does in a first-year calculus class). Investigating a circle magically leads to the trigonometric functions (which were already available), and they in turn supply the number π. However, as we have seen, the system already breaks down when one attempts the length of a part of an ellipse.

In 1880 a few elementary differential equations were solvable exactly, and there were some existence theorems that asserted the existence of solutions within a region but did not specify them. There were also procedures, usually of questionable validity, that allowed one to grind out an approximate solution in a given area using a series expansion of some kind. Some of these were clever and worked well in practice. Today one can buy software that will "solve" differential equations on a personal computer. These programs often make use of these old methods.

Poincaré set out to see what he could figure out about the global solutions of a general type of nonlinear differential equation, specifically, in Poincaré's terms, $dx/X = dy/Y$, where X and Y are polynomials and only real numbers are allowed. He started out by looking at the "critical" points of the equation, that is, when X and Y were both zero and the equation lost its meaning (no matter how one wrote it) because it involved dividing by zero. Cauchy and Charles Auguste Albert Briot (1817–1882) had already looked at these points and classified them as "nodes," "saddles," "spiral points," and "centers." However, Poincaré addressed the problem in wholly original terms. In order to get a picture of curves that ran off to infinity, he projected the plane onto a half-sphere sitting above it, then flattened the sphere so that everything could be drawn on a disc. Here we have one of those jumps so striking in the work of Poincaré. Suddenly the space one is working in is non-euclidean, and geometry and topology jump into a problem on differential equations. One is discovering how to fit things in and have the ends match up.

Poincaré generalizes the problem; the habitat for the curves being sought is a surface determined by a polynomial. Early topological invariants like genus and Betti numbers come into play. One way of viewing the problem when phrased in this way, and how geometry and topology get into it, is by viewing the vector field that is defined by these equations as hair growing on the surface that has to be combed and to lie flat. The vectors come out of every point on the surface (as the hair on one's head does not). The nodes and spirals, etc., represent different kinds of cowlicks. They are crucial in the behavior of any curve (or solution of the differential equation). In general, it is impossible to comb a hairy surface without getting some cowlicks. An exception is a torus.

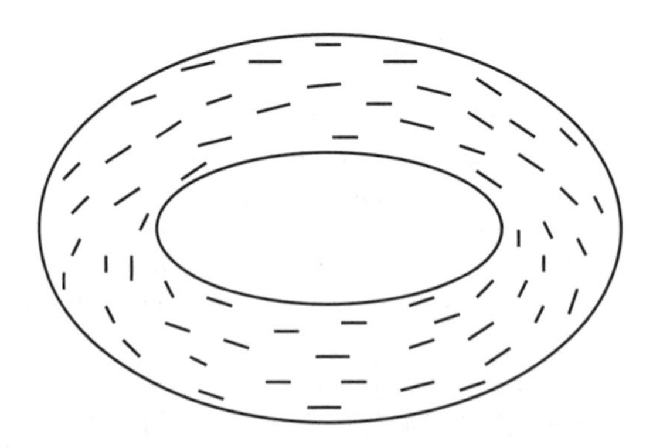

The hair is combed round and round in the same direction.

Types of critical points would include:

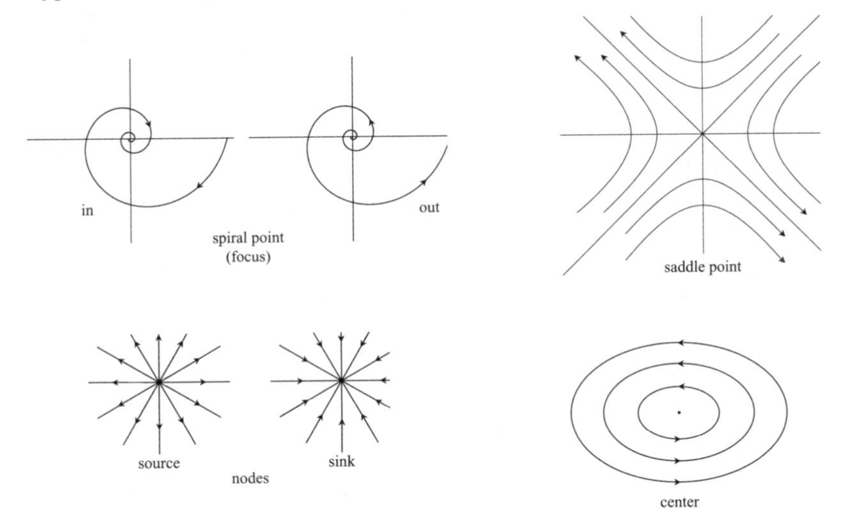

(You might get a saddle if you aimed two streams of water at each other head-on.)

Poincaré introduces the genus, or number of holes in the surface, and then proves a remarkable result:

$$\text{Genus} = (\text{Saddles} - \text{Spiral Points} - \text{Nodes} + 2)/2$$

He showed that centers, though common with very simple equations, occur only in exceptional cases when the equations are more complex. Something like what is now known as the ergodic hypothesis makes an appearance, though Poincaré was unable to prove it. If one looks at some of these illustrations Poincaré included after having followed the popularization of chaos theory and the many diagrams that have a similar feel, one is haunted by how deep Poincaré's vision into the future was.

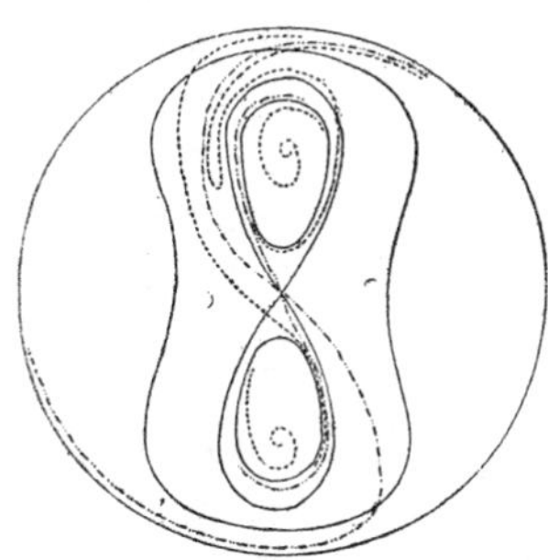

A qualitative picture of the solution to a differential equation. From *Ouevres de Henri Poincaré* (Vol. 2, p. 71).

In the last section of this paper, published in 1886, Poincaré looked at the problem in higher dimensions. He had already started looking at the three-body problem and the problems of celestial mechanics.

Poincaré was working on the qualitative theory of differential equations. The second half of Hilbert's sixteenth problem is a question in this area. The nodes (sources and sinks), spiral points, and saddles are all defined in terms of how solutions approach a singularity—a singularity, in this context, is where the function is not defined. Farther out in the region of a singularity one might find a limit cycle. A limit cycle is a periodic solution—a closed orbit—that has an "annulus-like" neighborhood in which there are no other periodic solutions. One example could be drawn:

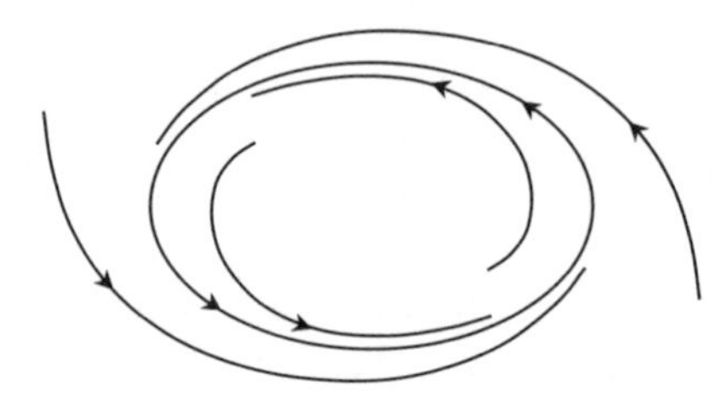

These are a bit like eddies. This particular limit cycle attracts nearby solutions, which approach it asymptotically. Hilbert simply wants to know how many limit cycles are possible for a given equation. (This is one of the unsolved problems.)

In 1923 the French mathematician Henri Dulac "proved" the number of limit cycles for these equations was finite. In 1957 Petrovsky and E. M. Landis (1921–1997) presented an estimate of how large this number was. The "solution" to the sixteenth problem Petrovsky and Landis published in 1957 was heralded. However, a gap in their argument was soon discovered, resulting in a retraction some years later. The correctness of the bound or limit Petrovsky and Landis found remained in a kind of limbo. In 1980 Shi Songling published a counterexample. Then in 1981 Yulij S. Ilyashenko found a gap in Dulac's proof, so that not only was the number of possible limit cycles not known, but it wasn't even certain that the number was finite. Finally, Ilyashenko (1991) and Jean Ecalle (1992) independently, and using different methods, gave rigorous proofs for Dulac's finiteness result, employing radically more sophisticated methods than Dulac had at his service.

This question is seen as important today. Smale included it on his list of problems in 1998, which implies some optimism that it might be solved. If we could better understand the second half of the sixteenth problem, it would be a step in the direction of better understanding the weather or other complex systems that can fall into repeating cycles. I asked Ilyashenko how things stood in late 2000. He was optimistic but didn't expect a solution soon. He said that a paper by S. Yakovenko and himself and subsequent work of V. Yu. Koloshin provide a "Hilbert type" bound to a special case of the sixteenth problem. This is significant because it is the first time any bound of this type has been found.

In 1885 King Oscar of Sweden proposed an international competition, with the goal of finding a convergent series solution to the *n*-body problem. With a good solution one might prove the stability of the solar system. This was a perfect opportunity for Poincaré. Mathematicians including Pierre-Simon de Laplace (1749–1827), Lagrange, Siméon-Denis Poisson (1781–1840), and Dirichlet all claimed to have proved the solar system stable. Dirichlet died before he could write down his proof, and the proofs that were written down were not convincing. These efforts mostly boiled down to approximate calculations of the movements of the planets based on infinite series that might not converge and trying to extend that indefinitely into the future. These expansions, though often useful for practical problems, invariably contained, sooner or later, terms with "small denominators" that could disturb whatever orderly pattern seemed to be unfolding in the lower-order approximations. However, these methods had been for years producing reliable calculations of where the planets should be.

Poincaré asked a great question: Why do these defective methods often work as well as they do? His answer was that many of the series involved were asymptotic. He showed that these series generally do not converge to any one answer, but they diverge so slowly that if we use them as approximations, the error, though it doesn't disappear in absolute terms, grows ever smaller on a percentage basis. Further, the percentage of error often becomes small after a small number of calculations. Poincaré—and independently Jan Stieltjes (1856–1894)—inaugurated the general theory of asymptotic series in 1886 papers.

Using a brilliant array of classical techniques with his clarifying theory of asymptotic series, as well as his new qualitative picture, Poincaré attacked the three-body and *n*-body problems. He proved rigorously that past "proofs" of stability in celestial mechanics were invalid. The series did not uniformly converge. Yet they were good mathematics. The three-body problem is complex, has many versions of exactly how it is stated, and, at best, starts from an approximation (point masses in Newton's equation). This qualification of the word "proof" is essential to understanding what has happened in the area. Poincaré's writings began a new era in celestial mechanics in which the statements of theorems and their proofs could be judged by more rigorous standards. In his first paper in this area in 1883, predating the prize, Poincaré had shown that there were an infinite number of solutions to a restricted three-body problem of a type G. W. Hill (1838–1914) discovered in 1878. Neither Hill nor Poincaré actually proved the convergence of their answers.[31] Poincaré returned to the restricted three-body problem and proved this time that if the mass of the second body was small enough, there were periodic solutions. Each member of the growing array of special solutions to the restricted three-body problem (including Poincaré's) was not proved stable if jostled in any tiny way.

Though not a solution to the question about the stability of the solar system, Poincaré's progress understanding the equations involved was substantial and he won King Oscar's Prize in 1889. Poincaré proved the existence of "homoclinic" points, points of stability that can be approached from more than one distinct direction and left in more than one direction. Disturbed in small ways, the system becomes wildly unstable—in modern terms, chaotic. The picture of the path of solution becomes a homoclinic tangle. Poincaré said he wouldn't try to draw the picture.

It has recently been found by scholars that the 273-page paper published as the prizewinner wasn't, in fact, exactly the one turned in. After the journal was already in print, errors were discovered. Poincaré went back and found that the error was deeper than he had initially thought. He ended up paying to have the issue reprinted, which exceeded the prize money he won. But the end result was good—the paper finally published was correct and Poincaré had discovered what we now call chaos theory in the course of his own somewhat chaotic process.

By the mid-1890s Poincaré had outrun the mathematical vocabulary of his time. For example, in attempting to formulate the question of stability, he was driven to talk about what we would now call measure theory. He ended up using the vocabulary of probability, which was scarcely more clearly grounded than the as-yet-nonexistent measure theory. Poincaré did not neglect calculation in his assault on celestial mechanics. His treatment of the classical methods of approximation in celestial mechanics is still being recommended, for example by V. I. Arnold in *Mathematical Methods of Classical Mechanics* (Russian 1974, English 1978). Poincaré wrote close to 100 papers on classical computational topics in the theory of the solar system.

Poincaré worked on many other physical problems, and he continued to be unafraid of difficult nonlinear problems. One of his most famous results was the stability of a rotating pear-shaped body of fluid held together only by gravity—this suggests the possibility of a pear-shaped star. He wrote and thought about the partial differential equations of physics—heat, fluid flow, elasticity, etc., which were at a very simple level of development in his time. Physicists today are likely to encounter methods associated with Poincaré in these areas. Writing about chaos theory, L. Galgani says:

> Obviously, no revolution is actually fully a revolution. Indeed, in a sense almost everything can already be found in Poincaré; the homoclinic phenomenon leading to chaos is explicitly described; the possibility of the existence of invariant tori with good frequencies satisfying suitable diophantine conditions is explicitly considered and it is shown that one cannot exclude it. . . .
>
> In this connection, it is of interest to remark that possibly the present state of awareness can be understood if one takes into account the role . . . [of high speed numerical computers]. Indeed the

> speculations of Poincaré on homoclinic points were probably too advanced, even for the community of pure mathematicians, and only after the exhibition of their effects by Hénon [1964] did people become acquainted with them.[32]

Topology particularly begged for development. Poincaré said, "Every problem I had attacked led me to *Analysis situs*."[33] The idea and words for an *analysis situs* or *geometria situs* go back at least as far as Leibniz, who wanted an inherently geometrical analysis of space that was as powerful as Descartes's analytic geometry, which used algebra. Despite Leibniz's early call for order, as of 1894 there was no organized discipline of *analysis situs* (or topology), just isolated and mostly intuitively derived special results. Riemann had introduced fundamental topological concepts, such as the degree of connectedness (how many cuts are required to cut a surface into separate, simple pieces) and the idea of a manifold (a surface that is locally similar to euclidean space). Enrico Betti (1823–1892) generalized the idea of connectivity to higher dimensions, but without a general foundation. Johann Listing (1806–1882), Augustus Ferdinand Möbius (1798–1868), and Klein discovered odd, one-sided surfaces. Poincaré had spent a major part of his career already dealing with topological questions and had had enough of this uncertainty. In 1895 he created algebraic topology—though it would not be commonly called topology until the 1920s and became heavily algebraic only with time.

As Poincaré himself described the subject:

> Imagine any sort of model and a copy of it done by an awkward artist: the proportions are altered, lines drawn by a trembling hand are subject to excessive deviation and go off in unexpected directions. From the point of view of metric or even projective geometry these figures are not equivalent but they appear as such from the point of view of geometry of position [that is, topology].[34]

Roughly, topology attempts to answer the question of which shapes are the same if we allow stretching but no tearing. Was his youthful effort a drawing of a horse? For example, the sphere, the torus, and the Klein bottle (these are surfaces, not solid objects):

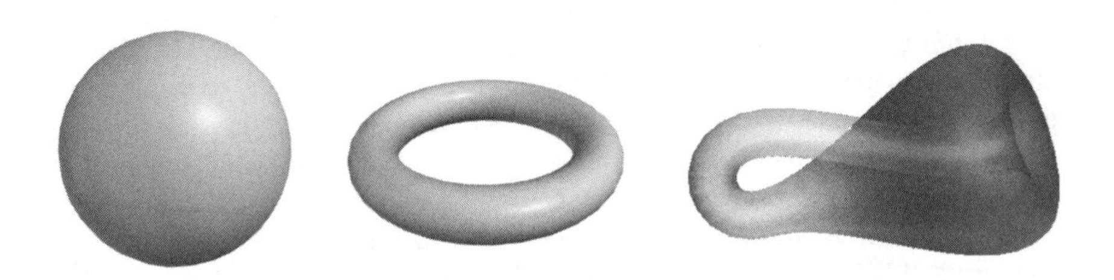

How are they different? The torus has a hole, while the sphere doesn't. That distinguishes them. Both the sphere and the torus have an inside and an outside to their surfaces. The Klein bottle doesn't. These ideas were known. The question facing Poincaré was how these insights could be made precise and generalized. He came up with two separate approaches—what we would today call homology theory and the first homotopy group (also known as the fundamental group). Dieudonné wrote: "It has been rightly said that until the discovery of the higher homotopy groups in 1933, the development of algebraic topology was entirely based on Poincaré's ideas and techniques."[35]

Poincaré had married Louise Poulain the year he was called to Paris, 1881. She was the great granddaughter of Étienne Geoffroy-Saint-Hilaire, who was the *antagoniste* of Cuvier, a naturalist famous for pathbreaking work on fossils and comparative anatomy. Despite their scientific disagreements, Saint-Hilaire was responsible for Cuvier's obtaining a position at the national museum. His descendants continued the association with the museum, and Darboux reports that Poincaré's mother-in-law actually lived at the museum for a long time. Poincaré and his wife had three girls and a boy. Much of Poincaré's extended family was in Paris either permanently or visiting. He saw much of his sister and her family, the Boutrouxs. Photographs of family gatherings or vacations are full of children. One imagines that the type of life Poincaré lived as a child continued into his adulthood. His fame and genius coupled with his mild, distracted disposition would have allowed him to be the focus as much as he wanted to be.

During the 1890s and the first decade of the 1900s Poincaré became increasingly lionized. He traveled widely and was invariably a featured speaker at any conference he attended, though Hilbert upstaged him a bit with his list of problems at the 1900 Paris Congress. He traveled to the U.S. more than once. Darboux says he was particularly excited by a trip to St. Louis, Missouri, in 1903. Toward the end of his life his opinion was sought on every conceivable subject, from politics to the effect of comets on the weather. In the words of Dieudonné, offering popular explanations of mathematics and science was "a chore for which he does not seem to have shown great reluctance."[36] He received honors and titles and served on committees for railroads, the post and telegraph, the national observatory, and educational institutions. Three times, in 1899, 1909, and 1910, he was president of the *Bureau des Longitudes*. He was a member of at least twenty-one foreign academies and many foreign societies, received at least nine honorary doctorates,[37] and was active in the cause of international bibliography. Poincaré never actually ceased to be a mining engineer, but remained under the control of the department of public works; he was just on assignment at the universities. In fact, in 1893 he became chief engineer of mines and in 1910 inspector general. Of course, he won many prizes.

Poincaré wrote three popular books on mathematics and science: *Science and Hypothesis* (1902), *The Value of Science* (1905), and *Science and Method* (1908), from which we quoted his account of his discovery of Fuchsian functions. The books mark a high point in writing about science and help explain his appointment to the French Academy as an arbiter of the language. Scholars in the history of science sometimes refer to these books as primary sources, in particular for the comments they contain on time, the aether, geometry, and the ideas behind chaos theory. According to Abraham Pais, Einstein had read parts of the first in the years preceding 1905 and the birth of the theory of special relativity.

There has been some debate whether Poincaré should receive equal credit with Einstein for the creation of special relativity, based on a paper he published in 1905 as well as for his clear, early statement of some of the basic principles in the popular books and other places. It is true that no one was more conversant with the mathematical theory of electromagnetism than Poincaré. The difficulties that require the special theory were known to him. The mathematics of the solution must have been child's play. But what did it mean physically? Poincaré had all the pieces, and in 1905 he did publish a theory of relativity that was similar to Einstein's. However, Pais makes a convincing argument that Poincaré didn't, in 1905 or ever, fully understand the way special relativity flows from two simple hypotheses, that light always travels at a constant speed and that the laws of physics remain the same in any inertial frame of reference. As late as 1909, in one of his lectures in Göttingen, Poincaré was still referring to a third hypothesis—length contraction, which follows from the other two. It therefore seems an exaggeration to give Poincaré equal credit with Einstein for special relativity.

Poincaré continuously wrote papers on ideas that occurred to him—on number theory, algebra, and myriad special topics from analysis. His work on abelian functions and algebraic geometry occupies almost as much space in his complete works as automorphic functions. However, he was not the clear leading figure in algebraic geometry in this period. (Since this is Poincaré, in another fifty years we might find out this is his deepest work.) His most important paper on number theory (1901), Dieudonné says, "was the first paper on what we now call 'algebraic geometry over the field of rationals.'"[38] This is ground zero for "arithmetic algebraic geometry."[39] Mordell's 1921 theorem and Siegel's 1929 results on diophantine equations start here. The mathematician Isabella G. Bashmakova notes: "It seems that Poincaré was completely unaware of the work of his predecessors bearing on the arithmetic of algebraic curves. He knew of Diophantus' procedures."[40]

The last of Poincaré's series of papers on *analysis situs*, in 1904, marks an end to the period of expansive creation that began on the omnibus in 1880. However, nothing is so cut and dried with Poincaré. The key in much of his last work is that by this time Poincaré and others had made progress toward creating techniques and vocabularies that allowed questions he might have groped after earlier to be addressed more precisely.

Having invented algebraic topology, Poincaré could now return to the topic of uniformization of surfaces defined in terms of algebraic and analytical equations. His original 1884 attempt at a proof hadn't been complete. Hilbert (though he seems to think Poincaré had earlier been more successful than he really had) called completing this project his twenty-second problem. Poincaré probably thought of it as unfinished business.

We can now express the problem simply, if we bear in mind that to uniformize is to parameterize a surface. The surfaces to be uniformized might be generated by polynomial equations or in some other way. Hilbert wants to prove there exists a parameterization using automorphic functions of any surface defined by "any analytic non-algebraic relations whatever." In particular, he wants to be sure "all regular points of the given analytic field are actually reached and represented." It is easy to parameterize a piece of such a surface, and Poincaré had already done this in the *Acta* papers. He proved the full result in 1907 using a process of exhaustion and a universal covering space (from algebraic topology), and thereby solved "Hilbert's" twenty-second problem at the age of fifty-three. Koebe solved the same problem independently and pursued his solution in its consequences with more thoroughness the same year.

Why is this seemingly technical result useful? To understand what a parameter is, consider the equation of a circle: $x^2 + y^2 = 1$ (We don't include complex numbers in this simple example). It is easy to specify both x and y as functions of a third variable, t (the parameter):

$$x = \frac{2t}{1+t^2} \qquad y = \frac{1-t^2}{1+t^2}$$

One of the central goals of number theory is to find rational solutions to polynomial equations. Once we have the parametric representation above, it is easy to see how to find a lot of rational solutions to the equation for a circle. Any rational number substituted for t yields a rational solution. Success in the twenty-second problem makes the automorphic functions a complete set of building blocks for parameterizing all of the large class of surfaces Hilbert asked about, despite the fact that they would seem to be a narrow subset of all possible analytic functions. Remember, it had been quite an achievement for Poincaré to prove the automorphic functions even existed. Now they could stand in for the functions using t in our simple example and provide a potential handhold on myriad new problems. Once again, Hilbert was concerned with stocking the tool chest of mathematics.

Poincaré received his first intimation of mortality in 1908 when he was attending the fourth International Congress of Mathematicians in Rome. He suffered from enlargement of the prostate, which prevented his delivery of a talk on the future of mathematics. He was operated on by Italian surgeons and seemed to recover. Tobias Dantzig gives a sense of his presence in this period:

> I saw him often between the years 1906 and 1910, when I was a student at the Sorbonne. I recall above all his unusual eyes: myopic,

Jules Henri Poincaré (courtesy of the Archives of the Mathematische Forschungsinstitut Oberwolfach).

> yet luminous and penetrating. Otherwise, my memory is that of a man small in stature, stooped and ill at ease, as it were, in limb and joint. This last impression was accentuated by his manner of writing on the blackboard. . . .[41]

In 1910 and 1911 Poincaré published in two parts what is probably his deepest paper on algebraic geometry. Then his difficulties returned, and in 1911 he wrote to the editor of *Rendiconti del Circolo Matematico di Palermo* that because of his age and health he was sending an incomplete work for publication. He had returned to the three-body problem. The work had given him great difficulties, but he thought it important and wanted his ideas published. This is sometimes known as "Poincaré's last theorem." He succeeded in reducing the problem of the existence of periodic solutions of the restricted three-body problem to a topological theorem about area-preserving mappings of an annulus onto itself, subject to certain boundary conditions. If all such maps had fixed points, that is, points that were mapped onto themselves, then there were periodic solutions to the restricted three-body problem. Dieudonné says this "was probably the first example of an existence proof in analysis based on algebraic topology."[42] In 1913 a young American, George D. Birkhoff (1884–1944), proved the theorem. Many more theorems of this type have followed. It was a new way to prove the existence of and understand periodic solutions, and a direct descendent is the Moser twist theorem in KAM theory. When one looks at

Jules Henri Poincaré (courtesy of the Musée de la Poste).

the theory of mechanics today one could mistakenly think one had wandered into an algebraic topology forum.

Poincaré did not see the proof of his theorem. On July 9, 1912, he had another surgery that seemed to be successful, but eight days later he died suddenly, probably because of an embolism. Great mourning ensued. Paul Painlevé—a mathematician turned premier/president of France—called him "the living brain of the rational sciences," which, though dramatic, was not far from the truth.

How could anyone be so productive? One answer comes from Poincaré's nephew Pierre Boutroux, in a letter to Magnus Mittag-Leffler:

> He thought in the street as he went to the Sorbonne, while he was attending some scientific meeting or while he was taking one of his habitual grand walks after lunch. He thought in his antechamber, or in the hall of meetings at the institute, while he walked with little steps, his physiognomy tense, shaking a bunch of keys. He thought at dinner table at family get-togethers, even in the sitting rooms, interrupting himself often brusquely in the midst of a conversation, and would leave his interlocutor planted, to follow a thought that crossed his spirit. All of his work of discovery was done mentally . . . without him, most often, having to check his calculations in writing.[43]

So the answer is: Think rapidly, think all the time, and write down everything only once, when you are going to publish it. Start with miraculous natural talent. One is drawn to compare Poincaré with Hilbert. Poincaré's relatively early death and Hilbert's long life tend to exaggerate their seven-year age difference. Both were universal mathematicians under E. T. Bell's definition. Both became mathematically productive relatively late—in their mid-twenties—and neither rushed his schooling. Each was viewed as the other's only real rival. Yet they were very different.

Hilbert was disciplined in his writing and, though he didn't always enjoy it, was systematic in his study of the literature, retaining assistants to do much of the slogging for him. He worked on one subject at a time, to the exclusion of others. Algebra, number theory, geometry, analysis (including analytic number theory), mathematical physics, foundations. Once done with a subject he seldom went back. In contrast, Poincaré wrote rapidly and seldom revised. When starting his career, he was almost unbelievably ignorant of the existing literature. He eventually came to know quite a lot, but less through systematic research than because he had an active mind, and once he glanced at something he understood and remembered it. Not only did Poincaré not work on only one subject at a time, the subjects he worked on were often hybrids.

We can see some of the point of view Poincaré wanted to leave in his paper "*L'avenir des mathématiques*" ("The Future of Mathematics"). This is the talk he didn't get to deliver at the 1908 International Congress in Rome because of the first flare-up of his illness—perhaps an answer to Hilbert's "Mathematical Problems." Its qualitative character stands out. Poincaré generally doesn't pose specific problems, but talks about directions. He discusses the importance of the invention of new words and the importance of words in mathematics. The word "energy" is an example, and he calls it prodigiously fecund, comparing it to a machine that economizes effort. He talks in terms of the elegance of methods of thought and of "the art of giving the same name to different things"[44] when what we say about one thing is equally true of the other. Where a mathematician less sensitive to language might think of dual theories, Poincaré uses the word "analogy, " and *analogie* or *analogue* appear again and again in the sixteen-page paper. As he thought of the equations needed for automorphic functions and stepped onto the bus in Coutances, he saw non-euclidean geometry. As he thought about finding the sometimes rational solutions to polynomial equations, he saw the separate rational solutions as part of a picture—an arithmetic algebraic geometry. When he thought of differential equations he saw one picture after another. In his work on dynamics, the three-body problem, and nonlinear differential equations, the specific results, important as they are, are probably not as important as the fact that Poincaré recast the mathematical language in which these problems are phrased. The coordinates of where something is located at a given moment supply the same information that a still photograph would and don't tell us the direction and speed things are going. Poincaré included velocities as coordinates of location in what is now called "phase space," and the still photo changes into a picture of a more complicated space that shows the system as flowing from one state into the next, into the next. A specific solution is a clean line or curve in this picture. The idea of hair and cowlicks, the qualitative theory, makes an appearance. This change of point of view paid huge dividends. Poincaré didn't originate all of the ideas of doing "mechanics on a differentiable manifold"—a string new of words—but he transformed bits and pieces into a systematic vision that he used to investigate hitherto intractable problems. His ideas are still being explored with great energy.

We have returned many times to the controversy over Cantor's set theory. Earlier, I glided over Poincaré's intuitionism and said that whatever reservations he may have had about Hilbert's program and set theory, Poincaré was hardly a partisan of a restricted mathematics. In "The Future of Mathematics" he puts his point of view concisely. He seems to be reacting almost instinctively to the limits of merely mechanical language and purely formal systems. For him a machine has levers and grease and pulleys and will let him ride down into a mine. In "The Future of Mathematics" he admits that certain results arising from Cantor's theory are useful. (After all, as we have seen, the words of set theory are a remarkably efficient machine for saying things about mathematics.) But then he says, "I think for my own account, and I am not alone, that the important point is to not introduce things that one cannot define completely in a finite number of words."[45] This almost seems to jump off the page and conjure Paul Cohen's minimal model, in which "every element can be 'named.' " In this model the continuum hypothesis is true, but Cohen then uses it to construct a second model in which the continuum hypothesis is false—proving it independent. Poincaré would have undoubtedly have seen Cohen's method as a more precise result, but in his time, more than fifty years earlier, his intuition told him that these questions were imprecise, and he had some intuition why.

This book celebrates Hilbert's problems. In his 1951 article, "A Half-Century of Mathematics," Hermann Weyl wrote:

> David Hilbert . . . formulated twenty-three unsolved problems which he expected to play an important role in the development of mathematics during the next era. How much better he predicted the future of mathematics than any politician foresaw the gifts of war and terror that the new century was about to lavish upon mankind! We mathematicians have often measured our progress by checking which of Hilbert's questions had been settled in the meantime. It would be tempting to use his list as a guide for a survey like the one attempted here.[46]

This chapter on Poincaré suggests some ways in which Hilbert's problems are incomplete. To cover more of the century, particularly the second half, we should discuss Poincaré's problems, though he didn't leave them in a tidy list. Here are nine such problems touched on in the previous pages and unsolved at the time of Poincaré's death. One could easily find others.

1. Describe the limit sets of Kleinian groups.

2. Investigate what are now called Teichmüller spaces—first suggested by Poincaré in the fourth *Acta* paper on automorphic functions, 1884.

3. Solve the second half of what came to be called Hilbert's sixteenth problem—and more generally extend Poincaré's qualitative methods in the study of differential equations.

4. Achieve further results on of the problem of small denominators in the n-body problem. (KAM theory did this, 1954–1963.)

5. What is the significance of homoclinic points and homoclinic tangles in the n-body problem? (Chaos theory addressed this in the second half of twentieth century.)

6. Put Poincaré's work in algebraic topology on a firm foundation–homotopy and homology. Explore the immediate consequences of these ideas. (Accomplished by many, 1900–1940.)

7. Prove or disprove the Poincaré conjecture for dimensions three and greater—a three-dimensional manifold is essentially the same as (homeomorphic to) a three-dimensional sphere if the manifold is closed, smooth, simply connected, and has the same homology. (Solved five or more dimensions, Smale, 1960–61; four dimensions, Michael Freedman, 1982; three dimensions, unsolved as of 2000.)

8. Explore the ideas in Poincaré's 1901 paper about algebraic geometry over the field of rational numbers to prove results about diophantine equations. This is now called arithmetic algebraic geometry. (Central to Wiles's 1994 proof of Fermat's last theorem.)

9. Prove Poincaré's last theorem—the theorem Poincaré published without proof just before he died. (George Birkhoff, 1913.) More generally, extend this topological method in the study of dynamics and stability. (Work is ongoing.)

Schools Amid Turbulence

The Thirteenth Problem

Andrei Nikolaevich Kolmogorov solved Hilbert's thirteenth problem, somewhat to his own surprise, at the age of fifty-four. He had recently made progress on one of Poincaré's favorite subjects, dynamical systems and the stability of the solar system. Always interested in irregular functions and the foundations of function theory, he decided to give a seminar, saying in his introductory lecture, "I have formulated the thirteenth Hilbert problem as a very remote and hardly realistic target."[1] Kolmogorov had already solved part of the sixth problem by successfully axiomatizing probability theory. Two years after he approached the thirteenth problem (the seminar was in 1955), it was solved, with the final steps taken by his student, V. I. Arnold. Kolmogorov then doubled back and proved a stronger version of the central theorem.

Kolmogorov's obituary by V. M. Tikhomirov in *Russian Mathematical Surveys (Uspehki Matematicheskikh Nauk)* begins: "On 20 October 1987 there ended the life of one of the greatest mathematicians in the history of the human race. . . . "[2] In the same issue Arnold says, "Kolmogorov, Poincaré, Gauss, Euler, Newton: five generations separate us from the origins of our science."[3]

A. N. Kolmogorov was born on April 25, 1903, when the Russians were still using the Julian calendar (on that calendar the date was April 12). His mother, Mariya Yakovlevna Kolmogorova, was the daughter of Yakov Stepanovich Kolmogorov, a member of the nobility who owned several large estates in the Yaroslavl and Vladimir regions, north and northeast of Moscow respectively. Describing Mariya and two of her sisters, the mathematician A. N. Shiryaev writes that they "were independent women with high social ideals."[4] Andrei's mother left home "against the wishes of her parents"[5] with Andrei's father, an agronomist and writer, Nikolai Matveevich Kataev. They neglected to marry. Kataev had been exiled to Yaroslavl, presumably for revolutionary activity, and that is how he came to meet Kolmogorova, according to D. G. Kendall, in his obituary on Kolmogorov for the Royal Society of London. She first gave birth to a girl, who soon died, then died herself giving birth to Andrei. The couple was in Tambov, a town about halfway back home from the Crimea. Kolmogorov was named after Prince Andrei Nikolaevich Bolkonski, the hero of *War and Peace*.

Kolmogorov was officially allowed to use the family name only after the Revolution, as he was illegitimate, but his mother's family assumed responsibility for the baby. Kataev accepted this and had no active role in his son's upbringing, though he visited him regularly. (After the Revolution, Kataev became a department head in the Agricultural Ministry. He "perished"[6] or "disappeared without a trace"[7] in 1919 in the civil war that followed the Revolution.) Brought back to the family estates after his birth, Andrei grew up at Tunoshna, on the banks of the Kotorosl River, which flows into the Volga. His mother's sister, Vera Yakovlevna, took charge of rearing the boy and adopted him.

Kolmogorov's aunt, parents, and grandparents were from the century of Gogol, Turgenev, Dostoyevsky, Tolstoy, and Chekhov, a time of extreme ferment in Russia. Sofya Kovalevskaya (1850–1891), like Kolmogorov, a member of a noble family, throws an interesting light on Kolmogorov's early milieu. Her beautiful memoir, *A Russian Childhood*, published in 1889, conveys the flavor of life on a Russian country estate, from the daily routine to the heady rush of mushroom hunting that overtook everyone in the fall. It gives a striking account of a child's first contact with advanced mathematics. When Kovalevskaya's family wallpapered the estate they came out one room short. Sending away for paper was prohibitive, so they decided to use whatever was at hand, which included a calculus textbook. The room was Sofya's nursery. In her words:

> These sheets, all speckled over with strange, unintelligible formulas, soon attracted my attention. I remember as a child standing for hours on end in front of this mysterious wall, trying to figure out at least some isolated sentences and to find the sequence in which the sheets should follow one another. From this protracted daily contemplation, the outer appearance of many of these formulas imprinted themselves in my memory: indeed, their very text left a deep trace in my brain, although they were incomprehensible to me while I was reading them.
>
> Many years later, when I was already fifteen, I took my first lesson in differential calculus from the eminent Petersburg professor Alexander Nikolayevich Strannolyubsky. He was amazed at the speed with which I grasped and assimilated the concepts of limit and of derivatives, "Exactly as if you knew them in advance." I recall that he expressed himself in just those words. And, as a mat-

> ter of fact, at the moment when he was explaining these concepts I suddenly had a vivid memory of all this, written on the memorable sheets of Ostrogradsky; and the concept of limit appeared to me as an old friend.[8]

The literary ferment of nineteenth-century Russia was not a distant rumor to Kovalevskaya. Her sister Anyuta wrote a story that Dostoyevsky accepted for *The Epoch*. An affair ensued, so young Sofya saw much of Dostoyevsky when the women in the family were in St. Petersburg every year for what amounted to a culture break. Kovalevskaya looked on as the power in the relationship gradually shifted from the great writer to her sister. A turning point was a scene at a party in which Anyuta humiliated Dostoyevsky. A fictionalized version of this appears in *The Idiot*, where Mishkin ends, after a tirade and a broken vase, in an epileptic seizure. Dostoyevsky himself used Sofya as a foil in his courtship of Anyuta, and Kovalevskaya, at fifteen, was infatuated with him. Kovalevskaya, who did fundamental work on the dynamics of a rotating rigid body, was consulted in the offering of King Oscar's Prize, which Poincaré won.

Tikhomirov describes Kolmogorov's childhood home:

> Photographs still exist of the grand old house with a second floor, constructed, most probably, at the beginning of the 19th century, on an old foundation, with four massive columns at the main entrance, with large windows on the first floor and terraces adjoining the house. In this house there were suites of rooms—large and small. The small rooms were named according to the color of the walls or of the curtains "green," "blue," "pink,". . . . Close by was a wing where the kitchens were. . . . Around the house and the wing were courtyards—front and back, flower gardens, a kitchen garden, a garden (so big that it could be called a park), barns, a cattle yard, a stable, a bathhouse—in a word, all those things infinitely remote from our present-day lives, and of which we now know only from books.[9]

There was a hidden press used for printing subversive material in this noble house, which was a mailing address for communications with revolutionaries living abroad. Illegal materials were hidden in Andrei's cradle when the house was searched when he was three months old. His aunt Vera was arrested on another occasion and spent some time under house arrest in St. Petersburg. Tikhomirov reports, " 'They took' her to the haymaking, where alongside the country women she combed and turned the hay (the Kolmogorov family followed the powerful ideas of Tolstoy's moral philosophy, committing man to work with the people and for his own good)."[10] It is not clear in what way haymaking would have been a real punishment for Vera, in that case. What is clearer is that Vera gave up her revolutionary activity to some extent to care for young Andrei.

Vera Yakovlevna showed Andrei the wonders of the physical world they lived in. She took him on walks through the fields and forests of the estate and talked to him about the profusion of life there—flowers, animals, herbs, trees. At night they looked at the sky and she taught him the constellations and bright stars. She read to him—Hans Christian Andersen and Selma Lagerlöf. When it came time for school, Vera and a friend, Matil'da Isadorovna Dubenskaya, the adoptive mother of Petr Savvich Kuznetsov (who ended up a famous philologist), started a school run on progressive principles. In the school journal, *Vesennie Lastochki*, "The Swallows of Spring,"[11] Andrei did the mathematical section. One of his first papers reported the discovery, at the age of five or six, that: $1 = 1^2$, $1 + 3 = 2^2$, $1 + 3 + 5 = 3^2$, $1 + 3 + 5 + 7 = 4^2$, etc. He also published an article analyzing all possible ways to tie on a button. As part of the family's Tolstoyan vision, young Andrei was required to sew on his own buttons, work in the garden, and help gather wood for the fires, though in the winter heating a house of that size north of Moscow must have involved the arrival of cartloads of wood, as well.

In 1910, when Andrei was six, Vera thought something more was needed, so they moved to Moscow and he was enrolled in the E. A. Repman Gymnasium. This was run by two women, Evgenia Al'bertovna Repman and Vera Fedorovna Fedorova, and was one of only two coeducational grammar schools in Moscow. They used experimental teaching methods. A bright child could study with older students. Any sign of talent was encouraged, rewarded, and fostered. There were no grades. All of this made the authorities suspicious, and everyone was perpetually worried that they would shut the school down. After the Revolution the school was not shut down but was taken over and renamed the Section Grade School No. 23. It was a small school and only open ten years, but in addition to Kolmogorov and Kuznetsov, it graduated two other academicians of the Soviet Academy of Sciences, a corresponding member, an "academician of the American Academy," and many other professors. In Kolmogorov's words: "The gymnasium rooms were small, each holding fifteen to twenty pupils. The teachers were enthusiasts of science, some of them being university lecturers. Our teacher of geography was involved in interesting expeditions. Many schoolchildren competed against each other in their private studies, sometimes even with the intention of shaming the less experienced teachers."[12] Kolmogorov states that he was "among" the best in mathematics, but that he was more interested in biology and history. By fourteen he was already studying higher mathematics out of the Brockhaus and Efron Encyclopedia. The articles in this encyclopedia were somewhat abbreviated, so he was forced to fill in the gaps.

At the school Andrei made friendships that would last all his life. He had lived mostly in the world of his aunt and her friends. Tikhomirov says he welcomed the "mischief, daring, bravery, dexterity [and] sportiveness"[13] that he encountered in the boys at the school. In this same period, in a parallel to Poincaré's Trinasie, Andrei dreamed of "organizing a state—a commune on a desert island—and he drafted a constitution for this state, where principles of higher justice had to be put into practice."[14] He also wanted to become a forester. One of his fellow students was Anna Dmitrievna Egorova. They later married, in 1942.

Kolmogorov's last years at the Repman Gymnasium coincided with the birth of the Soviet state. He said, "Together with other senior students I left Moscow for the construction of the railway from Kazan to Ekaterinburg. In addition to this work I continued with my independent studies preparing for the secondary school examination, and for taking the secondary school degree externally. Back in Moscow I felt a certain disappointment: They issued a certificate (the secondary school degree) without taking the trouble of testing my knowledge."[15]

In 1920 Kolmogorov enrolled simultaneously at Moscow State University and at the D. I. Mendeleev Chemical and Technological Institute, where he was intending to study metallurgy. He concentrated on studying mathematics at this time, but also studied history, participating in S. V. Bakhrushin's seminar—Bakhrushin was an expert in the history of Russian theater. Kolmogorov decided to analyze land registers of the fifteenth and sixteenth centuries from the Novgorod region, to the northwest of where he had spent his earliest years. These documents, called cadastres, contain listings of both settlements and inhabitants. Begun in the 1490s after Ivan III "joined" Novgorod to Moscow, kicked out the old landowners, and moved in his own people, the cadastres recorded a survey of the region for tax purposes and to formalize the new ownership. This was an important moment in the formation of the Russian state, and a six-volume edition of the records had been published in 1915. Much historical work had already been done on the cadastres when Kolmogorov set out to analyze the six volumes for his first college paper.

The documents were essentially lists of numbers. The books that had survived clearly had gaps. The Novgorod province was large and had substantial variation in types of land, climate, economic management, and density of population. Further, it was often not clear what was being counted, a fundamental uncertainty being exactly what was meant by "people." Heads of households? What was a household? Therefore it was difficult to arrive at any general descriptions, such as total population or total number of villages or settlements.

Kolmogorov waded into the land registers in a way that would become typical. Undaunted by complexity, he was downright interested in seeming incoherence and was capable of a large volume (or in this case volumes) of work. He pored over the numbers. Where he found either regularities or discrepancies he used probability, particularly Bayes's formula, to calculate the odds they were random or more likely the result of some hidden organizing principle. Typical of this work was his respect for detail. Through this analysis, Kolmogorov reached conclusions that differed from the contemporary wisdom. He concluded that the term "person" was more closely connected to the working population than to "heads of families," that the taxes were broken down often not in terms of individual farms but of entire estates. One of the most important results related to the term *obzha*. The land registers defined it: "One man on one horse ploughs an *obzha*, while he who has three horses, and himself ploughs on the third, ploughs a *sokha*."[16] It would appear that in practice an *obzha* was a measure of the total profitability of an estate rather than a measure of land or work, as the definition seems to say. The difficulty of counting serfs and the economic importance of this enterprise is something of a theme in Russian culture. Nikolai Gogol's *Dead Souls* (1842) recounts the adventures of a character called Chichikov as he travels through rural Russia attempting to buy dead serfs that still remain on the census rolls, so he can collect money from the czar for relocating them.

A new method in the study of history, "cliometrics," was heralded in *Time* and *Newsweek* in the U.S. in the early 1970s. The idea was that mathematical methods could be used, with the help of modern computers, to extract historical fact from disconnected, partial records. One can imagine the difficulty the seventeen-year-old Kolmogorov faced in 1921 presenting research of such originality, even though it was at least partially appreciated by his professor. However, when Kolmogorov asked if it could be published, Bakhrushin responded: "Not you, young man! You have found only one piece of evidence. That is very little for a historian. You need at least five pieces of evidence."[17] Kolmogorov liked to tell this story and is reported by the historian V. L. Yanin to have said: "I have decided to go into science, in which for a final conclusion one proof is sufficient."[18] Yanin, who also worked on the cadastres, expressed "great regret" that: "Had the work of Andrei Nikolaevich been published shortly after it was written, our knowledge today would be much greater and, most importantly, more exact."[19]

Kolmogorov turned his full attention to mathematics and at the same time strove to make his life a bit more bearable. He quickly tested out of the first year of the university, because second-year students received a ration of food—a *pood* of bread, about 16 kilograms, and a kilogram of butter each month. He made his own wooden-soled shoes. To earn extra money he taught mathematics and physics in a grammar school. He taught with enthusiasm and enjoyment.

Recognition of Kolmogorov's talents was not long in coming. Already he had drawn attention to himself. Luzin made a conjecture during a course on analytic functions. Kolmogorov supplied a counterexample, which he delivered in class. P. S. Urysohn (1898–1924) was present and asked Kolmogorov to study under him. Urysohn was further impressed when a subtle error in one of his lectures was spotted by his new pupil. Kolmogorov also started working with P. S. Aleksandrov (1896–1982), under whose direction he did his first serious research on the descriptive theory of sets, and in V. Stepanov's seminar on trigonometric series.

First Kolmogorov found an important result on Fourier series, that there is no slowest rate of convergence. Then in June 1922, at scarcely nineteen, he discovered a Lebesgue-integrable function whose Fourier series was divergent almost everywhere (which he soon extended to everywhere). Other questions in the general theory of Fourier series hinged on this result, so the function was more than the oddball exception physicists often dismiss as mere "pathology." It brought Kolmogorov international recognition, and from here the volume, quality, and variety of his research make even cursory summary difficult. His bibliography has more than 500 entries. Even when those that are not research papers are excluded, what remains is paper after paper of fundamental importance, as if Kolmogorov were actually trying to follow Gauss's dictum, "few but ripe," but the branches were heavy with fruit.

Starting when he was seventeen and through 1929, when he passed from student to faculty, Kolmogorov looked at a variety of things in the most fundamental ways. Cantor had been led to the creation of set theory through his study of trigonometric series. Now Kolmogorov's first papers were on the descriptive theory of sets (the question of what sets can be referred to), though they languished until 1928, and some considerably longer, in Professor Luzin's desk because Luzin believed incorrectly that there were problems with them. It would seem Kolmogorov did not think them important enough to fight over. Throughout his career he would emphasize that portion of set theory that is concretely constructible and has importance for concrete problems. As mentioned in the chapter on Gödel, Kolmogorov was sympathetic to the greater caution of intuitionist mathematics, which can also be traced back to Kronecker's objection to pure existence proofs.

Kolmogorov addressed the controversy with his 1925 paper, "On the Principle of Excluded Middle." Kendall says this and a later paper "are regarded with awe by specialists in the field."[20] Intuitionists held that the law of the excluded middle (either a statement is true or its negation is true) was invalid when applied to infinite processes. Stating, "We shall prove that all the finitary conclusions obtained by means of a transfinite use of the principle of excluded middle are correct and can be proved even without its help,"[21] Kolmogorov showed that no contradictions were introduced by using the law of the excluded middle in this way. The logical flow is the same as the proofs Hilbert used in his treatment of geometry. Just as non-euclidean geometries can be embedded in euclidean geometry and so are at least as consistent as euclidean geometry, Kolmogorov embeds standard, classical mathematics in the intuitionist system. To do this he had to formalize intuitionism, anticipating Heyting. While siding with the intuitionists and working from their viewpoint, Kolmogorov forwarded Hilbert's goal of avoiding expulsion from "Cantor's paradise," and set theory would be omnipresent in his work.

Still in his early twenties, Kolmogorov was also looking at the ideas of integration and differentiation in calculus in the most fundamental terms. What is the most general possible definition of derivative that can be applied to the most general possible type of function? In examining integration he was looking to formulate the idea of measure in the most general way possible. Measure theory extends our simpler notions of length or area in an attempt to obtain a "measure" of sets. The sets being "measured" could be made up of complicated things like functions and weren't limited to sets of points of a geometrical space. Another way to think about measure theory would be as an attempt to liberate the comparison or assessment of size from the shackles of one-to-one correspondence, or counting.

Beginning in 1924, Kolmogorov started working in the area of mathematics with which he is most closely associated, probability theory. The Russians have a long tradition in this area, starting with P. L. Chebyshev (1821–1894) and the St. Petersburg school, and including such exemplars as A. A. Markov (1856–1922) and A. M. Lyapunov (1857–1918), among many others. The point of entry was again through the theory of Fourier series. In this early period, sometimes in collaboration with A. Ya. Khinchin (1894–1959), Kolmogorov started looking at the foundations of classic probability theory. In his own words: "All my joint work in probability theory with Khinchin, as well as the whole initial period of my work in this area, were marked by the application of the methods developed in the metrical theory of functions. Such topics as conditions for the validity of the law of large numbers and conditions for convergence of series of independent random variables were actually tackled by methods developed by N. N. Luzin and his pupils in the general theory of trigonometric series."[22]

The foundations of probability theory were problematic. Von Mises and others, including Sergey Bernstein (1880–1968), had attempted or were attempting axiomatic treatments of what had historically developed in response to real-world problems. For von Mises, frequency was the central starting point, attractive because some believed that probability in nature arose from underlying processes that were harmonic and had frequencies. Here there is an obvious connection to Fourier series, which break a wide class of functions down into different frequencies. One way of understanding the importance of Kolmogorov's earlier creation of an integrable function that had no convergent Fourier series is that it thereby cast doubt on using frequency itself as a basis for probability theory. Frequency as a notion is not sufficiently fundamental to address all we want to address in probability. The function that Kolmogorov created would be left out. Probability theory needed to capture all possible random occurrences, and Kolmogorov's function, though highly irregular, was manifestly possible.

By looking fundamentally at the notions of differentiation, integration, and measure, Kolmogorov moved toward creating a mathematically precise set of axioms under which all of probability theory could be conducted. He saw that aspects of the theory of the measure of sets and of convergence founded on measure theory were structurally identical to the procedures and insights of probability theory that had grown up organically. Therefore the solution was to start with a simple group of axioms stated in the context of measure theory. This tended to separate vexing philosophical questions from the mathematical system. Once the mathematical structures were in place, the philosophy sublimed into questions of application to practical situations. Kolmogorov first presented a system of axioms in 1929, followed in 1933 by a more extensive monograph. Clear, simple, and convincing, these axioms became the model for how modern probability theory is set up. He looks at the set of all possible outcomes. The probability function is a function that assigns numbers between 0% and 100% to subsets of the set of possible outcomes. The set of all possible outcomes is assigned the probability of 100%. If two sets of outcomes are mutually exclusive—hot and dry all day or rainy and cold—the probability associated with the union of the two sets is the value of the first set of outcomes added to the value of the second—with a few more technical details added in. Not all subsets can always be assigned probabilities—some are too irregular. Any set that can be reasonably inquired about has a probability.

Kolmogorov's work amounts to a solution of the first underlined part of Hilbert's sixth problem, which reads: "To treat in the same manner [as Hilbert had treated geometry], by means of axioms, those physical sciences in which mathematics play an important part; in the first rank are the theory of probabilities and mechanics." However, the spirit of the sixth problem is broader, so that many treatments of the problem written recently don't even mention Kolmogorov. Kendall gives Kolmogorov credit for this not once but twice, including later work as a second solution.

In 1929 Kolmogorov completed his graduate education and became qualified to teach at the university level, having already published eighteen papers. Instead of taking exams he wrote papers, most of them now famous, in the areas in which he had to prove competence. At first there was some question whether he would be allowed to stay in Moscow. It was an issue of seniority. Good sense prevailed and he remained, teaching at the Institute of Mathematics and Mechanics at Moscow State University while also teaching at the K. Liebknecht Industrial and Teaching Institute from 1929 to 1931. In 1931 he became a professor at Moscow State University.

Aleksandrov, only a few years older and becoming a world leader in topology, was one of those instrumental in keeping Kolmogorov in Moscow. During the summer breaks Kolmogorov had started organizing boating or hiking trips with one to three companions, which would be a lifetime practice. These trips eventually covered a large part of the Soviet Union. In the summer of 1929 Aleksandrov was invited as the third traveler, almost as an afterthought. Up until this time contact between the two had been mainly professional. Kolmogorov wrote: "Pavel Sergeevich Aleksandrov and I date our friendship . . . which lasted 53 years [from the 1929 trip]. . . . For me these fifty-three years of close and indissoluble friendship were the basic reason why all my life turned out to be filled, on the whole, with happiness, and the basis of my good fortune was the unceasing thoughtfulness of Aleksandrov."[23]

The trip in question was organized through the Society for Proletarian Tourism and Excursions. They offered the rent of boats and tents for voyages on the Volga River (starting not far from where Kolmogorov had spent his first years). Kolmogorov, Aleksandrov, and Nikolai Dmitrievich Nyuberg set out from Yaroslavl on June 16 with no map or guidebook. The only book they took with them was Homer's *Odyssey*. They had no set idea of their destination or what they should do after turning the boat back in, wherever that might be. So they set out drifting down one of the world's great rivers. A striking account of this trip and the early years of this friendship appears in *Russian Mathematical Surveys*. Letters from Aleksandrov to Kolmogorov also appear. Here is an excerpt from Kolmogorov's account:

> The typical scenery on the banks of a big Central Russian river, be it the Oka, the Dnieper, the Don, or the Volga, is pretty much the same. The river is usually divided into several arms, which flow between green water meadows and round sandy islands, overgrown with osier beds. The sands on the Volga are washed away, leaving an almost pure whiteness (my description relates to the '20s and '30s, before the construction of large reservoirs on the Volga).
>
> This countryside is not without its own particular grandeur. Aleksandrov soon grew fond of it and in later years we have often sailed along the Belaya, the Kama, the Volga, and the Dnieper.
>
> When we sailed along the Volga we usually chose our route along some Volozhka (that is the name in the Volga dialect for the minor channels). As there was a strong enough current in the Volozhka, we could determine whether it was carrying us into the main stream.
>
> We usually pitched camp on a sandy island, on the upper or lower extremity, where the flow of the water is especially felt. During the first days of our journey we often swam at night; in the white summer nights, gliding along past the overgrown osier beds on the bank, the air filled with bird song, made a lasting impression on us. We wished that this could go on forever.[24]

They left the river to visit the Nizhegorodskii Kremlin in Simbirsk (Ulyanovsk). They climbed onto the high precipices along the banks of the Volga. "Dominating everything was the continuous movement downstream." They read the *Odyssey*. They were idle and had conversations "on every subject." After twenty-one days, they finally made it to Samara, 1300 kilometers from where they started, and turned in the boats. They continued by steamer until they arrived at Astrakhan on the Caspian Sea and then went to Baku, finally heading inland, arriving at Lake Sevan, not far from the border with Iran. There they went to an island, which they shared with the archimandrite of an empty monastery, his maid, the head of the meteorological station and his family, and the "captain" of the "Sevanski fleet"—one motorboat and a few pleasure boats. Aleksandrov stripped off all his clothes and started working on mathematics. Kolmogorov says, "To study I took refuge in the shade, while Aleksandrov lay for hours in full sunlight wearing only dark glasses and a white panama. He kept his habit of working completely naked under the burning sun well into his old age." They spent twenty days at Lake Sevan and then went to Yerevan (in Armenia). "The temperature was 40° [centigrade—104° Fahrenheit], the sky was a hazy blue, and only after sunset did there unexpectedly appear the peak of Ararat, suspended in this blue sky." On to Echmiadzin, then Alagez, where they stayed with a band of physicists who were "working on cosmic showers." They climbed Mount Alagez and continued on to Tbilisi. They went to the Orbelian bath house. Aleksandrov went on by train and steamer to Gagra on the coast of the Black Sea, while Kolmogorov set off on foot into the mountains, where it snowed.

Kolmogorov rejoined Aleksandrov and a group of mathematicians. "We all joined in bathing in a fairly strong surf, where simply mastering the exhausting waves requires a planned approach—one has to throw oneself under the wave with the head down." They stayed in a house with a balcony projecting over the crashing waves below. It was owned by a "former Abkhazian princess." Beset by bedbugs, they headed inland.

On their return, Kolmogorov and Aleksandrov rented half of a house together outside of Moscow. Vera Yakovlevna, Kolmogorov's aunt who raised him, moved in to run the house. The other half was occupied by the owners. Kolmogorov mentions that they had a cow "from which we got milk."

In 1930 and early 1931 Kolmogorov and Aleksandrov traveled abroad. They visited Berlin briefly and continued on to Göttingen, where Hilbert was still regnant with Courant, Noether, Landau, etc. Kolmogorov saw evidence of the turmoil overtaking Germany, with Nazis on one side and Communists on the other. Jokes were made about how it would turn out. Kolmogorov noted "crowds of unemployed people, often very badly dressed."[25]

No momentous events for Kolmogorov occurred in Göttingen. In Munich he met Carathéodory, one of the originators of the modern ideas of measure that he had been thinking about. Then Kolmogorov and Aleksandrov went to the Bavarian Alps, where they stayed in "a little inn by the small lake of Spitzensee." Kolmogorov says, "From there we did the 'ascent of the Rotwand.' Along the tourist path to the Rotwand were marching sturdy Bavarians wearing hats with a feather, Bavarian leather shorts, Alpine boots with knitted socks, and with Alpenstocks in their hands. Partly from a desire to tease these respectable Bavarians, we did all the ascent barefoot (Aleksandrov and I generally liked walking barefoot)."[26]

They toured Gothic cathedrals in Ulm and Freiburg. In Ulm they swam in the Danube, swimming naked downstream though the town and then running back along the banks of the river. (Germany had a nude "physical culture" movement at this time, and nude bathing is still fairly common in some parts of Germany, in fact.) France—Mulhouse, the "seven lakes," two days along the banks of the Rhône, a day on the shores of Lake Annecy, Marseille—and finally to Sanary-sur-Mer, near Toulon in the south of France, where the French mathematician Maurice Fréchet (1878–1973) was vacationing.

Next Aleksandrov and Kolmogorov traveled to Brittany where, on an earlier visit, Urysohn, one of Kolmogorov's first teachers, had drowned. Urysohn and Aleksandrov had gone for an ocean swim, as they had every day for a number of weeks. The North Atlantic off Brittany, near where the Loire empties into it at the village of Batz, is not the Black Sea. An unusually large wave hurled Aleksandrov entirely over a group of rocks and smashed Urysohn against them. Aleksandrov pulled him out, but he died on the beach. Aleksandrov wrote: "Some more time passed, and I went into my room and finally dressed. (Until then I had remained in my swimming clothes.) Pavel Urysohn lay on his bed, covered by a sheet; there were flowers at the head of the bed. It was here that I thought for the first time about what had happened. All my experiences, all my impressions of that summer, and indeed of the last two years, rose up in my consciousness, with such distinctness and clarity. All this merged into a single awareness of how good, how exceptionally good, things had been for each of us, only about an hour ago."[27] Kolmogorov and Aleksandrov visited Urysohn's grave and went for silent walks by the sea. After a few days in Brittany they left for Paris.

In Paris Kolmogorov saw the older French mathematicians, Émile Borel (1871–1956), Henri Lebesgue, and particularly Paul Lévy (1886–1971). He was too busy to see very much of the younger mathematicians. In the first volume of Kolmogorov's collected works there are sixty papers and only ninety-three other mathematicians are mentioned. Fifteen of the papers have no bibliography at all. About this, Tikhomirov reports Kolmogorov saying that "in those years he was so full of ideas that he could give some away rather than borrow any."[28]

Kolmogorov contracted a bad cold and returned to Göttingen a sick man. He was hospitalized and spent the next two months recuperating. In February Aleksandrov traveled to Princeton to talk to the algebraic topologists there. In the extracts of his letters to Kolmogorov printed in *Russian Mathematical Surveys*, Aleksandrov is impressed with the size and cleanliness of the Princeton swimming pool, where he swims every day, and particularly with the abundant, lather-producing soap that enables everyone to get clean before swimming. He is pleased that no one wears clothes at the pool and that the Princeton youth are "of excellent physique," with no unsightly hernias. This is a contrast to French beaches, where suits were worn but failed to entirely cover the hernias. Aleksandrov does object that clothes are worn on the running track and athletic field, though he admits that these locations are visible from the road. He queries Kolmogorov about his exercise program and offers suggestions of things he should purchase before returning to Russia, ranging from items of clothing to a large boat. In the first ominous reference to events at home he says: "Do not fail to buy yourself all these things, especially in case you have to return unexpectedly to Moscow. (Unfortunately I think this cannot be ruled out.)" A bit later he launches into an exhortation about nutrition, pleased that in the U.S. cream is available and that the prices of dairy products are fixed "almost entirely according to their nutritive value." He compliments American housewives for their openness to rational argument about nutrition and faults their "German counterpart(s)" for their dogged allegiance to milk over cream. He ends this section with the exhortation, "But you should buy both cream and milk (since you drink cream in your coffee, but milk by itself).[29] Of course the best of all is to drink cream by itself." (Aleksandrov lived well into his eighties.)

A later letter deals with how Aleksandrov is falling in love with the countryside around Princeton. He is entranced by "the Stony Brook." He swims in it—it is still winter. He approaches it from different directions. He describes it as something out of Turgenev. It seems enchanted and empty, but he recounts finding empty tins and the signs of a campfire. He has to walk half an hour through thickets and the thickets are beautiful. The cap of the letter is the following narration:

> Some days ago I went for a long four-hour walk and followed the course of the Stony Brook over a distance of 15 kilometers. This walk was often fraught with difficulties, even dangers, since I had not only to penetrate the thickets where there were no paths, but what was much worse—I had to go through private properties, fenced in with wire fences, on which there was every ten paces, instead of barbs, a printed notice stating that any stranger entering here was in breach of the laws of the USA, with all the ensuing consequences! But, thank God there weren't any consequences.[30]

Aleksandrov and Kolmogorov returned to Russia in June 1931 and the curtain soon descended. Aleksandrov traveled to Europe again in 1932, where he saw Hilbert, Weyl, Landau, Noether, Neugebauer, and others for the last time, but wrote: "Often I was woken in the morning by the sounds of '*Deutschland, erwache*'... sung by the young people of the '*Hitler-Jugend*,' as they marched up and down the streets."[31] After this, contact with the West, except for some journals and sometimes correspondence, was almost entirely cut off until the death of Stalin. In Russia, life for Kolmogorov and Aleksandrov appears to have been tolerable. In 1935 they bought an old manor house outside Moscow, which became a mathematical gathering place.

Kolmogorov had clearly moved into mathematical high gear in the 1930s, and a listing of his papers begins to seem like a Homeric catalogue. One can find entire issues of major scientific journals that merely attempt to give a census of his results to the community of mathematicians. An issue of *Russian Mathematical Surveys* is devoted to this task, as is an entire issue of *The Annals of Probability*, confined largely to his contributions to probability theory. Earlier, five issues of *Russian Mathematical Surveys* were dedicated to him on the occasions of his birthdays.

Kolmogorov was a rigorous mathematician. Like Weierstrass and Hilbert, he tried to make his definitions, his mathematical boundaries, and the philosophical underpinnings as clearly and completely understood as was possible—hence the papers on intuitionist logic, the investigations of the concepts of measure, differentiation, and integration, and the axiomatic approach to probability theory. His later work was less focused on basic concepts, but the attempt at mathematical rigor and logical clarity remained.

Another constant in Kolmogorov's work is its motivation. He was most interested in analyzing physical questions with mathematics and looked first to physics, returning to it again and again, but also to biology, geology, technology, history, and linguistics to find problems. Since the core of his work addressed difficult and complex problems where one never had a hope of possessing complete data, he tended to mount his offensives starting from ideas in probability and statistics, many of which he invented or significantly clarified.

Andrei Nikolaevich Kolmogorov.

In his autobiography, *Enigmas of Chance*, Mark Kac wrote, "Our work began [1935] at a time when probability theory was emerging from a century of neglect and was slowly gaining acceptance as a respectable branch of pure mathematics. This turnabout came as a result of a book by the great Soviet mathematician A. N. Kolmogorov on foundations of probability theory, published in 1933. It appeared to us awesomely abstract."[32] The American mathematician Norbert Wiener discovered many of the same results as Kolmogorov without ever talking to him or corresponding with him. In the second volume of his autobiography, *I Am a Mathematician*, Wiener wrote, "I had long had a peculiar sort of contact with the leading Russian mathematicians, although I had never met any of them. . . . For more than twenty years, we [referring here also to Khinchin, Kolmogorov's early collaborator] have been on one another's heels; either they had proved a theorem which I was about to prove, or I had been ahead of them by the narrowest of margins."[33] These great mathematicians were working where they were because there was almost-untouched, fertile ground.

In his obituary for Kolmogorov in *Physics Today*, Arnold stated: "His work on the foundations of probability theory, the theory of stochastic processes (including stationary processes), Markov chains and processes, the Fokker–Planck equation and Brownian trajectories forms the base of modern probability theory."[34] Definitions of some of this specialized vocabulary will clarify Kolmogorov's contribution. In the foundations, the law of large numbers gives mathematical assurance that probability does, in fact, work. If we do enough trials or throw a die enough times, we will almost surely obtain a result that matches what probability calculates it should be. The law of large numbers is a starting point for all that follows. In the 1920s Kolmogorov clarified exactly when the law applied, proved it rigorously for very general situations, and with Khinchin proved a strengthened version of the classical result, called the "strong law of large numbers."

In Markov chains, one thing follows another in discrete random jumps. There is no "memory" of what happened the time before last (which is close to the essence of probability). Brownian motion (or trajectories, to use Arnold's word) is a specific example. A small piece of pollen is knocked back and forth by random collisions with the water molecules it is floating in. How does it move? How far does it move in a given period of time? How big an area does it sweep out? This can also be treated as a continuously occurring process.

The word "stochastic" in "stochastic processes" simply means random. The word "processes" in this usage actually might contain more hidden meaning. The processes treated often vary continuously in time. Probability attempts to deal with different kinds of random processes, those that vary continuously as well as those that unfold as a series of discrete events.

Physicists arrived at various specialized equations in probability through physical intuition and trial and error. The Fokker–Planck equation describes diffusion processes, like the diffusion of scent when a rose is brought into a room or the spread of a genetic mutation in a biological population. Another equation, associated with Smoluchowski and Einstein, deals with types of Brownian motion. Kolmogorov analyzed this type of equation in a 1931 paper, one of his most important. Starting with probability theory, he rigorously derived the whole class of differential equations, making plain the nature of the connection between the probabilists' theory of continuous random processes and the differential equations that physicists possessed. These equations as a class are now often called Kolmogorov–Chapman equations by mathematicians. This is the kind of paper that causes buildings to be built on university campuses to house the mathematicians who follow and do the work that flows from it.

Kolmogorov indulged in a wide-ranging group of what looked like mathematical side-interests, enough to have established his reputation if he had worked only on them. He wrote two foundational papers on geometry—classical geometries of constant curvature and projective geometry—in 1930 and 1932, respectively. In the mid-1930s, in the area of algebraic topology, he introduced the idea of cohomology groups and defined a multiplication on the cohomology group to create cohomology rings. He did this simultaneously with and independently of J. W. Alexander at Princeton. (Arnold says Kolmogorov was a bit earlier.) Poincaré had introduced the idea of homology groups in an attempt to tell one kind of manifold (surface) from another, cutting the manifold into pieces and calculating the homology group from the relation the pieces have with one another. Cohomology is done in a different way, but homology and cohomology are often "dual" (generate the same algebraic groups) and so give alternative methods of finding the same thing.

It seems astounding that an outsider would be able to introduce an idea this important to a discipline as "hard" and well-staffed as algebraic topology. It certainly surprised the topologists who were gathered for an international conference in Moscow in 1935 when a probabilist showed up and presented these important results. Hassler Whitney wrote that Kolmogorov was "an unlikely person at the conference."[35] However, Kolmogorov's friendship with Aleksandrov kept him well-versed in what was happening in topology. Aleksandrov is credited as the originator of one of the main species of homology theory in the 1920s. Arnold says Kolmogorov got the idea for cohomology theory while thinking about the distributions of electric charges and currents in space. Tikhomirov says that the idea came from attempts to describe "the flow of liquid over a manifold."[36] It seems Kolmogorov was trying to "fit" classical things like the energy of a fluid flow or electric and magnetic fields into a geometrically phrased problem.

Also spurred on by Aleksandrov, and reminiscent of his early construction of an integrable function whose Fourier series diverges almost everywhere, Kolmogorov constructed a compact one-dimensional topological object and a continuous topological mapping of it onto a compact two-dimensional object (which he also constructed) that took open sets to open sets. These objects are the limits of constructions and can be embedded in three- and four-dimensional spaces respectively. There is a lot of cutting out holes and gluing möbius strips back in their places. Again, Kolmogorov showed that he had an unsurpassed ability in constructing examples of things that people thought ought not to exist.

Kolmogorov also did work on approximation theory and other subjects related to his probability work, applying it to concrete scientific problems. There is work on the crystallization of metals (he'd originally signed on to study metallurgy) and on the reversibility of certain physical processes. There is rather extensive application of the diffusion theory that we mentioned under probability to biology. This was done independently of R. A. Fisher (1890–1962) in England. Kolmogorov examines the spread of genes in a habitat and calculates extinction probabilities. In 1939–40 he engaged in a public controversy over the proper interpretation of some biological data with the ideological and scientifically corrupt biologist Lysenko. Looking at the data of Lysenko's student, N. I. Ermolayeva, Kolmogorov explained an apparent discrepancy between the data and Mendel's laws by examining what kind of statistical error from the predicted norm ought to be expected in this sort of situation. Since the spread or error in the data was exactly what would be expected in a carefully done experiment where there was no fudging of the data (consciously or unconsciously) to make it look prettier, he complimented Ermolayeva's care in doing the experiment—perhaps ironically, with the implication was that Ermolayeva and Lysenko were not up to a statistical analysis of the data sufficient to fool him about its authenticity. In these dangerous times one must admire Kolmogorov's dexterity in arguing his position.

The Second World War began, and Kolmogorov worked on the sort of applied problems that mathematicians and physicists turned to in every country. The problem of firing or aiming cannons accurately was important, if seemingly pedestrian, and it attracted Wiener in the U.S. It is, in fact, a deep problem. Because of wartime security, neither knew what the other was doing, and after the war much of the material remained classified for many years. Wiener gives a balanced account of who did what, in what order, in his autobiography, stating with disappointment still evident: "Nevertheless, all my really deep ideas were in Kolmogorov's work before they were in my own, although it took me some time to become aware of this."[37] The issue of priority, in these circumstances, seems less important than independence. Kolmogorov introduced to the subject of firing the notion that in certain real-world situations, better results will be obtained by occasionally misaiming. If you control your aim too closely and your analysis is off you will miss every shot. He analyzed when and how to misaim.

If the study of firing and prediction theory allowed Kolmogorov to pursue ideas that were of natural interest to him during the war, his other big wartime pursuit was even more fortunate. He worked on turbulence, and in this very difficult area he achieved some of his most important results. In the 1930s Kolmogorov had been connecting measure theory to probability, probability to differential and integral equations, and all of this to complex physical situations through treatment of statistical ensembles using the new methods. In the late 1930s he had added to ergodic theory, in which progress had recently been made in treating complex but not inherently statistical situations in dynamics. Now, in turbulence, Kolmogorov applied some of his growing arsenal of techniques to a physical situation that was classical and deterministic but so complex that statistical methods and insights were appropriate.

When a fluid, air for example, flows past something reasonably smooth, such as an airplane wing, it can do so with laminar flow or turbulence. An image for laminar flow might be to see it as a bending freeway moving around the airplane wing with many lanes (hence the word "laminar," as in lamination). As long as all the "cars" in this metaphor can negotiate the curves and stay in their lanes, everything is OK. Slow cars stay out of the lanes with fast cars. If the overall speed of cars gets high enough so that a significant number of cars bounce off the railings or can't stay in their lanes, all hell breaks loose.

Turbulence is chaotic and is not easy to describe accurately. There are many different situations where it can occur—winds hit a mountain, there is a thunderstorm, or convection currents meet in the ocean or air. Water flowing rapidly in a pipe can be turbulent. Turbulence can be quite violent or more smooth. One of the main questions, again in the example of an airplane wing, is how all the macroscopic swirls and eddies that are visible with smoke in a wind tunnel end up dispersed and dissipated, persisting as a slight increase in the temperature of the air. Kolmogorov advanced notions of self-similarity and scaling that later had far-reaching effects in statistical physics and, more remarkably, in quantum field theory.

If we look at a system of eddies in a turbulent medium and then magnify a small section of it, we see a statistically identical picture of smaller eddies. If we magnify again we see yet smaller eddies, but statistically the picture is the same. Obviously this can't go on forever. How does the energy of the macroscopic movements cascade down into the atomic scale? Kolmogorov found a way to analyze the problem if there are enough different scales of similar eddies. He made some assumptions motivated by physical intuition and dimensional analysis and derived what is known as "Kolmogorov's rule of two-thirds." It states, if certain assumptions are met: "The average (mean) square of the difference of the velocities at two points, located at distance r (neither too big nor too small), is proportional to $r^{2/3}$."[38]

Many types of physicists and engineers use this result. Astronomers are interested because it can be extended to predict a similar pattern of variations in the refractive index in the path of light making its way through the earth's atmosphere. Each of the swirls within swirls in the atmosphere acts as a very weak, unfocused lens, spreading the light—corrugating the incoming wave front—leading to the twinkling we see in stars and the blurring of the image a telescope finally focuses. Nontrivial, but straightforward, calculations from classical optics applied to the pattern that the Kolmogorov rule presents lead to a formula for the angular resolution of an otherwise idealized telescope. No (passive) earthbound telescope can escape this blurring even if the optics are perfect.

Kolmogorov's work on turbulence reads as if it were written by a mathematically inclined physicist. He was thinking about a problem, and his solution was pragmatic. Kolmogorov is always efficient. If "awesome abstractness" is helpful, it appears. If something as ad hoc (though deeply motivated) as dimensional analysis is needed, he uses it.

Readers interested in science will be aware of all the excitement about chaos theory not long ago and might have read James Gleick's *Chaos: Making a New Science* (1987), in which the ideas of self-similarity and scaling are central. Gleick says in the book that the "blossoming" of chaos science in the West "inspired considerable bewilderment [in Russia], because much of the new science was not so new in Moscow."[39] He ascribes the lack of communication between the Soviet Union and the West to a combination of language differences and the travel restrictions of the Cold War. This is partly true, but it was also due to the fact that the creators of chaos theory in the United States were a heterogeneous bunch, including population biologists, meteorologists, and relatively low-to-the-ground experimental physicists. These people wouldn't necessarily have known much about the purely mathematical end of the theory. Much as small swirls don't "know" what big swirls are doing, there seems to be self-similarity within chaos theory itself. One advantage the Russian system had was a historical bias toward applying mathematics to grungy and impure scientific questions, a bias that goes back to the St. Petersburg school in the time of Kovalevskaya and well pre-dates the Bolshevik Revolution with its concomitant, often false, scientization of everything. If we look at a theoretical mathematician like Smale, who carried the topological/analytic mathematical assault in the U.S. to the next stage, we see more awareness of the whole picture. Smale wrote about a 1961 visit to the Soviet Union: "After Kiev

I went back to Moscow where Anosov introduced me to Arnold, Novikov, and Sinai. I must say I was extraordinarily impressed to meet such a powerful group of four young mathematicians. In the following years, I often said there was nothing like that in the West."[40]

A second way in which the Russian system might have had an advantage was that historically, and certainly after the collectivization of Soviet society, the Russians showed a tendency to organize themselves into schools and identify themselves with schools. With something like turbulence this is a distinct advantage. Kolmogorov headed a laboratory from 1946–1949. Even later, when he is in his late sixties, we find him out on a boat, the Dmitrii Mendeleev, making sure the right things are being measured. This tradition of applying theory to gritty practice and working together in groups goes a long way toward explaining how a country where animals were still being used (in significant numbers) to pull things could build atomic weapons with surprising rapidity and then develop rockets that in many ways were superior to those of the U.S. Certainly the Soviets captured some German know-how, as well as some Germans, after the Second World War, but their achievements were still amazing because of the lack of infrastructure on which to build.

By the early 1950s, Kolmogorov was beginning to show signs that the Stalin years had worn him down. Perhaps only because he was middle-aged, his mathematical production went into a relative lull. It is likely that he had been spending time on classified work. As a younger man, in 1936, he had been drawn into speaking against Luzin at Luzin's hearing at the Academy of Sciences (see the section on Gelfond). Though Kolmogorov's denunciation had been unenthusiastic, Luzin could not have regarded it positively. Luzin had not retaliated against his attackers, with the exception of blocking Aleksandrov's admission to the Academy of Sciences. There were a lot of near misses in this harrowing period. For example, in the summer of 1937, Kolmogorov's student B. V. Gnedenko (1912–1995) went on a hiking expedition to the Caucasus with other mathematicians, including, for part of the trip, Kolmogorov. People talked freely. Gnedenko was denounced by someone who went on the trip. In December he was arrested, and considerable pressure was put on him for six months in prison to implicate Kolmogorov as an "enemy of the people." He didn't succumb, reasoning that if he did, they both would be finished. Then suddenly he was released. Kolmogorov and Khinchin, at considerable risk to themselves, managed to get him reinstated to his position at the university.[41]

In 1946 Luzin seemed to agree to support Aleksandrov's admission to the Academy, but then withdrew. Kolmogorov slapped Luzin in the face at the Academy. The president of the Academy, S. I. Vavilov, said to Kolmogorov, in A. P. Yushkevich's account:

> "Well, old man, you sure started something; the Academy has seen no such thing since Lomonosov's time." Vavilov told Stalin what

> had happened. Stalin said, "Well, such things occur with us as well." That was the end of it. I am sure that Kolmogorov took the incident very much to heart.[42]

Vavilov's brother, N. I. Vavilov (1887-1943) had traveled to Iran, Afghanistan, Ethiopia, China, and South America collecting specimens for experiments in plant breeding. Between 1916 and 1933 he had brought back 19,000 different kinds of wheat alone and an additional 50,000 varieties of wild plants. He was first denounced by Lysenko in 1934 for his Mendelianism, was arrested in 1940 and died in the camps in 1943.

In the late 1940s, Kolmogorov had to retreat on his earlier work on genetics. Lysenko had taken nearly total control over Soviet agriculture. There was perpetual pressure to betray others. Sometimes, as with Gnedenko, Kolmogorov was able to help people, but the difficulty of the situation comes through in the following quote from Arnold:

> "Sometime I will explain everything to you," Andrei Nikolaevich would say to me each time he would act in a way that clearly contradicted his principles. Apparently he was experiencing pressure from some kind of evil genius whose influence was enormous (well-known mathematicians had the role of a link transmitting the pressure). Andrei Nikolaevich did not quite live until the time when it became possible to talk about these pressures, and like almost all the people of his generation who lived through the 1930s and 1940s, he feared "them" to his last day. We must not forget that for professors of that time not to report on seditious speeches of an undergraduate or postgraduate student quite often meant being charged the next day with sympathizing with seditious ideas (following denunciation by this undergraduate or postgraduate provocateur).[43]

The lull in Kolmogorov's creative life came to an end when Stalin died. In 1954 truly major papers begin to appear, first among them papers on the three-body problem. Arnold, curious about where the new ideas came from, hypothesized a theory of the chain of ideas in Kolmogorov's work that led to his progress on this problem. When he asked him about it, Kolmogorov replied, "I didn't think that at all. The main thing was that in 1953 there appeared some hope. I became somehow quite enthusiastic, I had thought about problems of celestial mechanics ever since childhood. . . . Several times I tried but it didn't turn out. Now I was able to make some progress."[44, 45] Kolmogorov's heretofore wide-ranging work on a dizzying variety of subjects begins to show a constant focus.

If Kolmogorov had always been interested in finding ways he could use theoretical mathematics to understand the world of his direct experience—classical science in complex situations—then by the mid-to-late 1950s that interest was clearly central. One can begin to see almost every piece of mathematics Kolmogorov ever did as being somehow tied to these human-scale problems. The work in Fourier series, descriptive set theory, the foundations of differentiation and integration, and even intuitionist logic begins to seem a setting of the stage. As noted earlier, physicists historically tended to dismiss as "pathological" examples of odd functions constructed by mathematicians that are constantly changing directions. The assertion was that they were of interest only to mathematicians—that real physical processes don't generate pathological functions. We have already seen how Brownian motion required the generation of a complex and abstract probability theory for an accurate treatment. In Brownian motion taken to its mathematical limit, the piece of pollen jostled by random collisions with water molecules is constantly changing direction. Its path is continuous but nowhere differentiable, just as in the historic pathological examples.

Kolmogorov's work on dynamical systems (in Shiryaev's words, "the quasiperiodical motions of Hamiltonian systems under small perturbations"[46] and what Poincaré had called "the basic problem of mechanics") can be said to begin this last great period of his career. The orbits of planets and their stability is a concrete example. After twenty years, Kolmogorov could again leave Russia. Just as Hilbert gave focus and direction to mathematical research in 1900, Kolmogorov gave an address to the International Congress of Mathematicians in 1954 in Amsterdam that would have wide-ranging effects. He invoked Hilbert's lecture in his introduction, caused a sensation, and had to deliver the address twice, once in German and once in French. His results, in the words of Arnold, "imply the stability of the motion of a planetoid in an elliptical orbit, perturbed by a 'Jupiter.'. . . This was the first rigorous result regarding typical motions in the three-body problem, whose formation goes back to Newton."[47]

Some of the flavor of the address comes through in Kolmogorov's description of the problem:

> In conservative systems, asymptotically stable motions are impossible. Therefore, for instance, the determination of individual periodic motions, however interesting it may be from the viewpoint of mathematics, has only a rather restricted real physical significance in the case of conservative systems. For conservative systems, the metrical approach is of basic importance, making it possible to study properties of a major part of motions. For this purpose, contemporary general ergodic theory has elaborated a system of notions whose conception is highly convincing from the viewpoint of physics. However, up to now the progress made towards the application of these modern approaches to the analysis of specific problems of classical mechanics has been more than limited.[48]

Then he sets out to make some progress.

Until this time, complex classical problems that did not lose energy (were "conservative") had few points of entry—and as soon as three things were interacting, the problem was complex. One could measure where the system was today and by some approximation or brute calculation attempt to calculate it a relatively small distance into the future. Another way was to find single and exactly specified initial conditions or orbits where everything was balanced perfectly, hence Kolmogorov's phrase "rather restricted physical significance." In practical terms, a system would never chance to fall into these not just narrow but infinitely exact conditions, and even if it did, how could we ever measure things accurately enough to tell if it had?

Another approach was to follow the ergodic hypothesis: In phase space—absent some reason to the contrary flowing from a particular problem—no state should be preferred. Therefore, except when everything balances and the system repeats perfectly, a system wanders over time into or near every possible state. More than one mathematician came upon this general idea and translated it into specific mathematics. We have seen that Poincaré formulated something close to the ergodic hypothesis in his studies of differential equations and celestial mechanics. Perhaps Ludwig Boltzmann (1844–1906), one of the originators of statistical mechanics, was the first to clearly state the hypothesis. He doesn't appear to have coined the name, which is based on Greek roots for "path of energy." In thermodynamics, entropy, or the degree of disorder, always increases. Boltzmann thought of the ergodic hypothesis in order to explain why entropy increases. If we look at something like the air in a room, one possible state is to have all the air on one side of the room. There are few states that achieve this, however, and they are unthinkably unlikely compared to a vastly larger number of states where the air is spread out all over the room. One finds the likelihood of something occurring by looking at the measure of the points that correspond to it in phase space and then comparing that to the measure of all of phase space. Here is a connection to Kolmogorov's measure-theoretic formulation of probability.

Over time, ergodic theorems were proved, starting with a 1931 theorem by Birkhoff commonly known as "the ergodic theorem." Kolmogorov wrote a paper in 1938 offering a simplified proof. However, as he remarked in his talk, the method had not been applied to concrete problems in classical mechanics. The qualitative hypothesis that a general physical system would wander everywhere it could was something like a mystical belief. Not only had it not been proven in general, it was becoming clear that it would probably have to be weakened in some way.

In his address Kolmogorov outlined a program that used deep insights and techniques from differential geometry, point set topology, differential equations, measure theory, number theory, and probability and applied them to an analysis of phase space considered in a way similar to ergodic theory. His result is that in certain cases of real physical significance, there are solutions that are, in fact, stable within a range of starting points or perturbations. He showed that the ergodic hypothesis is probably false in its most general form. Instead of wandering everywhere, a system could get stuck in stable bands that could persist forever. Within a region of phase space the system might wander, but now there were regions of phase space that might be separate dynamically from one another. In a way that was typical, Kolmogorov didn't apply the program himself. He didn't fully prove his assertions but said he saw a way how to. This was accomplished by Arnold and Jürgen Moser, and the development of what is now called KAM (Kolmogorov–Arnold–Moser) theory extended over a period of years. Ya. G. Sinai says that in 1957 he was a member of a class in which Kolmogorov gave a complete proof of the main theorem.

We have seen that Poincaré proved that "the problem of small denominators" caused any series approximation of the solution of the three-body problem to not converge uniformly. In 1941 Siegel proved that the points that had this problem were not only numerous but everywhere dense. However, Siegel also wrote: "It would be important to obtain also some information about the distribution of the regular points [where there wasn't a problem] . . ., but this seems to be rather a difficult problem."[49] Here we come to a marvelous example of how seemingly esoteric work in mathematics can be applicable to a problem as basic as the stability of the solar system. The rational numbers are everywhere dense; so are the points where small denominators are a problem. Set theory supplied language that allowed mathematicians to see that despite the omnipresence of rational numbers, almost all numbers were transcendental. Measure theory allowed us to see that the measure of the rationals in any line segment was 0 while the measure of the transcendentals was the length of the line segment. Poincaré, struggling after a definition of stability without the help of these tools, said that an outcome or set of outcomes whose probability was 0 could be ignored. He phrased the problem in terms of differential geometry in phase space. Kolmogorov had used set theory and measure theory to make probability a rigorous part of mathematics. In his 1954 address he connected all these things.

Kolmogorov said the problem of small denominators was not as omnipresent as it seemed. Sometimes the set of points that led to the problem of small denominators was, in fact, negligible in a region, when one asked what would really happen in the n-body problem—just as the set of rational numbers is negligible when talking about length. It is easy to find rational numbers and hard to find specific numbers that you can prove are transcendental, yet just as the curtain of rational numbers turned out to be diaphanous, hanging in front of the real numbers, the points where the small denominators swamped the ever-smaller coefficients in the approximation were like a fine haze that did not obscure what really happened. Like Siegel, Kolmogorov saw that good frequencies existed where the small denominators were overwhelmed by a sort of superconvergence of the series in which they appear. He proved that in many cases the good frequencies were actually the rule, not the exception. Siegel also had shown that there were points where the small denominators were not a problem. However, he proved his result by calculation using specific numbers, so his methods provided no direct way of making contact with a more general result.

Kolmogorov also observed that there were truly bad frequencies that did cause instability in their neighborhood. There are small gaps in Saturn's rings and gaps in the asteroid belts. Debris in these areas couldn't orbit stably and flew off. KAM theory is more a method than a single theory. Many believe that Kolmogorov's 1954 paper and computational results done in the same year by Enrico Fermi, J. R. Pasta, and Stanislaw Ulam, which found empirical computational evidence of chaos in simple systems and surprising regularities in chaotic systems, mark the beginning of chaos theory. (A 1945 paper by Mary Cartwright [1900–1998] and J. E. Littlewood also contains a clear recognition of an aspect of chaos—they had been doing wartime work on radar.)

Kolmogorov hadn't been only thinking of dynamics. He investigated the idea of computable functions in its Russian guise, "algorithms," supervising V. A. Uspensky's 1952 diploma thesis and publishing his own "On the Concept of an Algorithm" in 1953. He had discovered Claude E. Shannon's (1916–2001) revolutionary 1948 paper, "A Mathematical Theory of Communication," and lectured on Shannon's information theory in late 1953. In 1954 a young A. G. Vitushkin published a paper looking at superpositions of smooth functions. If we have two functions of two variables, $f(x, y)$ and $g(x, y)$, we can build a function of three variables $f(x, g(y, z))$—this is called superposition. Vitushkin used the topological machinery developed by Petrovsky and Oleinik mentioned in the chapter on the sixteenth problem. Kolmogorov's interest was piqued—this kind of jump in category is always interesting. Vitushkin's results were clearly related to Hilbert's thirteenth problem.

Kolmogorov started thinking about the problem himself. Though the thirteenth problem talks about algebraic equations and approximate "nomographic" solutions, what it really asks is whether there exist continuous functions that are inherently three-variable functions and can't be represented as superpositions of continuous two-variable functions.* Kendall captures some of the flavor of what Kolmogorov was up against in this enterprise: "Of course careful note must be taken of the fact that continuous functions form a wide class, and can have horrendous properties from a practical point of view (compare Brownian paths)."[51] But Kolmogorov already had a lot of experience with Brownian paths.

Kolmogorov decided to hold a seminar on, in his words, "the theory of approximate representation of functions of several variables including problems of approximate nomography." He had published on the theory of approximation in 1936 and 1948. He did not think a solution of the entire thirteenth problem was approachable when he set out to investigate it in 1955, but the problem opened up

*Some people say that if Hilbert were really interested in algebra in this problem, he must have meant to ask the question about analytic or even algebraic functions, but he clearly says "continuous"—a broader class—in the highlighted part of the problem. In the words of Arnold, "It is strange . . . but he has done it."[50]

surprisingly. If we look at superposition of two functions, $f(g(x))$, only part of the domain of the second function f is used—the range of g. The game becomes building a function with many different pieces packed in and functions that send the variable to the right place. Kolmogorov began to think of the "trees" and "universal trees" in the work of A. S. Kronrod to do the packing. By 1956 he succeeded in proving that no matter how many variables a continuous function has, it can be constructed using continuous functions of three variables, using a finite number of substitutions. Kolmogorov said that this proof required the longest period of concentration on a single problem in his career.[52] He started thinking of other things again and left the problem to his students.

In 1957 Arnold, then nineteen and in his third year, finished the solution of the thirteenth problem by proving that two variables were enough, solving the general problem. Kolmogorov had been thinking about his first method of proof, what he felt was its abstractness and lack of intuitive clarity. In 1957 he saw a way to improve the result and proved a theorem that other mathematicians tend to characterize with words like "remarkable" and "astounding." Using his first proof as a way in, Kolmogorov broke down what it was doing into a method of explicit construction. If one admits the operation of addition, which is an inherently two-variable function, one can construct all multivariable continuous functions with superpositions of continuous functions of one variable in a surprisingly straightforward way. This paper is very dense, even "intricate" (Kaplansky's word).[53] Kolmogorov published it in *Doklady,* the premier Soviet journal for reporting new results, and this journal had a rule that no article could be longer than four pages!

There are a lot of constants with four indices as in $\lambda^{p,q}_{k,i}$. The article is perhaps easier to understand if you think of it as describing an algorithm or computer program that (in the case of three dimensions) breaks the unit cube into smaller cubes that have gaps between them. The gaps allow you to avoid tearing as you send each of them to its own address somewhere on a line segment. Then you do it again and shift each of them a bit—then again and again. This way every point of the cube has at least four addresses on the line. Next do the same thing with a finer grain. You end with a series of approximations that converge to the functions that are needed—this is where the analysis fits in. Otherwise the proof is discrete and elementary. These addresses can be added together in a way that doesn't lose the information they contain. If Hilbert felt that new mathematical ideas were so difficult there had to be a simpler way of looking at them, Kolmogorov's point of view seems to have been, well, I can understand everything, but I bet I can solve the problem using calculus (just as Siegel got a lot of mileage out of algebra).

Kolmogorov began to focus on the idea of complexity itself. Shannon had not framed information theory in mathematically rigorous terms. He worked for Bell Labs and was interested in how much information can be packed into a signal and how that information is corrupted by the injection of random error (or noise) during transmission and reception. Shannon talked about the "entropy"—also here a measure of disorder—of an encoded message (or sequence of numbers), "ergodic sources," and probability. Kolmogorov—as did his early collaborator in probability theory, Khinchin—worked to make information theory mathematically precise and extended it. In the late 1950s, Kolmogorov took the ideas of information theory that dealt with discrete series of numbers and extended them to continuous strings of numbers. Compare a message sent in Morse code—a discrete sequence of dots and dashes—to a message sent by a human voice. In this generalization Shannon's definition of entropy didn't work. All natural generalizations said just about every continuous string of numbers or function of a real variable was infinitely disordered—yet the human voice is not disordered. Kolmogorov, at this point, had also been using epsilon nets in his approach to the thirteenth problem and to the idea of approximating functions—in 1955 and 1956 he had published papers on these topics.

Now Kolmogorov connected his new ideas about information theory and his new-and-improved ergodic techniques to hydrodynamical instability. He had never really dropped turbulence—for example, in 1949 he published "On the Breakdown of Drops in a Turbulent Flow" and in 1952 "On the Velocity Profile Drag in Turbulent Flow in Tubes." His new ideas amounted to a program and involved many of his best students, like Arnold and Sinai. The key published result by Kolmogorov here was the introduction of the concept of entropy into this area through the path of information theory. He was trying to find descriptions of what we would now call chaotic behavior. Kolmogorov had spent much of his life working in terms of probability, investigating systems that are inherently random in the subprocesses. Now he was trying to study random behavior in classical systems governed by differential equations that are so complex that their behavior is in practice random. Moving through the door opened by KAM theory, the goal was to find patterns that a chaotic system could fall into, in its wandering path, that would hold the system quasi-stable. The red spot on Jupiter has stayed in one place for a long time. Kolmogorov didn't succeed in achieving what he took to be hard results, so he didn't publish his conjectures, except for his introduction of entropy. Arnold says, "His conjectures differ from the modern 'strange attractors' program only in terminology."[54]

In 1961, Kolmogorov was involved with his last major investigation—the theory of algorithms used as a basis on which to analyze complexity. An algorithm is any effective method of calculating a solution to a problem, the same idea as effectively computable functions. Kolmogorov focused on algorithms for generating strings of counting numbers. He and Uspensky had worked on the subject earlier. Now they refined their ideas and, like Turing, ended with a type of ideal computer—a Kolmogorov–Uspensky machine. Turing machines were designed to be maximally simple, and so it is not always easy to see how to persuade a Turing machine to actually calculate something. Given a method of calculating in the real world, it is more clearly efficient to program a Kolmogorov machine to do it. Kolmogorov showed that it was "easy" to program a Kolmogorov machine to compute a series of numbers that was intuitively regular and "harder" to program it to calculate sequences that seemed more intuitively random. (It is "easy" to calculate a sequence if we can write a short program to do it.) Therefore, Kolmogorov said, a sound measure of the randomness or complexity of a sequence is the length of the shortest possible program that produces it. He proved that this measure is not essentially dependent on the quirks of any given scheme for setting up the machines. Similar definitions of algorithmic complexity were also made independently by R. J. Solomonoff and Gregory Chaitin. Kolmogorov began to demonstrate that one could construct both information theory and probability starting in a natural way from the concept of algorithms.

In 1962 Kolmogorov began publishing papers analyzing the metrical structure of Russian poetry, which he loved, using statistics. He eventually published eleven papers; it was more than just a passing fancy. He examined poetry by Eduard Bagritsky, Anna Akhmatova, Pushkin and Vladimir Mayakovsky. He also wrote on meter and the connection between the rhythms of Russian speech and classical meters. There were precedents in Russian mathematical history for this. A. A. Markov (senior), of the Markov chains, asserted that the string of vowels and consonants in *Eugene Onegin* was as pure an example of a Markov chain as he could find.

Kolmogorov believed that the school years were crucial in the creation and development of mathematicians. He had seen older mathematicians continuing at the university as if nothing had changed, as they aged and their work declined. So, in large part, he changed his emphasis to the creation of new mathematicians. He continued to attend and contribute ideas in seminars at the university, to head departments—he headed the logic department at MSU after Markov's death—and was instrumental in establishing the Laboratory of Statistical Methods at MSU. He still did creative mathematical work but published at a slower rate. However, the largest part of Kolmogorov's time from 1963 on was spent establishing Physics and Mathematics School No. 18 and then contributing to it as a teacher and in other ways. School No. 18 (often referred to as the Kolmogorov School) accepted students with high mathematical and scientific aptitude coming from many places, but it focused on students from the provinces with poor access to first-class mathematics education. Matiyasevich attended this school—though he came from Leningrad. In 1970 a documentary film on the school was made with the title "Ask Your Questions Boys." Yet the school was more inclusive than that title suggests. There is a beautiful photograph on the back cover of *Kolmogorov in Perspective* of Kolmogorov surrounded by school-age pupils, some of whom are certainly girls.

Kolmogorov threw himself into the school for fifteen years, never receiving any compensation—teaching older students, going on school camping trips, telling the students about music and literature. Many mathematicians close to Kolmogorov taught there, likewise for the pleasure of teaching. He worked on curricula and texts for high school mathematics. He worked on *Quantum*, a science magazine for high school students also distributed in the U.S. For years, he had been writing articles for the *Large Soviet Encyclopedia*, possibly remembering how he had first

Kolmogorov with his students at School No. 18.

studied mathematics out of an encyclopedia. In the 1980s, he was editor-in-chief for *Russian Mathematical Surveys*.

Kolmogorov's physical energy waned and toward the end of his life he was severely limited by a neurological disorder some have called Parkinson's disease. However, the symptoms sometimes mentioned as the worst, muscular weakness and blindness, don't match Parkinson's. Toward the end he was bedridden and could not speak, though during earlier periods of the disease there were stories of him escaping his caregiver to go bathing in one more cold Russian river or pond. Aleksandrov continued also to seek out physical challenges. B. A. Rosenfeld says: "I last saw him on the Courland Spit (Kursu Neringa) in Lithuania, where he was recovering from injuries suffered when a motorboat passed over him while he was swimming under water."[55] At the end, Kolmogorov's students and former students helped his wife attend him twenty-four hours a day.

In April 1986, on his eighty-third birthday, a group gathered at the country home Kolmogorov had purchased with Aleksandrov, now dead. Kolmogorov's students—presumably some getting older themselves—delivered testimonials. Kolmogorov couldn't speak that day, so the next day dictated the following:

> There was talk about my allegedly inexhaustible youth. I am grateful for such an appraisal, but I'd better introduce certain limits to it. Age is anyway objective and one cannot escape it. Happy age. . . . How can it be realized? Either by the refusal to produce any new results, or by the tolerance of an actually shallow existence. Leaving that aside, the old man can view this period as bright and happy,

> but it will be inevitably combined with sad feelings about whether I can do this or that. . . .[56]

No matter what he looked at, be it as incomplete and misleading as the Novgorod cadastres or as hard and clearly drawn as Hilbert's thirteenth problem, Kolmogorov was not a man who fooled himself. Confronting physical decay and death, there is not a hint of dishonesty. However, neither is there defeat. He goes on to talk a bit about what he still can do.

Kolmogorov died October 20, 1987, about a year and a half after that day with his students.

> *We usually pitched camp on a sandy island, on the upper or lower extremity, where the flow of the water is especially felt. During the first days of our journey we often swam at night; in the white summer nights, gliding along past the overgrown osier beds on the bank, the air filled with bird song, made a lasting impression on us. We wished that this could go on forever.*

It is strange that the greatest flowering of Russian mathematics occurred during a period of repression and murder. V. I. Arnold, co-solver of the thirteenth problem and a contributor to the sixteenth, has been an honest presence during this period. We have quoted him several times already, as he is prone to saying interesting things in interesting ways. If one were to form a club of great living mathematicians, Arnold would be a charter member. He is still active, and appreciations and surveys of his life and work by his contemporaries are just beginning to appear.

Vladimir Igorevich Arnold was born June 12, 1937, in Odessa. His father, Igor Vladimirovich Arnold (1900–1948), was a well-known number theorist. He died when Arnold was eleven. Arnold says that the physicist M. A. Leontovich "helped our family to survive when my father died."[57] In 1954 Arnold began at Moscow State University. He has described Kolmogorov as a teacher:

> He never explained anything, just posed problems, and didn't chew them over. He gave the student complete independence and never forced one to do anything, always waiting to hear from the student something remarkable.[58]

Arnold thrived in this setting and in 1957, at nineteen, brought Kolmogorov his solution to Hilbert's thirteenth problem, finishing what Kolmogorov himself had begun. He received his *kandidat* degree in 1961 with a thesis on the solution.

Kolmogorov then suggested that Arnold choose a problem on his own. Arnold has written that next he "wanted to choose something completely orthogonal to all the works of Kolmogorov. This was difficult, because he was working on so many subjects, but still I tried to invent my own problem. I had the list of Hilbert problems, written down one by one in my notebook. (I. M. Gelfand once saw it and laughed a lot)."[59] In his work on the thirteenth problem, Arnold had been thinking about functions that lived on trees. It seemed natural to study functions that lived on more complicated curves, which included curves with cycles. This led him to Poincaré and the topic of repetition in dynamical problems like the three-body problem. He proved some theorems and brought them to Kolmogorov. He was young and not at all widely read. Kolmogorov told him to read his 1954 paper on the problem of small denominators in the three-body problem. When he did this, Arnold realized he had not escaped Kolmogorov's orbit. Kolmogorov had not actually published a full proof of his major theorem. Moser, in one of the more difficult-to-write less-than-a-page reviews ever recorded, made the case that the way to the proof was not contained in Kolmogorov's paper and that, as it stood, there were substantial difficulties with the brief outline Kolmogorov had presented. Arnold was drawn to this problem and in 1963 published the first full proof of the 1954 theorem. In a series of papers published 1961-63 Arnold substantially broadened and deepened KAM theory. He received his Russian doctoral degree for this work.

In 1964 Arnold found instances of instability within the larger pattern of KAM theory, where some initial configurations drifted or leaked substantially away from those of the majority. This is called "Arnold diffusion" now. In 1966 he published a paper on hydrodynamics, analyzing the shape of solutions that explained mathematically why fluid flow tends to be so unstable. (He looked at it in the context of infinite-dimensional Lie groups.) This leads to a deeper understanding of why the weather is so difficult to predict, for deep reasons residing in the equations, even with all the improvements of data collection and computing power that have occurred. In 1971 Arnold published his work on the sixteenth problem and inaugurated a new period in the subject of real algebraic geometry. He has published important work in a wide range of areas—well over 300 publications so far.

The word "singularity" (for a place where a function is undefined or goes to infinity) appears in Arnold's work again and again, in questions of plane curves and caustics formed from plane curves, in wave propagation and reflections, in hydrodynamics, and generally in studies trying to make some sense of complex systems. Trying to understand where the mathematical system that gives you a value breaks down has amounted to a program. (Crystallographic groups are prominent in one approach.) Arnold has written many textbooks and has had many students.

Arnold was made a full professor at Moscow State University in 1965. In 1986 he severed ties with the university, saying that after Petrovsky's death in 1973 the university had been ruined by "party rectors." Since then, he has worked at the Steklov Institute in Moscow. Since 1993 he has been spending half the year at the University of Paris. Arnold could have worked anywhere he wanted after the dissolution of the Soviet Union but continued to spend half the year in Moscow.

Though Arnold is drawn to "difficult" subjects, he wants things explained in as simple language as is possible. This comes through in the following:

> It is almost impossible for me to read contemporary mathematicians who, instead of saying "Petya washed his hands," write simply: "There is a $t_1 < 0$ such that the image of t_1 under the natural mapping $t_1 \rightarrow$ Petya (t_1) belongs to the set of dirty hands, and a t_2, $t_1 < t_2 \leq 0$ such that the image of t_2 under that above-mentioned mapping belongs to the complement of the set defined in the preceding sentence."[60]

His own writing usually follows this advice. He is opinionated, referring to "criminal bourbakizers,"[61] "some criminal algebraist,"[62] and "administrators in most countries, like pigs under an oak tree"[63] and stating opinions such as:

> By the way, the 200-year interval from Huygens and Newton to Riemann and Poincaré seems to me to be a mathematical desert filled only with calculations."[64]

or

> Bizarre questions like Fermat's problem or problems on sums of prime numbers were elevated to supposedly central problems in mathematics. ("Why *add* prime numbers?" marveled the great physicist Lev Landau. "Prime numbers are made to be multiplied, not added!")[65]

Arnold loves vigorous physical culture in the Russian tradition. Smilka Zdravkovska, who interviewed Arnold, refers to "almost daily marathon hiking, biking, swimming or skiing trips."[66] One story that I have heard conveys these enthusiasms. When Arnold was visiting Berkeley in summer of 1989, he told Marina Ratner that he wanted to swim across the bay at the Golden Gate Bridge. They picked a day and went down to the bridge. Arnold gave her his video camera, and she went up on the bridge to film his progress. The wind was blowing so hard that she had trouble holding on to the camera, and as she looked down at about the first pier she realized that though he was swimming very hard he was not moving. She began to worry. "After some time I saw that he finally was turning back and, knowing that Arnold would never turn back unless there was an extraordinary reason, I realized that something went very wrong. When he returned, he was very angry and told me that he could not overcome an extreme force that was holding him back."[67] Yakov Eliashberg says that the next day Arnold appeared at UC Santa Cruz to give a talk and when someone asked, "How are you?" he replied, "Terrible, yesterday I failed to cross Golden Gate."[68] Richard Montgomery says that Arnold told him about his exploit and says, "I can say with some confidence that it is lucky that the tide felt like a wall to him. If he had made it, say, 1/3 of the way across into an ebb tide, he would have been swept miles out to sea. The ebb tide there is about 6 knots. A kayak can barely keep even against it. A person probably swims, at a full sprint,

Vladimir Igorevich Arnold on the Golden Gate Bridge.

at a speed of around 4 knots."[69] Of course, it seems odd that someone who has done so much work in the mathematics of physics wouldn't have thought about this. Montgomery says that, the same month, when he and Arnold were kayaking to Angel Island, they wandered into a sailboat race "with 40 ft. boats whizzing past us." He had some difficulty persuading Arnold to turn back.

Then there are his work habits:

> When I cannot prove something, I put on my skis and ski 40 or 60 kilometers (usually in bathing trunks). During that time, the difficulty usually resolves itself, and I return with a ready solution. . . .[70]

He then comments that the solution he has just thought of is usually wrong, but he is now thinking about the problem in a new way.

Arnold's interests are remarkably broad. There may not be any universal mathematicians anymore, but on Arnold's Web site I found the following:

> Research Interests:
>
> Dynamical Systems, Differential Equations, Hydrodynamics, Magnetohydrodynamics, Classical and Celestial Mechanics, Geometry,

> Topology, Algebraic Geometry, Symplectic Geometry, Singularity theory[71]

Well, there is no number theory on the list (though Arnold has published several papers on number theory). In 1997 *Russian Mathematical Surveys* published a sketch of Arnold's career in honor of his sixtieth birthday. It was signed by twenty-two mathematicians from an array of specialties. A small planet, about six kilometers in length, has been named after him. It is called Vladarnolda. As this manuscript was going to press, he won a Wolf Prize.

After the 1994 International Congress of Mathematicians in Zurich, Arnold published "Will Mathematics Survive? Report on the Zurich Congress." It contained some of the opinions quoted above, but ended on the note:

> After the tiring Congress, I spent a day at the home of my old friend A. Haefliger near Geneva. We climbed from 1500m to 3000m in the mountains near the Rhône valley, about halfway between the Jungfrau and the Matterhorn, and I got to swim in a glacial lake. On the return I picked mushrooms, sorrel, blueberries, and wild strawberries, and made my hosts a dinner. . . .[72]

Past Chernaya Rechka, to 61 Savushkina Street

The Twenty-First Problem

« Plemelj, Bolibruch »

Hilbert asks for "Proof of the Existence of Linear Differential Equations Having a Prescribed Monodromic Group." This problem is about linear differential equations that have points where the coefficients (rational functions of a variable) have singularities, that is, become infinite. When attempting to solve such an equation (or a collection of equations with coefficients that have a finite collection of singularities), one can form what is called the "monodromy group." The solutions to this kind of problem, as with many things in complex analysis, are not single-valued. One circles a singularity (like a horse in a hippodrome), jumping from one local value of the solution to another. The monodromy group captures this. An element of the monodromy group takes one solution and gives you another, and so the group is like a map that shows how the solutions connect to each other. Riemann mentioned going the other direction—starting with the monodromy and singularities and finding equations that fit them. He assumed this could be done. Hilbert wanted proof.

This is often referred to as the Riemann–Hilbert problem. The italicized portion is "*To show that there always exists a linear differential equation of the Fuchsian class, with given singular points and monodromic group.*" (Mathematicians believe Hilbert must have meant "system of equations" rather than a single equation, because it was known at the time that the problem was not true if one is limited to a single equation—further, by using flexible notation, one can write a system of equations as a single equation.) In the words of D. V. Anosov and A. A. Bolibruch: "This is a distinctly formulated problem which has to be answered 'yes' or 'no' (whereas some of Hilbert's problems are formulated not so distinctly . . .)."[1] Hilbert is asking if, given a monodromy group and set of singularities, one can always find a Fuchsian system of equations to match.

In 1905 Hilbert himself published a complicated paper that purported to give at least a partial solution to the twenty-first problem. (Freudenthal credited this work as the solution.) However, just a year or two later, Josip Plemelj (1873–1967) offered a simpler proof of a more general theorem based on his own methods, which were of value in and of themselves. This work was expanded and published in 1908. Mathematicians didn't bother to try to get through Hilbert's proof, which claimed less, with this new, clear result available. Plemelj was widely credited with solving the twenty-first problem in the affirmative.

Plemelj was born on December 11, 1873, in a small village, Grad, on the shores of a small lake in the Julian Alps in what today would be the northwest corner of Slovenia. His father was a "joiner," probably a cabinetmaker, who died when Josip was two, leaving behind his wife and five children. Josip received an education, however, finishing all his grammar school mathematics in four years and then tutoring his classmates. He moved on at twelve to the gymnasium in Lyublyana and then went to the University of Vienna intending to study astronomy but gravitating to mathematics. In 1898 Plemelj graduated with a dissertation on differential equations that mostly duplicated work already done but accomplished the results using simpler techniques. He then traveled to Berlin and Göttingen, where he stayed from 1899 to 1902, studying with Klein and Hilbert. He was already working on differential equations and monodromy groups. He used this work for his *venia legendi*, which allowed him to become a privatdozent at the University of Vienna.

After he solved the Hilbert problem (though he thought of it more as the Riemann problem), Plemelj was appointed by imperial decree first extraordinarius

Josip Plemelj (courtesy of Bojanka Pust).

and then ordinarius professor at Chernovitz University at the eastern edge of the Austro-Hungarian Empire (today Chernovsty in the Ukraine).[2] He moved back to Lyublyana in 1919, where he was appointed professor at the newly established university, in just-created Yugoslavia. He stayed in this position until 1957, when he was appointed professor at the Slovene Academy of Science and Arts at the age of eighty-three. Plemelj published a book in English in 1964, *Problems in the Sense of Riemann and Klein*, that is in part about his solution of Hilbert's twenty-first problem. He was 91. He died in 1967 thinking he had solved the problem.

Andrei Bolibruch was born on January 30, 1950, in Moscow. His father, Andrei Vlasovich Bolibruch, had been a gold medal student in his village in the Ukraine who qualified for Kiev State University but never attended, as he was drafted into the army when the Second World War began. Bolibruch's father was wounded seven times, rising to the rank of major. After the war he continued in the army and attended the military Academy of Frunze and then the Academy of General Headquarters. He spoke French and a little German and was knowledgeable in literature and history. Andrei's mother, Tatyana Ivanovna (Pimanikhina), was good at mathematics and entered college. She married at nineteen, though, and led an itinerant life following her husband's military postings. Bolibruch remembers that when he was six they lived in a forest in Belorussia where there were only two buildings. He wrote:

> When I was 6 years old my father decided to prepare me for school. I could read, write and add numbers. My father decided to teach me math. In particular he tried to explain to me how to calculate the area of a circle!! And I remember very well that I could not understand him. He told me, "The area of the circle equals pi r squared. Do you understand?" No, I answered, you say to write numbers, but what is an area? . . . Now I understand that he was not a mathematician and he could not say that area is a functional on the set of measurable figures with the following properties, etc.[3]

He says that later in school he became quite good at manipulating the formula for the area of the circle but regrets that he didn't ask again for a long time the "childish question," what is area?

Andrei's father was very busy, but vacation times were good. Traveling by car or train, the family visited interesting places in the Baltic republics, in the area around Kaliningrad (formerly Königsberg, Hilbert's home town), and in the Crimea. Bolibruch particularly remembers what he calls "fishing with disks" with his father, for pike.[4] They would put a small fish on the hook and throw it in the lake with the line up over a disk of about eight inches diameter as a float. The underside of the disk was painted red, so when a pike took the hook, the disk flipped over showing the red. They slept in a tent and left the lines in the water overnight. He says:

"This kind of fishing is not usual and I don't remember why and how we began to use it. It is impossible to use it in a river, just in a lake, and very often you lose disks. But when you see a red disk it is really very impressive."[5] He remembers one particular morning when they saw a red disk. He and his father got the fish into a boat and transported it to shore, still alive. It was almost as long as Andrei was tall. On the bank the fish slipped its hook and there was a wrestling match. Bolibruch has a picture of himself holding the successfully subdued fish.

Bolibruch (like Matiyasevich) particularly remembers the *kruzhoks*, study groups that were varied in their focus, especially "*kruzki*" on "radio, math, electronics, sport (swimming, chess, fencing)." He also remembers the "pioneer camps." He wrote to me:

> I spent a lot of time in such camps. They also were almost free and they were located in excellent places. Usually the time of one's childhood is the best time one's life, but my childhood also coincided with the time of Khrushchev's thaw. Of course, I could not understand things concerning the political or economic situation in the country . . . but it was, or may have been, the best period of the Soviet Union. I think the atmosphere of the (relative) prosperity of the country influenced positively the whole life of that time.[6]

Bolibruch spent most of his childhood in Moscow; in Tallin, Estonia; and then in Kaliningrad. He was active in school theater in Kaliningrad when he was eleven to fourteen years old and won prizes in recitation. He liked literature. At twelve he had a good mathematics teacher; she gave many interesting problems, particular in geometry. He dates his interest in mathematics, which he started studying intensely when he was fourteen, from that time. That year he won the mathematics Olympiad in Kaliningrad and went on to place second in the all-Soviet Olympiad, sharing that spot with A. Suslin and I. Krichever—both well-known mathematicians now. This won Bolibruch an invitation to Boarding School No. 45 in Leningrad, a specialized mathematics and physics school (similar to Boarding School No. 18, in Moscow). A Web page for alumni of School No. 45 shows them to be scattered all over the former Soviet Union, Israel, and the United States. In 1994 a reunion of the mathematics alumni resulted in a volume of mostly research papers. Bolibruch contributed "Some Memories of Boarding School #45."

He wrote, "In those days the boarding school was located . . . on Savushkina Street, beyond Chernaya Rechka," only 30–45 minutes from the center of town. Every week of winter 1966 the students were taken to the Hermitage Museum and shown especially the modern art, "from the impressionists to Picasso," with good commentary.[7] Bolibruch became a passionate theatergoer. There were regular concerts, readings by "unofficial poets" with discussions following, and presentations of what was, in effect, underground theater. Bolibruch said, "Every other student of our school wrote poetry."[8] He read unapproved poets: Velimir

Khlebnikov, Boris Pasternak, and Marina Tsvetaeva. They were not actually forbidden, but seldom published and hard to come by. He never got into trouble for reading them.

He says that "the peak of this poetic activity was reached in winter of 1966."[9] A school newspaper appeared with "optimistic" poems by students and was tacked up all over the school. A parallel set of literary journals was prepared in the school's dormitories containing less optimistic poems. Bolibruch's dormitory was known as "Uncle Toms' Cabin" in honor of the room's president, Victor Toms. Bolibruch was editor and editorial board rolled into one for their journal, *Quiet Pond. Quiet Pond* contained "dismal words" and was posted inside the clothes closet. At just this time the Minister of Education came to visit. Inspecting the school, he opened the closet and came upon the journal. He said: "Now I see what is really going on here," and departed for Moscow.[10] This minister was opposed to elite, specialized schools, because they picked their students on merit and gave them a better education. The irreverent poetry was ammunition for him—it was possible he might succeed in closing the school down. "At the time we were still in the classrooms, but 'good news spreads quickly,' and we returned to the dorm with our heads down, in the worst mood imaginable."[11] The teachers, instead of punishing them, came in and apologized "for the behavior of some grown-ups incapable of understanding modern poetry. They asked us not to keep a grudge against such people, who did not know any better." They invited the dormitory poets to read at an upcoming reading given by the school poets. The school survived, and Bolibruch won the gold medal at graduation. He closes his article:

> Today our boarding school is housed in a different part of the city, closer to the new university campus. But each time I come to St. Petersburg (which does not happen as often as I would like), I get on the #80 bus and ride all the way past Chernaya Rechka, to 61 Savushkina Street.[12]

Bolibruch was accepted at Moscow State University. "It was maybe the best place to study math throughout all the world," he wrote to me. "Some of the names of mathematicians who worked at that time in MSU: Kolmogorov, Aleksandrov, Arnold, Novikov, Sinai, Anosov, Manin, etc. . . . My mathematical life was extremely interesting."[13] During this period there were many "very good" theaters in Moscow with inexpensive student tickets. He also mentions the libraries.

Bolibruch was first interested in algebraic topology and studied under M. M. Postnikov. As a graduate student he worked on Pfaffian systems on a complex manifold, because this subject mixed the methods of algebraic topology and analysis. He was drawn into a seminar given by V. A. Golubeva and A. V. Chernavskii on Fuchsian differential equations on a complex manifold, a related topic. They tried to investigate a generalization of Hilbert's twenty-first problem to a multidimensional case, following the ideas of Gerard, Deligne, Röhrl, and Hain. Bolibruch's

1977 thesis was on this topic. Everyone involved thought the classical case, as stated in Hilbert's twenty-first problem, was solved in the affirmative.

After completing his thesis, Bolibruch worked for thirteen years at the Moscow Institute of Physics and Technology, working on what he calls "elastohydrodynamics" or "tribology" to earn money. These topics are part of the mathematical theory of lubrication. He wasn't working primarily as a pure mathematician. In these years, Armando Treibich Kohn (published 1983) and Arnold and Ilyashenko (published 1988) pointed out that there was a gap in Plemelj's proof. Bolibruch had returned to the subject on his own and was again working on a more modern generalization of Plemelj's result. Bolibruch's flow back and forth illustrates the connectedness between relatively abstract mathematical topics in analysis and applied results. He turned more attention to work on Fuchsian systems, again with Golubeva, Chernavskii, and V. P. Lexine. They felt the ground under them unsure and so looked at a variant of the problem and got nowhere. In 1987 Bolibruch returned to the twenty-first problem in its original form and in 1989 found a counterexample to Plemelj's result. Not only was there a gap in the proof, but the result was false. The counterexample can be stated simply, but proving that it is a counterexample is less simple.

Anosov and Bolibruch, in a book on the problem, comment dryly, "Of course, misunderstandings are common for human activity, but it is not so common that a misunderstanding in mathematics remains for more than 70 years."[14] This is doubly or even triply embarrassing for mathematics, because the problem had been "re-solved" a number of times over the years. For example, for the 1974 AMS conference on Hilbert's problems, Nicholas Katz wrote: "Traditionally, it was viewed as a problem in 'function theory,' and in that setting has been solved repeatedly, by such men as George Birkhoff, Plemelj, and most recently by Röhrl. But as Lipman Bers remarked in his talk on uniformization, if a problem is worth solving, it's worth solving several times."[15]

The confusion hangs on the gap between a Fuchsian singularity and a "regular" singularity. A Fuchsian singularity is a "simple pole." A regular singularity is one where the function doesn't grow too rapidly as one approaches the pole. A simple pole does not grow too rapidly, so it is easy to show that a Fuchsian singularity is regular. Plemelj proved that a regular system of the desired type exists and then more as an afterthought—the heavy lifting was done—showed that he could use this to build "another system with the same monodromy and the same singular points, which is now a Fuchsian system for all except perhaps one of the points."[16] He never quite got around to that one possibly suspect point.

Plemelj's result and his methods were important in their own right. Bolibruch wrote, "After the publication of Plemelj's result, the subject matter of articles connected with the Riemann–Hilbert problem moved, basically, towards the effective construction of a Fuchsian system with given monodromy. . . ."[17] The basic existence theorem wasn't questioned. The positive result that Plemelj had proved would probably be accepted as a solution in a positive sense if he had also found the counterexample to the more general result. During the years that followed Plemelj's paper, the language of mathematics changed drastically. By the time Helmut Röhrl offered his modern solution (and solution to a broader and different problem), contact had been effectively lost with the original problem and language. Therefore, Röhrl proved something that was true but was not actually a solution to Hilbert's problem. The abstract of Katz's article, "An Overview of Deligne's Work on Hilbert's Twenty-First Problem" is:

> Hilbert's twenty-first problem on the existence of differential equations with *regular* singular points and prescribed monodromy is interpreted as a . . . problem of algebraic-analytic comparison. Deligne's solution is outlined. Several open questions are raised.[18] [my italics]

This is a correct statement of what can be proved in a positive direction—a theorem with *regular* singularities. Only when Kohn, Arnold, and Ilyashenko went back and looked at the historical record was it observed that Hilbert's problem was not actually solved! The received assumption of what the problem is dies hard, and in *The Ball and some Hilbert Problems* (1995), Rolf-Peter Holzapfel writes in passing, "The final solution of this Hilbert Problem has been given by H. Röhrl."[19]

How much work has been done based on an affirmative solution to Hilbert's twenty-first problem? Bolibruch wrote to me:

> I cannot estimate the number of wrong works based on the wrong result on the twenty-first Hilbert problem, I just know that this number was large enough. Maybe a more correct question here could be the following: Are there important results which turned out to be wrong because of using the wrong result of Plemelj? I know at least one such work: it was published in 1970 by a well-known Japanese mathematician, where he used the "positive solvability of the 21st Hilbert Problem" to prove a very important but wrong result that the Schlesinger equation has no movable singular points at all![20]

It could have been worse. Often, later results that are correct were used; though they were not actually solutions to Hilbert's problem, they were still correct.

Unsolving or solving Hilbert's twenty-first problem had a good effect on Bolibruch's life. In 1990 he was invited to work at the Steklov Institute and then, in 1991, he used his result for the Russian doctoral degree. In 1994 he was elected a corresponding member of the Russian Academy of Sciences and was made a full member in 1997. In 1994 he was made vice director of the Steklov Mathematical Institute and also became vice president of the Moscow Mathematical Society with Arnold as president. In 1994 he gave an invited talk at the International Congress of Mathematicians in Zurich.

Andrei Bolibruch (courtesy of Andrei Bolibruch).

Bolibruch's work schedule and pattern of employment tell a lot about life for mathematicians in Russia at the end of the twentieth century. Salaries were low, about $80 a month, so "almost everybody" had two or three additional jobs. The Soros grants—Matiyasevich has one—were, and are, important. There are grants from the Russian Foundation of Basic Research, the European INTAS, the U.S. National Science Foundation, and joint grants with Germany and France. Apartments were cheap in Moscow—$10 to $15 a month—and so was public transportation, but everything else was almost as high as in France. Bolibruch is able to earn money outside of Russia. In 1993–1997 he had a temporary position in Nice, France, for three months a year. As of 1999 he had a similar position in Strasbourg. He has spent time at the Max Planck Institute in Bonn and has visited Mexico, the U.S. (Princeton, Penn State, Berkeley), Brazil, Holland, and Italy, etc. Like Arnold, he tries to stay in Moscow about half the year. On April 2, 1999, Bolibruch wrote:

> [These travels abroad] are a good possibility to do mathematics, to be in contact with my colleagues in the West and of course to earn money for life in Russia. And this life is complicated enough, but extremely interesting: a lot of changes, good and not so good. The situation changes rapidly.[21]

Work on It

The Nineteenth, Twentieth, and Twenty-Third Problems

The Nineteenth Problem

The calculus of variations was an early but elegant arrival at the party presided over by post-Newtonian physics. It is based on the assumption that nature prefers the world arranged so that certain mathematical functions are minimized. The dancers at the party will follow a path that minimizes the integral of the kinetic minus the potential energy (the lagrangian). The goal is to find the minimum: the most effortless dance, the most efficient configuration, the path that "costs" the least energy. For example, soap bubbles always arrange themselves so that a specific mathematical function is minimized. In the eighteenth problem we saw that the possible shapes for crystals could be understood by looking at geometrical symmetries and analyzing those symmetries algebraically. Crystals can also be approached through the calculus of variations. They are the shapes that minimize certain functions. Newton's equation can be rewritten using the calculus of variations in a way that makes some problems easier to solve. There is a fine explanation and history of this subject, *The Parsimonious Universe* by Stefan Hildebrandt and Anthony Tromba.

In the nineteenth problem Hilbert asks, "Are the Solutions of Regular Problems in the Calculus of Variations Always Necessarily Analytic?" The highlighted portion of the problem is, "*Does every lagrangian partial differential equation of a regular variation problem have the property of admitting analytic integrals exclusively?*" Hilbert uses general terms—"lagrangian partial differential equation," "a regular variation problem," and "calculus of variations"—that can mean different things to different people. He talked about "this sort of problem" rather than using more precise language. The problem is phrased in the context of the calculus of variations, but historically the answers have been given as statements about classes of partial differential equations.

This problem has considerable importance outside of pure mathematics. If a scientist or engineer has an equation that comes from the calculus of variations, knows that there is a solution, and knows that the solution is "reasonably" smooth, then he or she can try to write a computer program that grinds out an answer based on the assumption that the answer is in the form of a power series. On a practical level this has meant that workers in this area focused more on the word "regular," as it is usually understood, than on the details of how Hilbert defined "regular" mathematically.

In 1904, Sergey Natanovich Bernstein of Russia defended a thesis at the Sorbonne that contained a general result about a class of "elliptical partial differential equations," a big step forward. Bernstein's father was a lecturer in anatomy and physiology at the university in Odessa. Bernstein was in Göttingen for two terms in 1902–1903 and studied under Hilbert, among others. Despite his degree from Paris and his brilliant thesis, he was not allowed to teach when he returned to Russia until he got a Russian "master's degree" (not the highest degree granted), in part because he was Jewish. He had defended his doctoral thesis in Paris before Poincaré, Hadamard, and Picard—not good enough! He passed his master's examination in 1906 in St. Petersburg and then taught high school for a time. Back in Kharkov, in

Sergey Natanovich Bernstein (courtesy of the Archives of the Mathematische Forschungsinstitut Oberwolfach).

his native Ukraine, he passed his master's exam a second time, in 1908, and defended his thesis again. Only then was he allowed to lecture at this provincial university. In 1913 Bernstein received his Russian doctoral degree from Kharkov, this time for a thesis that "solved" Hilbert's twentieth problem—more on this shortly. Only after the Revolution was he appointed professor and allowed to participate fully in the university.

In 1917 Bernstein offered the first axiomatic treatment of probability theory and wrote many fundamental papers on this subject. Many honors came to him, including memberships in the USSR Academy of Sciences and the Paris Academy of Sciences. He was one of two mathematicians—the other was A. N. Krylov—to try to defend Luzin at his 1936 trial and, not long after this incident, at fifty-eight, published one of his most important papers on the constructive theory of functions. Bernstein continued publishing papers into his seventies and died at eighty-eight.

There is a history of disagreement over whom to credit as solver of the nineteenth problem. Enrico Bombieri wrote for the AMS conference on Hilbert's problems that Bernstein's theorem "was considered at the time a solution to Hilbert's 19th problem." In an obituary for Bernstein, Aleksandrov, N. I. Akhiezer, Gnedenko and Kolmogorov wrote: "He proved and generalized the theorem, conjectured by Hilbert in his Nineteenth Problem. . . ."[1] Aleksandrov, writing a few years later with Oleinik, however, referred to work Petrovsky did in the 1940s, saying, "This result was a complete solution of Hilbert's nineteenth problem and a broad generalization of Bernstein's theorem."[2] Michael Monastyrsky gives credit to Petrovsky in an aside in his book on the Fields Medals. Petrovsky did important work on this problem; however, I see no worldwide consensus that gives him specific credit. The generality of the way the problem is phrased makes it difficult to decide such an issue.

Ivan Georgievich Petrovsky, who it is safe to say did important work on the nineteenth problem and played a role in both halves of the sixteenth, was born in 1901 into a prosperous family in Sevsk, south-southwest of Moscow, almost in the Ukraine. His grandfather, a merchant, expected him to take over the family business. Petrovsky enrolled in the university in Moscow in 1917, however, intending to study chemistry. His studies were immediately interrupted by the October Revolution. He realized that his family was in danger, returned home, and persuaded them to abandon their property and flee to Elizavetburg, in what is now Azerbaijan. There he worked at odd jobs and after a few months entered a local technical school to learn machine construction. At the school he encountered a good mathematics book and found his lifelong interest. Petrovsky returned to Moscow, working as a janitor, and in 1922 enrolled in Moscow State University, this time as a proletarian.

Petrovsky was part of the constellation around Luzin. Landis writes, "Unlike many of Luzin's students, he was not brilliant."[3] Petrovsky studied under Egorov, reading mathematics with great thoroughness and, as it turned out, generated original and penetrating ideas despite his deliberate style. In 1933, 1938, and 1949 he published significant partial solutions to part one of the sixteenth problem. He published many papers on partial differential equations and did important work on probability theory, collaborating with Kolmogorov.

Petrovsky was elected dean of the Mechanics and Mathematics faculty at Moscow in 1940 and in 1951 became rector of the university. He loved books and, despite the collectivization of Soviet society, managed to accumulate a personal library of 30,000 volumes on a wide range of subjects (5,000 on mathematics) and a space large enough to hold them. There was no catalog, but it is said that he could find any book that he wanted. His wife donated these books to the university after his death, and in 1976 a Petrovsky library was opened. Landis says that Petrovsky was interested in a wide variety of the subjects he was in charge of as rector and even participated in archeological digs near Novgorod. According to Landis: "He knew a great deal about painting. This was all the more remarkable because he was colorblind."[4]

Paolo Marcellini published an article in Italian in 1997 that explicitly deals with progress on the nineteenth and twentieth problems while actually referring to their text. The nineteenth problem seems to be widely identified with the question, "Do all partial differential equations of the *elliptic* type admit only analytic solutions?" Hilbert's attempt at pinning down regularity is technical. The definition of "elliptic partial differential equation" has changed historically, as the initial concept was enlarged and generalized. Today things are often phrased in terms of "elliptic operators." However, one thing Hilbert's definition of "regular" and the various versions of "elliptic" tend to agree on is that the character of the equations as measured by an inequality doesn't change from place to place—it doesn't go to zero or change signs. Marcellini says that when speaking of regularity "in modern terms (see for example L. Nirenberg [1955]) one says the condition of ellipticity (or strong ellipticity)."[5]

According to mathematician and historian Morris Kline, the first time anyone called a partial differential equation elliptic was in 1889, when Paul David Gustav Du Bois-Reymond (1831–1889) published a method of classifying general homogeneous linear equations of the second order using "projections of the characteristic curves."[6] These projections were of the form $Tdx^2 + Sdxdy + Rdy^2 = 0$ (T, S, and R are smooth functions of x and y), and this equation approximates the behavior of the partial differential equation close to a given point. If $TR - S^2 > 0$, this is similar in form to the equation of an ellipse; if $= 0$, a parabola; and if < 0, a hyperbola. An equation that is elliptic (at least for smooth linear equations of the second order) can be transformed by a change of variable into the form of Laplace's equation—describing a conservative force field in the plane. Likewise, a parabolic equation is essentially like the equation for the flow of heat, and a hyperbolic equation is like the equation of a vibrating membrane—a wave equation. One talks about an equation being elliptic in a region. If the sign of $TR - S^2$ changes from place to place (is not regular), then the equation can't be classified in this way. This classification system was generalized, expanded, and adjusted to apply to more complicated equations with more variables, possibly with higher-order derivatives. Today's breakdown into elliptic, parabolic, and hyperbolic is often all covered under a more general concept of being "elliptic" (e.g., Gerald Folland's *Introduction to Partial Differential Equations*). Not all books agree on the definition, but all good books are consistent in their usage and what is proved matches the definition used. The concept bifurcates into locally elliptic, globally elliptic, strongly elliptic, properly elliptic, pseudoelliptic, elliptic of order . . . even hypoelliptic—presaging the onset of turbulence for the reader. The ways smoothness is measured become increasingly general and sensitive.

Why should solutions to all elliptic partial differential equations be analytic? This is a strong condition and Hilbert comments, "One of the most remarkable facts in the elements of the theory of analytic functions appears to me to be this: That there exist partial differential equations . . ." that only allow analytic solutions. We can get a hint of why this might be if we think of the Cauchy–Riemann condition/equation in complex analysis. This is a simple partial differential equation. The central and amazing fact of complex analysis is that if a function is simply differentiable (smooth) at a point, it is analytic (smooth an infinity of derivative layers deep). A function that satisfies the Cauchy–Riemann equation is differentiable and hence analytic. We can think of the results of the nineteenth problem as a somewhat amazing generalization of this.

Marcellini gives a concise, historical rundown of the central developments of this problem that I will take as my answer, as of this writing, to the question of who solved Hilbert's nineteenth problem. "S. Bernstein (1904) has furnished the first response," which deals with second-order equations in two variables.[7] Bernstein's theorem says that if a solution to a partial differential equation with two variables is smooth measured three layers deep, then that solution is analytic—smooth an infinite number of layers deep. After briefly mentioning Bernstein's methods, Marcellini continues: "L. Lichtenstein (1912) improved the result of Bernstein." Solutions that are smooth two layers deep are smooth three layers deep and are therefore analytic. In 1929 E. Hopf improved the result further in a more technical way. Marcellini continues:

> With contributions from many authors such results were extended to the general case $n \geq 2$, $m \geq 2$ [two or more coupled equations and two or more variables]. We can cite for example: Leray-Schauder (1934), Petrovsky (1939), Caccioppoli (1934, 1950-51), Morrey (1958), etc.
>
> Returning to an imprecise, but effective mode of speaking, in the '50s one could affirm that *every sufficiently smooth solution of a regular variational problem is an analytic function.*[8]

This is essentially what Hilbert asked for.

The question now naturally turns to defining more precisely what it means to be "sufficiently smooth." One chain of results talked about functions that were smooth in the sense that they belonged to a specified "Sobolev space." Hopf's 1929 result talked about functions that were "Hölder continuous." In 1938 C. B. Morrey proved a very strong result for equations with two variables.

For the next twenty years, after Morrey's paper, no one managed to extend the result to equations with three or more variables. Then in 1957 and 1958 Ennio De Giorgi and John Nash, working independently, each did this using different methods. In 1960 Moser found a much more elegant and simple proof of these results that has been fundamental to later work. These ideas have been further explored by Morrey (1966), Guido Stampacchia (1966), O. A. Ladyzhenskaya, and N. N. Uraltseva (1968), etc.

Knowing when regularity begins also brings knowledge of when it breaks down, which then brings efforts both at salvage and to look for unusual things in nature that correspond to the mathematical breakdown. Peter Lax told me that he thought this area of breakdown was one of the most interesting things about this problem. In 1968 De Giorgi showed that his and Nash's earlier results cannot be extended to systems of equations (with higher numbers of variables), by finding an example of an elliptic variational problem whose solution is not analytic—it was not analytic at one point. So the question becomes, when are the solutions to variational problems guaranteed to be almost analytic?

De Giorgi and Nash are an interesting pair. De Giorgi published first, in 1957, but in an obscure place, and so Nash wasn't aware of this work. No one would have been aware of it quickly if Paul Garabedian of Stanford, working in 1957 as a naval attaché in London, hadn't taken a long car trip. He visited mathematicians in Rome and had some good, long lunches. Nobody mentioned De Giorgi's new result. In Naples someone did mention De Giorgi. According to Sylvia Nasar in her biography of Nash, *A Beautiful Mind*:

> Garabedian looked De Giorgi up on his way back through Rome. "He was this bedraggled, skinny little starved-looking guy. But I found out he'd written this paper."
>
> De Giorgi, who died in 1996, came from a very poor family in Lecce in Southern Italy [the heel of the boot]. Later he would become an idol to the younger generation. He had no life outside mathematics, no family of his own or other close relationships, and, even later, literally lived in his office. Despite occupying the most prestigious mathematical chair in Italy, he lived a life of ascetic poverty, completely devoted to his research, teaching, and, as time went on, a growing preoccupation with mysticism that led him to attempt to prove the existence of God through mathematics.[9]

In the year he died, De Giorgi stated one of his central beliefs: "I see this as a journey throughout which, until the end, one must love knowledge completely, expecting that this love will continue in another form even after death."[10]

Nash makes De Giorgi seem conventional. In 1958, after achieving the big result and then having the shock of finding he had not made the discovery alone, Nash began to think that the staff of MIT, where he taught, was behaving strangely toward him. He noticed men in red neckties and decided this meant they were crypto-communists. He said that "abstract powers from outer space . . . were communicating with him through *The New York Times*."[11] His delusions grew. He interrupted a seminar to tell Al Vasquez that there was a photo of him on the cover of *Life* magazine. Vasquez might not have recognized Nash because the photo was disguised to be a picture of Pope John XXIII. Nash knew it was of himself because 23 was his favorite prime number. He rejected an appointment to a prestigious chair at the University of Chicago, apologizing that he couldn't take it because he was scheduled to become the Emperor of Antarctica. He was hospitalized at different times and early on spent time in a locked ward with residents who included the poet Robert Lowell. He became obsessed with the idea of world government and left for Paris, traveling also to Switzerland and Luxembourg, trying to renounce his U.S. citizenship and become *Citoyen du Monde*.

Nash gradually became interested in mathematics again. He played on a computer at Princeton, became a very good programmer, and became known as "the ghost of Fine Hall," who had once been a very good mathematician. By 1990 Nash's obsession with the deep significance of numbers in a kind of code—for example in the *New York Times*—had shifted to questions that looked like number theory. He asked Peter Sarnak after a talk for a copy of a paper Sarnak had written on the zeta function. Wanting to be kind, Sarnak gave him a copy. Nasar describes what happened next:

> A few days later, at teatime, Nash approached him again. He had a few questions, he said, avoiding looking Sarnak in the face. At first, Sarnak just listened politely. But within a few minutes, Sarnak found himself having to concentrate quite hard. Later, as he turned the conversation over in his mind, he felt rather astonished. Nash had spotted a real problem in one of Sarnak's arguments. What's more, he also suggested a way around it. "The way he views things is very different from other people," Sarnak said later. "He comes up with instant insights I don't know I'd ever get to. Very, very outstanding insights. Very unusual insights."[12]

Nash was back—at least to a significant extent. Nasar says the new anti-psychotic medications being developed were not responsible and that Nash's improvement appeared to come from changes within him.

A number of Nash's early ideas survived the passage of time and seemed increasingly important. In 1994 he was awarded the Nobel Prize in Economics—despite some nervousness about what would happen when they actually presented him the award. His private audience with the King of Sweden lasted longer than most; the two were having a conversation on the difficulty of driving on the left side of the road. Nash published new mathematical papers in the 1990s.

Mathematicians are judged for a few lucid moments, and if a mathematician has these moments, he or she might be judged great. In short, eccentricity up to and including madness is not as great a hindrance to career advancement as it would be in some other professions. Some mathematicians cultivate an image of eccentricity—and mathematicians garner no praise for accepting received wisdom. However, if we looked at the rate of occurrence of severe mental illness in mathematicians (as in the case of Nash), I suspect that it wouldn't be much different from that of the general population. It does seem, however, that more members of this minority of mathematicians progress far enough to enter the history of mathematics and receive prizes. In this they would be similar to poets, like Lowell, mentioned above, who need only a handful of truly good poems.

The Twentieth Problem

Hilbert calls the twentieth problem "The General Problem of Boundary Values." Where the nineteenth problem asks if we can ensure that the only solutions we have to look for are analytic (if there be solutions at all), the twentieth problem asks, "Has not every regular variation problem a solution . . .?" This is too big to pin down. It asks if we can ensure that any problem of this type is solvable by restricting conditions at the boundary. Hilbert's salvage of Dirichlet's principle in the year before he delivered his problems was the model for what he wanted done with more general problems here. Despite its generality, the twentieth has been seen as a good problem. James Serrin wrote for the AMS conference:

> Among the prophetic problems in Hilbert's famous list one must surely include the twentieth, the general problem of boundary values for elliptic partial differential equations. This subject, only a seedling in the year 1900, has burst into flower during our century, has developed in directions surely Hilbert never imagined, and today encompasses a vast area of work which to a mathematician of seventy-five years ago would seem little short of astonishing. A list of the mathematicians who have significantly developed the theory is itself impressive: beginning with the work which Hilbert himself contributed in 1900 in the form of a legitimization of Riemann's version of the Dirichlet principle, other important advances through the first half-century were supplied by Serge Bernstein, Jacques Hadamard, Henri Lebesgue, Eberhard Hopf, Richard Courant, O. Perron, Norbert Wiener, Jules Schauder, Jean Leray, K. O. Friedrichs, G. Giraud, C. B. Morrey, Jr., Lars Garding, and others. An even longer list would be required for the twenty-five years since 1950, while a complete bibliography would be monumental.[13]

A good source for detail on work on this problem is the second half of D. Gilbarg and N. S. Trudinger's classic monograph, *Elliptic Partial Differential Equations of Second Order*. Another is an extensive 1969 paper of Serrin's, "On a Problem of Dirichlet . . . ," where, in addition to a detailed study of boundary solvability conditions, there is an extensive historical discussion of previous work, particularly that of Leray and Schauder, Bernstein, and Robert Finn.

The key first steps were supplied by Bernstein again. He approached the problem using the idea of *a priori* estimates—now a bread-and-butter technique in texts on partial differential equations. He showed in a series of papers, 1910–1912, that a well-specified class of equations had solutions. His methods were complicated and difficult, so that the "brilliant underlying idea" was not recognized even by Bernstein himself for twenty years.[14] Once articulated, a flood of new results followed, some of them Bernstein's.

This has some of the feel of a foundation problem. Hilbert wants to be sure that all these problems do have solutions so that we can look for them, secure in the knowledge that they exist, if we are only clever enough to find them. Since a huge number—I am tempted to say "most"—of the questions in physics and other hard sciences can be phrased in this way, the equations at issue in this problem are then, essentially, the equations of the physical world.

The Twenty-Third Problem

Hilbert begins the twenty-third problem, "Further Development of the Methods of the Calculus of Variations":

> So far, I have generally mentioned problems as definite and special as possible, in the opinion that it is just such definite and special problems that attract us the most and from which the most lasting influence is often exerted upon science. Nevertheless, I should like to close with a general problem, namely with the indication of a branch of mathematics repeatedly mentioned in this lecture—which in spite of the considerable advancement lately given it by Weierstrass, does not receive the general appreciation which in my opinion, is its due—I mean the calculus of variations.

The calculus of variations has already been the subject of the nineteenth and twentieth problems—both already general. Still not satisfied, here Hilbert points to the entire subject and says, essentially, work on it. There follows an essay of some substance—much longer than the text of any other problem. The focus is on developing actual techniques that can be used to solve specific equations.

It doesn't make sense to talk of this problem being solved, much less to talk about a specific solver. The relevant question is whether mathematicians responded to Hilbert's suggestion. The answer is: Yes, mathematicians have worked fruitfully on the calculus of variations along the lines Hilbert suggested.

To see what Hilbert himself did on this problem, look at *Methods of Mathematical Physics* by Hilbert and Courant in the 1924 German edition. *The Calculus of Variations* by Gilbert Ames Bliss (1876–1951) also displays early work. For more recent perspectives, look at Stampacchia's article delivered at the AMS conference and Chern's article in the *Mathematical Intelligencer* from the point of view of global geometry.

We Come to Our Census

Census

So, how many of Hilbert's problems have been solved?

Sixteen problems have been solved in a discrete way. The core of the question Hilbert asked received an answer that is complete with respect to the original question (within reason), mathematically rigorous, and unlikely to be dramatically improved upon in the future. These problems are 1, 2, 3, 4, 5, 7, 9, 10, 11, 13, 14, 15, 17, 18, 21, and 22.

Four problems—12, 19, 20, and 23—are what amount to programs for research and are somewhat vague in their statement. These have prompted a great deal of successful research, and I think it is time to count them as "solved," as problems of Hilbert's. Their concerns can be embodied in new problems written in new language.

Three problems—6, 8, and 16—have not been solved. The sixth problem, axiomatizing physics, seems to be a program, but since today there exist theories that purport to cover all physical phenomena in a single mathematical treatment, it is possible that physics will, in fact, be axiomatized. We don't know that the sixth problem is unsolvable. It is possible that it will be solved in a decisive way based on approaches that exist today. The algebraic half of the sixteenth problem is a program that is still vital. The eighth problem and the analysis half of the sixteenth problem are both clear and unsolved. They are hard problems. Stephen Smale included both on his 1998 list of problems.

It would seem natural to conclude this book with a substantial summing up. However, I think this goes against the spirit of David Hilbert and mathematics. Hilbert composed a list of good problems, so good that it has been said that they have distorted mathematics somewhat with their attractiveness. Most of them have now been solved. I hope mathematics in the next hundred years finds an organizing principle that is as straightforward and compelling.

Appendix

Mathematical Problems

« David Hilbert »

Lecture delivered before the International Congress of Mathematicians at Paris in 1900.

Who of us would not be glad to lift the veil behind which the future lies hidden; to cast a glance at the next advances of our science and at the secrets of its development during future centuries? What particular goals will there be toward which the leading mathematical spirits of coming generations will strive? What new methods and new facts in the wide and rich field of mathematical thought will the new centuries disclose?

History teaches the continuity of the development of science. We know that every age has its own problems, which the following age either solves or casts aside as profitless and replaces by new ones. If we would obtain an idea of the probable development of mathematical knowledge in the immediate future, we must let the unsettled questions pass before our minds and look over the problems which the science of today sets and whose solution we expect from the future. To such a review of problems the present day, lying at the meeting of the centuries, seems to me well adapted. For the close of a great epoch not only invites us to look back into the past but also directs our thoughts to the unknown future.

The deep significance of certain problems for the advance of mathematical science in general and the important rôle which they play in the work of the individual investigator are not to be denied. As long as a branch of science offers an abundance of problems, so long is it alive; a lack of problems foreshadows extinction or the cessation of independent development. Just as every human undertaking

Reprinted from Bulletin of the American Mathematical Society **8** (July 1902), 437–479.
Originally published as *Mathematische Probleme. Vortrag, gehalten auf dem internationalen Mathematike-Congress zu Paris 1900*, Gött. Nachr. 1900, 253–297, Vandenhoeck & Ruprecht, Göttingen.
Translated for the *Bulletin*, with the author's permission, by Dr. Mary Winston Newson, 1902.

pursues certain objects, so also mathematical research requires its problems. It is by the solution of problems that the investigator tests the temper of his steel; he finds new methods and new outlooks, and gains a wider and freer horizon.

It is difficult and often impossible to judge the value of a problem correctly in advance; for the final award depends upon the gain which science obtains from the problem. Nevertheless we can ask whether there are general criteria which mark a good mathematical problem. An old French mathematician said: "A mathematical theory is not to be considered complete until you have made it so clear that you can explain it to the first man whom you meet on the street." This clearness and ease of comprehension, here insisted on for a mathematical theory, I should still more demand for a mathematical problem if it is to be perfect; for what is clear and easily comprehended attracts, the complicated repels us.

Moreover a mathematical problem should be difficult in order to entice us, yet not completely inaccessible, lest it mock at our efforts. It should be to us a guide post on the mazy paths to hidden truths, and ultimately a reminder of our pleasure in the successful solution.

The mathematicians of past centuries were accustomed to devote themselves to the solution of difficult particular problems with passionate zeal. They knew the value of difficult problems. I remind you only of the "problem of the line of quickest descent," proposed by John Bernoulli. Experience teaches, explains Bernoulli in the public announcement of this problem, that lofty minds are led to strive for the advance of science by nothing more than by laying before them difficult and at the same time useful problems, and he therefore hopes to earn the thanks of the mathematical world by following the example of men like Mersenne, Pascal, Fermat, Viviani and others and laying before the distinguished analysts of his time a problem by which, as a touchstone, they may test the value of their methods and measure their strength. The calculus of variations owes its origin to this problem of Bernoulli and to similar problems.

Fermat had asserted, as is well known, that the diophantine equation

$$x^n + y^n = z^n$$

(x, y and z integers) is unsolvable—except in certain self-evident cases. The attempt to prove this impossibility offers a striking example of the inspiring effect which such a very special and apparently unimportant problem may have upon science. For Kummer, incited by Fermat's problem, was led to the introduction of ideal numbers and to the discovery of the law of the unique decomposition of the numbers of a circular field into ideal prime factors—a law which to-day in its generalization to any algebraic field by Dedekind and Kronecker, stands at the center of the modern theory of numbers and whose significance extends far beyond the boundaries of number theory into the realm of algebra and the theory of functions.

To speak of a very different region of research, I remind you of the problem of three bodies. The fruitful methods and the far-reaching principles which Poincaré has brought into celestial mechanics and which are to-day recognized and applied in practical astronomy are due to the circumstance that he undertook to treat anew that difficult problem and to approach nearer a solution.

The two last mentioned problems—that of Fermat and the problem of the three bodies—seem to us almost like opposite poles—the former a free invention of pure reason, belonging to the region of abstract number theory, the latter forced upon us by astronomy and necessary to an understanding of the simplest fundamental phenomena of nature.

But it often happens also that the same special problem finds application in the most unlike branches of mathematical knowledge. So, for example, the problem of the shortest line plays a chief and historically important part in the foundations of geometry, in the theory of curved lines and surfaces, in mechanics and in the calculus of variations. And how convincingly has F. Klein, in his work on the icosahedron, pictured the significance which attaches to the problem of the regular polyhedra in elementary geometry, in group theory, in the theory of equations and in that of linear differential equations.

In order to throw light on the importance of certain problems, I may also refer to Weierstrass, who spoke of it as his happy fortune that he found at the outset of his scientific career a problem so important as Jacobi's problem of inversion on which to work.

Having now recalled to mind the general importance of problems in mathematics, let us turn to the question from what sources this science derives its problems. Surely the first and oldest problems in every branch of mathematics spring from experience and are suggested by the world of external phenomena. Even the rules of calculation with integers must have been discovered in this fashion in a lower stage of human civilization, just as the child of today learns the application of these laws by empirical methods. The same is true of the first problems of geometry, the problems bequeathed us by antiquity, such as the duplication of the cube, the squaring of the circle; also the oldest problems in the theory of the solution of numerical equations, in the theory of curves and the differential and integral calculus, in the calculus of variations, the theory of Fourier series and the theory of potential—to say noting of the further abundance of problems properly belonging to mechanics, astronomy and physics.

But, in the further development of a branch of mathematics, the human mind, encouraged by the success of its solutions, becomes conscious of its independence. It evolves from itself alone, often without appreciable influence from without, by means of logical combination, generalization, specialization, by separating and collecting ideas in fortunate ways, new and fruitful problems, and appears then itself as the real questioner. Thus arose the problem of prime numbers and the other problems of number theory, Galois's theory of equations, the theory of algebraic invariants, the theory of abelian and automorphic functions; indeed almost all the nicer questions of modern arithmetic and function theory arise in this way.

In the meantime, while the creative power of pure reason is at work, the outer world again comes into play, forces upon us new questions from actual experience, opens up new branches of mathematics, and while we seek to conquer these new fields of knowledge for the realm of pure thought, we often find the answers to old unsolved problems and thus at the same time advance most successfully the old theories. And it seems to me that the numerous and surprising analogies and that apparently prearranged harmony which the mathematician so often perceives in the questions, methods and ideas of the various branches of his science, have their origin in this ever-recurring interplay between thought and experience.

It remains to discuss briefly what general requirements may be justly laid down for the solution of a mathematical problem. I should say first of all, this: that it shall be possible to establish the correctness of the solution by means of a finite number of steps based upon a finite number of hypotheses which are implied in the statement of the problem and which must always be exactly formulated. This requirement of logical deduction by means of a finite number of processes is simply the requirement of rigor in reasoning. Indeed the requirement of rigor, which has become proverbial in mathematics, corresponds to a universal philosophical necessity of our understanding; and, on the other hand, only by satisfying this requirement do the thought content and the suggestiveness of the problem attain their full effect. A new problem, especially when it comes from the world of outer experience, is like a young twig, which thrives and bears fruit only when it is grafted carefully and in accordance with strict horticultural rules upon the old stem, the established achievements of our mathematical science.

Besides it is an error to believe that rigor in the proof is the enemy of simplicity. On the contrary we find it confirmed by numerous examples that the rigorous method is at the same time the simpler and the more easily comprehended. The very effort for rigor forces us to find out simpler methods of proof. It also frequently leads the way to methods which are more capable of development than the old methods of less rigor. Thus the theory of algebraic curves experienced a considerable simplification and attained greater unity by means of the more rigorous function-theoretical methods and the consistent introduction of transcendental devices. Further, the proof that the power series permits the application of the four elementary arithmetical operations as well as the term by term differentiation and integration, and the recognition of the utility of the power series depending upon this proof contributed materially to the simplification of all analysis, particularly of the theory of elimination and the theory of differential equations, and also of the existence proofs demanded in those theories. But the most striking example for my statement is the calculus of variations. The treatment of the first and second variations of definite integrals required in part extremely complicated calculations, and the processes applied by the old mathematicians had not the needful rigor. Weierstrass showed us the way to a new and sure foundation of the calculus of variations. By the examples of the simple and double integral I will show briefly, at the close of my lecture, how this way leads at once to a surprising simplification of the calculus of variations. For in the demonstration of the necessary and sufficient criteria for the occurrence of a maximum and minimum, the calculation of the second variation and in part, indeed, the wearisome reasoning connected with the first variation may be completely dispensed with—to say nothing of the advance which is involved in the removal of the restriction to variations for which the differential coefficients of the function vary but slightly.

While insisting on rigor in the proof as a requirement for a perfect solution of a problem, I should like, on the other hand, to oppose the opinion that only the concepts of analysis, or even those of arithmetic alone, are susceptible of a fully rigorous treatment. This opinion, occasionally advocated by eminent men, I consider entirely erroneous. Such a one-sided interpretation of the requirement of rigor would soon lead to the ignoring of all concepts arising from geometry, mechanics and physics, to a stoppage of the flow of new material from the outside world, and finally, indeed, as a last consequence, to the rejection of the ideas of the continuum and of the irrational number. But what an important nerve, vital to mathematical science, would be cut by the extirpation of geometry and mathematical physics! On the contrary I think that wherever, from the side of the theory of knowledge or in geometry, or from the theories of natural or physical science, mathematical ideas come up, the problem arises for mathematical science to investigate the principles underlying these ideas and so to establish them upon a simple and complete system of axioms, that the exactness of the new ideas and their applicability to deduction shall be in no respect inferior to those of the old arithmetical concepts.

To new concepts correspond, necessarily, new signs. These we choose in such a way that they remind us of the phenomena which were the occasion for the formation of the new concepts. So the geometrical figures are signs or mnemonic symbols of space intuition and are used as such by all mathematicians. Who does not always use along with the double inequality $a > b > c$ the picture of three points following one another on a straight line as the geometrical picture of the idea "between"? Who does not make use of drawings of segments and rectangles enclosed in one another, when it is required to prove with perfect rigor a difficult theorem on the continuity of functions or the existence of points of condensation? Who could dispense with the figure of the triangle, the circle with its center, or with the cross of three perpendicular axes? Or who would give up the representation of the vector field, or the picture of a family of curves or surfaces with its envelope which plays so important a part in differential geometry, in the theory of differential equations, in the foundation of the calculus of variations and in other purely mathematical sciences?

The arithmetical symbols are written diagrams and the geometrical figures are graphic formulas; and no mathematician could spare these graphic formulas, any more than in calculation the insertion and removal of parentheses or the use of other analytical signs.

The use of geometrical signs as a means of strict proof presupposes the exact knowledge and complete mastery of the axioms which underlie those figures; and in order that these geometrical figures may be incorporated in the general treasure of mathematical signs, there is necessary a rigorous axiomatic investigation of their conceptual content. Just as in adding two numbers, one must place the digits under each other in the right order, so that only the rules of calculation, *i.e.*, the axioms of arithmetic, determine the correct use of the digits, so the use of geometrical signs is determined by the axioms of geometrical concepts and their combinations.

The agreement between geometrical and arithmetical thought is shown also in that we do not habitually follow the chain of reasoning back to the axioms in arithmetical, any more than in geometrical discussions. On the contrary we apply, especially in first attacking a problem, a rapid, unconscious, not absolutely sure combination, trusting to a certain arithmetical feeling for the behavior of the arithmetical symbols, which we could dispense with as little in arithmetic as with the geometrical imagination in geometry. As an example of an arithmetical theory operating rigorously with geometrical ideas and signs, I may mention Minkowski's work, Die Geometrie der Zahlen.[1]

Some remarks upon the difficulties which mathematical problems may offer, and the means of surmounting them, may be in place here.

If we do not succeed in solving a mathematical problem, the reason frequently consists in our failure to recognize the more general standpoint from which the problem before us appears only as a single link in a chain of related problems. After finding this standpoint, not only is this problem frequently more accessible to our investigation, but at the same time we come into possession of a method which is applicable also to related problems. The introduction of complex paths of integration by Cauchy and of the notion of the IDEALS in number theory by Kummer may serve as examples. This way for finding general methods is certainly the most practicable and the most certain; for he who seeks for methods without having a definite problem in mind seeks for the most part in vain.

In dealing with mathematical problems, specialization plays, as I believe, a still more important part than generalization. Perhaps in most cases where we seek in vain the answer to a question, the cause of the failure lies in the fact that problems simpler and easier than the one in hand have been either not at all or incompletely solved. All depends, then, on finding out these easier problems, and on solving them by means of devices as perfect as possible and of concepts capable of generalization. This rule is one of the most important levers for overcoming mathematical difficulties and it seems to me that it is used almost always, though perhaps unconsciously.

Occasionally it happens that we seek the solution under insufficient hypotheses or in an incorrect sense, and for this reason do not succeed. The problem then arises: to show the impossibility of the solution under the given hypotheses, or in the sense contemplated. Such proofs of impossibility were effected by the ancients, for instance when they showed that the ratio of the hypotenuse to the side of an

[1] Leipzig, 1896.

isosceles right triangle is irrational. In later mathematics, the question as to the impossibility of certain solutions plays a preëminent part, and we perceive in this way that old and difficult problems, such as the proof of the axiom of parallels, the squaring of the circle, or the solution of equations of the fifth degree by radicals have finally found fully satisfactory and rigorous solutions, although in another sense than that originally intended. It is probably this important fact along with other philosophical reasons that gives rise to the conviction (which every mathematician shares, but which no one has as yet supported by a proof) that every definite mathematical problem must necessarily be susceptible of an exact settlement, either in the form of an actual answer to the question asked, or by the proof of the impossibility of its solution and therewith the necessary failure of all attempts. Take any definite unsolved problem, such as the question as to the irrationality of the Euler-Mascheroni constant C, or the existence of an infinite number of prime numbers of the form $2^n + 1$. However unapproachable these problems may seem to us and however helpless we stand before them, we have, nevertheless, the firm conviction that their solution must follow by a finite number of purely logical processes.

Is this axiom of the solvability of every problem a peculiarity characteristic of mathematical thought alone, or is it possibly a general law inherent in the nature of the mind, that all questions which it asks must be answerable? For in other sciences also one meets old problems which have been settled in a manner most satisfactory and most useful to science by the proof of their impossibility. I instance the problem of perpetual motion. After seeking in vain for the construction of a perpetual motion machine, the relations were investigated which must subsist between the forces of nature if such a machine is to be impossible;[2] and this inverted question led to the discovery of the law of the conservation of energy, which, again, explained the impossibility of perpetual motion in the sense originally intended.

This conviction of the solvability of every mathematical problem is a powerful incentive to the worker. We hear within us the perpetual call: There is the problem. Seek its solution. You can find it by pure reason, for in mathematics there is no *ignorabimus*.

The supply of problems in mathematics is inexhaustible, and as soon as one problem is solved numerous others come forth in its place. Permit me in the following, tentatively as it were, to mention particular definite problems, drawn from various branches of mathematics, from the discussion of which an advancement of science may be expected.

Let us look at the principles of analysis and geometry. The most suggestive and notable achievements of the last century in this field are, as it seems to me, the arithmetical formulation of the concept of the continuum in the works of Cauchy, Bolzano and Cantor, and the discovery of non-euclidean geometry by Gauss, Bolyai,

---

[2] See Helmholtz, "Ueber die Wechselwirkung der Naturkräefte und die darauf bezüglichen neuesten Ermittelungen der Physik"; Vortrag, gehalten in Königsberg, 1854.

and Lobachevsky. I therefore first direct your attention to some problems belonging to these fields.

1. Cantor's Problem of the Cardinal Number of the Continuum

Two systems, *i.e.*, two assemblages of ordinary real numbers or points, are said to be (according to Cantor) equivalent or of equal *cardinal number*, if they can be brought into a relation to one another such that to every number of the one assemblage corresponds one and only one definite number of the other. The investigations of Cantor on such assemblages of points suggest a very plausible theorem, which nevertheless, in spite of the most strenuous efforts, no one has succeeded in proving. This is the theorem:

Every system of infinitely many real numbers, *i.e.*, every assemblage of numbers (or points), is either equivalent to the assemblage of natural integers, 1, 2, 3, ... or to the assemblage of all real numbers and therefore to the continuum, that is, to the points of a line; *as regards equivalence there are, therefore, only two assemblages of numbers, the countable assemblage and the continuum.*

From this theorem it would follow at once that the continuum has the next cardinal number beyond that of the countable assemblage; the proof of this theorem would, therefore, form a new bridge between the countable assemblage and the continuum.

Let me mention another very remarkable statement of Cantor's which stands in the closest connection with the theorem mentioned and which, perhaps, offers the key to its proof. Any system of real numbers is said to be ordered, if for every two numbers of the system it is determined which one is the earlier and which the later, and if at the same time this determination is of such a kind that, if a is before b and b is before c, then a always comes before c. The natural arrangement of numbers of a system is defined to be that in which the smaller precedes the larger. But there are, as is easily seen, infinitely many other ways in which the numbers of a system may be arranged.

If we think of a definite arrangement of numbers and select from them a particular system of these numbers, a so-called partial system or assemblage, this partial system will also prove to be ordered. Now Cantor considers a particular kind of ordered assemblage which he designates as a well ordered assemblage and which is characterized in this way, that not only in the assemblage itself but also in every partial assemblage there exists a first number. The system of integers 1, 2, 3, ... in their natural order is evidently a well ordered assemblage. On the other hand the system of all real numbers, *i.e.*, the continuum in its natural order, is evidently not well ordered. For, if we think of the points of a segment of a straight line, with its initial point excluded, as our partial assemblage, it will have no first element.

The question now arises whether the totality of all numbers may not be arranged in another manner so that every partial assemblage may have a first element, *i.e.*, whether the continuum cannot be considered as a well ordered assemblage—a question which Cantor thinks must be answered in the affirmative. It appears to me most desirable to obtain a direct proof of this remarkable statement of Cantor's, perhaps by actually giving an arrangement of numbers such that in every partial system a first number can be pointed out.

2. The Compatibility of the Arithmetical Axioms

When we are engaged in investigating the foundations of a science, we must set up a system of axioms which contains an exact and complete description of the relations subsisting between the elementary ideas of that science. The axioms so set up are at the same time the definitions of those elementary ideas; and no statement within the realm of the science whose foundation we are testing is held to be correct unless it can be derived from those axioms by means of a finite number of logical steps. Upon closer consideration the question arises: *Whether, in any way, certain statements of single axioms depend upon one another, and whether the axioms may not therefore contain certain parts in common, which must be isolated if one wishes to arrive at a system of axioms that shall be altogether independent of one another.*

But above all I wish to designate the following as the most important among the numerous questions which can be asked with regard to the axioms: *To prove that they are not contradictory, that is, that a finite number of logical steps based upon them can never lead to contradictory results.*

In geometry, the proof of the compatibility of the axioms can be effected by constructing a suitable field of numbers, such that analogous relations between the numbers of this field correspond to the geometrical axioms. Any contradiction in the deductions from the geometrical axioms must thereupon be recognizable in the arithmetic of this field of numbers. In this way the desired proof for the compatibility of the geometrical axioms is made to depend upon the theorem of the compatibility of the arithmetical axioms.

On the other hand a direct method is needed for the proof of the compatibility of the arithmetical axioms. The axioms of arithmetic are essentially nothing else than the known rules of calculation, with the addition of the axiom of continuity. I recently collected them[3] and in so doing replaced the axiom of continuity by two simpler axioms, namely, the well-known axiom of Archimedes, and a new axiom essentially as follows: that numbers form a system of things which is capable of no further extension, as long as all the other axioms hold (axiom of completeness). I am convinced that it must be possible to find a direct proof for the compatibility of the arithmetical axioms, by means of a careful study and suitable modification of the known methods of reasoning in the theory of irrational numbers.

To show the significance of the problem from another point of view, I add the following observation: If contradictory attributes be assigned to a concept, I say, that *mathematically the concept does not exist*. So, for example, a real number

[3] *Jahresbericht der Deutschen Mathematiker-Vereinigung*, vol. 8 (1900), p. 180.

whose square is -1 does not exist mathematically. But if it can be proved that the attributes assigned to the concept can never lead to a contradiction by the application of a finite number of logical processes, I say that the mathematical existence of the concept (for example, of a number or a function which satisfies certain conditions) is thereby proved. In the case before us, where we are concerned with the axioms of real numbers in arithmetic, the proof of the compatibility of the axioms is at the same time the proof of the mathematical existence of the complete system of real numbers or of the continuum. Indeed, when the proof for the compatibility of the axioms shall be fully accomplished, the doubts which have been expressed occasionally as to the existence of the complete system of real numbers will become totally groundless. The totality of real numbers, *i.e.*, the continuum according to the point of view just indicated, is not the totality of all possible series in decimal fractions, or of all possible laws according to which the elements of a fundamental sequence may proceed. It is rather a system of things whose mutual relations are governed by the axioms set up and for which all propositions, and only those, are true which can be derived from the axioms by a finite number of logical processes. In my opinion, the concept of the continuum is strictly logically tenable in this sense only. It seems to me, indeed, that this corresponds best also to what experience and intuition tell us. The concept of the continuum or even that of the system of all functions exists, then, in exactly the same sense as the system of integral, rational numbers, for example, or as Cantor's higher classes of numbers and cardinal numbers. For I am convinced that the existence of the latter, just as that of the continuum, can be proved in the sense I have described; unlike the system of *all* cardinal numbers or of *all* Cantor's alephs, for which, as may be shown, a system of axioms, compatible in my sense, cannot be set up. Either of these systems is, therefore, according to my terminology, mathematically non-existent.

From the field of the foundations of geometry I should like to mention the following problem:

3. The Equality of the Volumes of Two Tetrahedra of Equal Bases and Equal Altitudes

In two letters to Gerling, Gauss[4] expresses his regret that certain theorems of solid geometry depend upon the method of exhaustion, *i.e.*, in modern phraseology, upon the axiom of continuity (or upon the axiom of Archimedes). Gauss mentions in particular the theorem of Euclid, that triangular pyramids of equal altitudes are to each other as their bases. Now the analogous problem in the plane has been solved.[5] Gerling also succeeded in proving the equality of volume of symmetrical polyhedra by dividing them into congruent parts. Nevertheless, it seems to me probable

[4] Werke, vol. 8, pp. 241 and 244.

[5] Cf., beside earlier literature, Hilbert, Grundlagen der Geometrie, Leipzig, 1899, ch. 4. [Translation by Townsend, Chicago, 1902.]

that a general proof of this kind for the theorem of Euclid just mentioned is impossible, and it should be our task to give a rigorous proof of its impossibility. This would be obtained, as soon as we succeeded in *specifying two tetrahedra of equal bases and equal altitudes which can in no way be split up into congruent tetrahedra, and which cannot be combined with congruent tetrahedra to form two polyhedra which themselves could be split up into congruent tetrahedra.*[6]

4. Problem of the Straight Line as the Shortest Distance Between Two Points

Another problem relating to the foundations of geometry is this: If from among the axioms necessary to establish ordinary euclidean geometry, we exclude the axiom of parallels, or assume it as not satisfied, but retain all other axioms, we obtain, as is well known, the geometry of Lobachevsky (hyperbolic geometry). We may therefore say that this is a geometry standing next to euclidean geometry. If we require further that that axiom be not satisfied whereby, of three points of a straight line, one and only one lies between the other two, we obtain Riemann's (elliptic) geometry, so that this geometry appears to be the next after Lobachevsky's. If we wish to carry out a similar investigation with respect to the axiom of Archimedes, we must look upon this as not satisfied, and we arrive thereby at the non-archimedean geometries which have been investigated by Veronese and myself. The more general question now arises: Whether from other suggestive standpoints geometries may not be devised which, with equal right, stand next to euclidean geometry. Here I should like to direct your attention to a theorem which has, indeed, been employed by many authors as a definition of a straight line, viz., that the straight line is the shortest distance between two points. The essential content of this statement reduces to the theorem of Euclid that in a triangle the sum of two sides is always greater than the third side—a theorem which, as is easily seen, deals solely with elementary concepts, *i.e.*, with such as are derived directly from the axioms, and is therefore more accessible to logical investigation. Euclid proved this theorem, with the help of the theorem of the exterior angle, on the basis of the congruence theorems. Now it is readily shown that this theorem of Euclid cannot be proved solely on the basis of those congruence theorems which relate to the application of segments and angles, but that one of the theorems on the congruence of triangles is necessary. We are asking, then, for a geometry in which all the axioms of ordinary euclidean geometry hold, and in particular all the congruence axioms except the one of the congruence of triangles (or all except the theorem of the equality of the base angles in the isosceles triangle), and in which, besides, the proposition that in every triangle the sum of two sides is greater than the third is assumed as a particular axiom.

---

[6] Since this was written Herr Dehn has succeeded in proving this impossibility. See his note: "Ueber raumgleiche Polyeder", in *Nachrichten d. K. Geselsch. d. Wiss. zu Göttingen*, 1900, and a paper soon to appear in the *Math. Annalen* [vol. 55, pp. 465–478].

One finds that such a geometry really exists and is no other than that which Minkowski constructed in his book, Geometrie der Zahlen,[7] and made the basis of his arithmetical investigations. Minkowski's is therefore also a geometry standing next to the ordinary euclidean geometry; it is essentially characterized by the following stipulations:

1. The points which are at equal distances from a fixed point O lie on a convex closed surface of the ordinary euclidean space with O as a center.

2. Two segments are said to be equal when one can be carried into the other by a translation of the ordinary euclidean space.

In Minkowski's geometry the axiom of parallels also holds. By studying the theorem of the straight line as the shortest distance between two points, I arrived[8] at a geometry in which the parallel axiom does not hold, while all other axioms of Minkowski's geometry are satisfied. The theorem of the straight line as the shortest distance between two points and the essentially equivalent theorem of Euclid about the sides of a triangle, play an important part not only in number theory but also in the theory of surfaces and in the calculus of variations. For this reason, and because I believe that the thorough investigation of the conditions for the validity of this theorem will throw a new light upon the idea of distance, as well as upon other elementary ideas, *e.g.*, upon the idea of the plane, and the possibility of its definition by means of the idea of the straight line, *the construction and systematic treatment of the geometries here possible seem to me desirable.*

5. Lie's Concept of a Continuous Group of Transformations Without the Assumption of the Differentiability of the Functions Defining the Group

It is well known that Lie, with the aid of the concept of continuous groups of transformations, has set up a system of geometrical axioms and, from the standpoint of his theory of groups, has proved that this system of axioms suffices for geometry. But since Lie assumes, in the very foundation of his theory, that the functions defining his group can be differentiated, it remains undecided in Lie's development, whether the assumption of the differentiability in connection with the question as to the axioms of geometry is actually unavoidable, or whether it may not appear rather as a consequence of the group concept and the other geometrical axioms. This consideration, as well as certain other problems in connection with the arithmetical axioms, brings before us the more general question: *How far Lie's concept of continuous groups of transformations is approachable in our investigations without the assumption of the differentiability of the functions.*

Lie defines a finite continuous group of transformations as a system of transformations

[7] Leipzig, 1896.

[8] *Math. Annalen*, vol. 46, p. 91.

$$x_i' = f_i(x_1, \dots, x_n; a_1, \dots, a_r) \qquad (i = 1, \dots, n)$$

having the property that any two arbitrarily chosen transformations of the system, as

$$x_i' = f_i(x_1, \dots, x_n; a_1, \dots, a_r)$$
$$x_i'' = f_i(x_1', \dots, x_n'; b_1, \dots, b_r)$$

applied successively result in a transformation which also belongs to the system, and which is therefore expressible in the form

$$x_i'' = f_i\{f_i(x, a), \dots, f_n(x, a); b_1, \dots, b_r\} = f_i(x_1, \dots, x_n; c_1, \dots, c_r),$$

where $c_1, \dots, c_r$ are certain functions of $a_1, \dots, a_r$ and $b_1, \dots, b_r$. The group property thus finds its full expression in a system of functional equations and of itself imposes no additional restrictions upon the functions $f_1, \dots, f_n; c_1, \dots, c_r$. Yet Lie's further treatment of these functional equations, viz., the derivation of the well-known fundamental differential equations, assumes necessarily the continuity and differentiability of the functions defining the group.

As regards continuity: this postulate will certainly be retained for the present—if only with a view to the geometrical and arithmetical applications, in which the continuity of the functions in question appears as a consequence of the axiom of continuity. On the other hand the differentiability of the functions defining the group contains a postulate which, in the geometrical axioms, can be expressed only in a rather forced and complicated manner. Hence there arises the question whether, through the introduction of suitable new variables and parameters, the group can always be transformed into one whose defining functions are differentiable; or whether, at least with the help of certain simple assumptions, a transformation is possible into groups admitting Lie's methods. A reduction to analytic groups is, according to a theorem announced by Lie[9] but first proved by Schur,[10] always possible when the group is transitive and the existence of the first and certain second derivatives of the functions defining the group is assumed.

For infinite groups the investigation of the corresponding question is, I believe, also of interest. Moreover we are thus led to the wide and interesting field of functional equations which have been heretofore investigated usually only under the assumption of the differentiability of the functions involved. In particular the functional equations treated by Abel[11] with so much ingenuity, the difference equations, and other equations occurring in the literature of mathematics, do not directly involve anything which necessitates the requirement of the differentiability of the accompanying functions. In the search for certain existence proofs in the

[9] Lie-Engel, Theorie der Transformationsgruppen, vol. 3, Leipzig, 1893, §§82, 144.

[10] "Ueber den analytischen Charakter der eine endliche Kontinuierliche Transformationsgruppen darstellenden Funktionen", *Math. Annalen*, vol. 41.

[11] Werke, vol. 1, pp. 1, 61, 389.

calculus of variations I came directly upon the problem: To prove the differentiability of the function under consideration from the existence of a difference equation. In all these cases, then, the problem arises: *In how far are the assertions which we can make in the case of differentiable functions true under proper modifications without this assumption?*

It may be further remarked that H. Minkowski in his above-mentioned Geometrie der Zahlen starts with the functional equation

$$f(x_1+y_1, \ldots, x_n+y_n) \leq f(x_1, \ldots, x_n) + f(y_1, \ldots, y_n)$$

and from this actually succeeds in proving the existence of certain differential quotients for the function in question.

On the other hand I wish to emphasize the fact that there certainly exist analytical functional equations whose sole solutions are non-differentiable functions. For example a uniform continuous non-differentiable function $\varphi(x)$ can be constructed which represents the only solution of the two functional equations

$$\varphi(x+\alpha) - \varphi(x) = f(x), \qquad \varphi(x+\beta) - \varphi(x) = 0,$$

where α and β are two real numbers, and $f(x)$ denotes, for all the real values of x, a regular analytic uniform function. Such functions are obtained in the simplest manner by means of trigonometrical series by a process similar to that used by Borel (according to a recent announcement of Picard)[12] for the construction of a doubly periodic, non-analytic solution of a certain analytic partial differential equation.

6. Mathematical Treatment of the Axioms of Physics

The investigations on the foundations of geometry suggest the problem: *To treat in the same manner, by means of axioms, those physical sciences in which mathematics plays an important part; in the first rank are the theory of probabilities and mechanics.*

As to the axioms of the theory of probabilities,[13] it seems to me desirable that their logical investigation should be accompanied by a rigorous and satisfactory development of the method of mean values in mathematical physics, and in particular in the kinetic theory of gases.

Important investigations by physicists on the foundations of mechanics are at hand; I refer to the writings of Mach,[14] Hertz,[15] Boltzmann[16] and Volkmann.[17] It

[12] "Quelques théories fondamentales dans l'analyse mathématique", Conférences faites à Clark University, *Revue générale des Sciences*, 1900, p. 22.

[13] Cf. Bohlmann, "Ueber Versicherungsmathematik", from the collection: Klein and Riecke, Ueber angewandte Mathematik und Physik, Leipzig, 1900.

[14] Die Mechanik in ihrer Entwickelung, Leipzig, 4th edition, 1901.

[15] Die Prinzipien der Mechanik, Leipzig, 1894.

[16] Vorlesungen über die Principe der Mechanik, Leipzig, 1897.

[17] Einführung in das Studium der theoretischen Physik, Leipzig, 1900.

is therefore very desirable that the discussion of the foundations of mechanics be taken up by mathematicians also. Thus Boltzmann's work on the principles of mechanics suggests the problem of developing mathematically the limiting processes, there merely indicated, which lead from the atomistic view to the laws of motion of continua. Conversely one might try to derive the laws of the motion of rigid bodies by a limiting process from a system of axioms depending upon the idea of continuously varying conditions of a material filling all space continuously, these conditions being defined by parameters. For the question as to the equivalence of different systems of axioms is always of great theoretical interest.

If geometry is to serve as a model for the treatment of physical axioms, we shall try first by a small number of axioms to include as large a class as possible of physical phenomena, and then by adjoining new axioms to arrive gradually at the more special theories. At the same time Lie's principle of subdivision can perhaps be derived from a profound theory of infinite transformation groups. The mathematician will have also to take account not only of those theories coming near to reality, but also, as in geometry, of all logically possible theories. He must be always alert to obtain a complete survey of all conclusions derivable from the system of axioms assumed.

Further, the mathematician has the duty to test exactly in each instance whether the new axioms are compatible with the previous ones. The physicist, as his theories develop, often finds himself forced by the results of his experiments to make new hypotheses, while he depends, with respect to the compatibility of the new hypotheses with the old axioms, solely upon these experiments or upon a certain physical intuition, a practice which in the rigorously logical building up of a theory is not admissible. The desired proof of the compatibility of all assumptions seems to me also of importance, because the effort to obtain each proof always forces us most effectually to an exact formulation of the axioms.

So far we have considered only questions concerning the foundations of the mathematical sciences. Indeed, the study of the foundations of a science is always particularly attractive, and the testing of these foundations will always be among the foremost problems of the investigator. Weierstrass once said, "The final object always to be kept in mind is to arrive at a correct understanding of the foundations of the science....But to make any progress in the sciences the study of particular problems is, of course, indispensable." In fact, a thorough understanding of its special theories is necessary to the successful treatment of the foundations of the science. Only that architect is in the position to lay a sure foundation for a structure who knows its purpose thoroughly and in detail. So we turn now to the special problems of the separate branches of mathematics and consider first arithmetic and algebra.

7. Irrationality and Transcendence of Certain Numbers

Hermite's arithmetical theorems on the exponential function and their extension by Lindemann are certain of the admiration of all generations of mathematicians. Thus the task at once presents itself to penetrate further along the path here entered, as A. Hurwitz has already done in two interesting papers,[18] "Ueber arithmetische Eigenschaften gewisser transzendenter Funktionen". I should like, therefore, to sketch a class of problems which, in my opinion, should be attacked as here next in order. That certain special transcendental functions, important in analysis, take algebraic values for certain algebraic arguments, seems to us particularly remarkable and worthy of thorough investigation. Indeed, we expect transcendental functions to assume, in general, transcendental values for even algebraic arguments; and, although it is well known that there exist integral transcendental functions which even have rational values for all algebraic arguments, we shall still consider it highly probable that the exponential functions $e^{i\pi z}$, for example, which evidently has algebraic values for all rational arguments z, will on the other hand always take transcendental values for irrational algebraic values of the argument z. We can also give this statement a geometrical form, as follows:

If, in an isosceles triangle, the ratio of the base angle to the angle at the vertex be algebraic but not rational, the ratio between base and side is always transcendental.

In spite of the simplicity of this statement and of its similarity to the problems solved by Hermite and Lindemann, I consider the proof of this theorem very difficult; as also the proof that

The expression α^β, for an algebraic base α and an irrational algebraic exponent β, e.g., the number $2^{\sqrt{2}}$ or $e^\pi = i^{-2i}$, always represents a transcendental or at least an irrational number.

It is certain that the solution of these and similar problems must lead us to entirely new methods and to a new insight into the nature of special irrational and transcendental numbers.

8. Problems of Prime Numbers

Essential progress in the theory of the distribution of prime numbers has lately been made by Hadamard, de la Vallée-Poussin, Von Mangoldt and others. For the complete solution, however, of the problems set us by Riemann's paper "Ueber die Anzahl der Primzahlen unter einer gegebenen Grösse", it still remains to prove the correctness of an exceedingly important statement of Riemann, viz., *that the zero points of the function $\zeta(s)$ defined by the series*

$$\zeta(s) = 1 + \frac{1}{2^s} + \frac{1}{3^s} + \frac{1}{4^s} + \ldots$$

[18] *Math. Annalen*, vols. 22, 32 (1883, 1888).

all have the real part ½, except the well-known negative integral real zeros. As soon as this proof has been successfully established, the next problem would consist in testing more exactly Riemann's infinite series for the number of primes below a given number and, especially, *to decide whether the difference between the number of primes below a number x and the integral logarithm of x does in fact become infinite of an order not greater than ½ in x.*[19] Further, we should determine whether the occasional condensation of prime numbers which has been noticed in counting primes is really due to those terms of Riemann's formula which depend upon the first complex zeros of the function $\zeta(s)$.

After an exhaustive discussion of Riemann's prime number formula, perhaps we may sometime be in a position to attempt the rigorous solution of Goldbach's problem,[20] viz., whether every [even] integer is expressible as the sum of two positive prime numbers; and further to attack the well-known question, whether there are an infinite number of pairs of prime numbers with the difference 2, or even the more general problem, whether the linear diophantine equation

$$ax + by + c = 0$$

(with given integral coefficients each prime to the others) is always solvable in prime numbers x and y.

But the following problem seems to me of no less interest and perhaps of still wider range: *To apply the results obtained for the distribution of rational prime numbers to the theory of the distribution of ideal primes in a given number-field k*—a problem which looks toward the study of the function $\zeta_k(s)$ belonging to the field and defined by the series

$$\zeta_k(s) = \sum \frac{1}{n(j)^s},$$

where the sum extends over all ideals j of the given realm k, and $n(j)$ denotes the norm of the ideal j.

I may mention three more special problems in number theory: one on the laws of reciprocity, one on diophantine equations, and a third from the realm of quadratic forms.

9. Proof of the Most General Law of Reciprocity in any Number Field

For any field of numbers the law of reciprocity is to be proved for the residues of the lth power, when l denotes an odd prime, and further when l is a power of 2 or a power of an odd prime.

[19] Cf. an article by H. von Koch, which is soon to appear in the *Math. Annalen* [Vol. 55, p. 441].
[20] Cf. P. Stäckel: "Über Goldbach's empirisches Theorem", *Nachrichten d. K. Ges. d. Wiss. zu Göttingen*, 1896, and Landau, *ibid.*, 1900.

The law, as well as the means essential to its proof, will, I believe, result by suitably generalizing the theory of the field of the lth roots of unity,[21] developed by me, and my theory of relative quadratic fields.[22]

10. Determination of the Solvability of a Diophantine Equation

Given a diophantine equation with any number of unknown quantities and with rational integral numerical coefficients: *To devise a process according to which it can be determined by a finite number of operations whether the equation is solvable in rational integers.*

11. Quadratic Forms with any Algebraic Numerical Coefficients

Our present knowledge of the theory of quadratic number fields[23] puts us in a position *to attack successfully the theory of quadratic forms with any number of variables and with any algebraic numerical coefficients.* This leads in particular to the interesting problem: to solve a given quadratic equation with algebraic numerical coefficients in any number of variables by integral or fractional numbers belonging to the algebraic realm of rationality determined by the coefficients.

The following important problem may form a transition to algebra and the theory of functions:

12. Extension of Kronecker's Theorem on Abelian Fields to Any Algebraic Realm of Rationality

The theorem that every abelian number field arises from the realm of rational numbers by the composition of fields of roots of unity is due to Kronecker. This fundamental theorem in the theory of integral equations contains two statements, namely:

First. It answers the question as to the number and existence of those equations which have a given degree, a given abelian group and a given discriminant with respect to the realm of rational numbers.

Second. It states that the roots of such equations form a realm of algebraic numbers which coincides with the realm obtained by assigning to the argument z in the exponential function $e^{i\pi z}$ all rational numerical values in succession.

The first statement is concerned with the question of the determination of certain algebraic numbers by their groups and their branching. This question corresponds, therefore, to the known problem of the determination of algebraic func-

[21] *Jahresber. d. Deutschen Math.-Vereinigung*, "Ueber die Theorie der algebraischen Zahl-körper", vol. 4 (1897), Part V.

[22] *Math. Annalen*, vol. 51 and *Nachrichten d. K. Ges. d. Wiss. zu Göttingen*, 1898.

[23] Hilbert, "Ueber den Dirichlet'schen biquadratischen Zahlenkörper", *Math. Annalen*, vol. 45; "Ueber die Theorie der relativquadratischen Zahlkörper", *Jahresber. d. Deutschen Mathematiker-Vereinigung*, 1897, and *Math. Annalen*, vol. 51; "Ueber die Theorie der relativ-Abelschen Körper", *Nachrichten d. K. Ges. d. Wiss. zu Göttingen*, 1898; Grundlagen der Geometrie, Leipzig, 1899, Chap. VIII, §83 [Translation by Townsend, Chicago, 1902]. Cf. also the dissertation of G. Rückle, Göttingen, 1901.

tions corresponding to given Riemann surfaces. The second statement furnishes the required numbers by transcendental means, namely, by the exponential function $e^{i\pi z}$.

Since the realm of the imaginary quadratic number fields is the simplest after the realm of rational numbers, the problem arises, to extend Kronecker's theorem to this case. Kronecker himself has made the assertion that the abelian equations in the realm of a quadratic field are given by the equations of transformation of elliptic functions with singular moduli, so that the elliptic function assumes here the same rôle as the exponential function in the former case. The proof of Kronecker's conjecture has not yet been furnished; but I believe that it must be obtainable without very great difficulty on the basis of the theory of complex multiplication developed by H. Weber[24] with the help of the purely arithmetical theorems on class fields which I have established.

Finally, the extension of Kronecker's theorem to the case that, *in place of the realm of rational numbers or of the imaginary quadratic field, any algebraic field whatever is laid down as realm of rationality*, seems to me of the greatest importance. I regard this problem as one of the most profound and far-reaching in the theory of numbers and of functions.

The problem is found to be accessible from many standpoints. I regard as the most important key to the arithmetical part of this problem the general law of reciprocity for residues of lth powers within any given number field.

As to the function-theoretical part of the problem, the investigator in this attractive region will be guided by the remarkable analogies which are noticeable between the theory of algebraic functions of one variable and the theory of algebraic numbers. Hensel[25] has proposed and investigated the analogue in the theory of algebraic numbers to the development in power series of an algebraic function; and Landsberg[26] has treated the analogue of the Riemann-Roch theorem. The analogy between the deficiency of a Riemann surface and that of the class number of a field of numbers is also evident. Consider a Riemann surface of deficiency $p = 1$ (to touch on the simplest case only) and on the other hand a number field of class $h = 2$. To the proof of the existence of an integral everywhere finite on the Riemann surface, corresponds the proof of the existence of an integer α in the number field such that the number $\sqrt{\alpha}$ represents a quadratic field, relatively unbranched with respect to the fundamental field. In the theory of algebraic functions, the method of boundary values (*Randwerthaufgabe*) serves, as is well known, for the proof of Riemann's existence theorem. In the theory of number fields also, the proof of the existence of just this number α offers the greatest difficulty. This proof succeeds with indispensable assistance from the theorem that in the number field there are always prime ideals corresponding to

[24] Elliptische Functionen und algebraische Zahlen, Braunschweig, 1891.

[25] Jahresber. d. Deutschen Math.-Vereinigung, vol. 6, and an article soon to appear in the *Math. Annalen* [Vol. 55, p. 301]: "Ueber die Entwickelung der algebraischen Zahlen in Potenzreihen".

[26] *Math. Annalen*, vol. 50 (1898).

given residual properties. This latter fact is therefore the analogue in number theory to the problem of boundary values.

The equation of Abel's theorem in the theory of algebraic functions expresses, as is well known, the necessary and sufficient condition that the points in question on the Riemann surface are the zero points of an algebraic function belonging to the surface. The exact analogue of Abel's theorem, in the theory of the number field of class $h = 2$, is the equation of the law of quadratic reciprocity[27]

$$\left(\frac{\alpha}{j}\right) = +1,$$

which declares that the ideal j is then and only then a principal ideal of the number field when the quadratic residue of the number α with respect to the ideal j is positive.

It will be seen that in the problem just sketched the three fundamental branches of mathematics, number theory, algebra and function theory, come into closest touch with one another, and I am certain that the theory of analytical functions of several variables in particular would be notably enriched if one should succeed *in finding and discussing those functions which play the part for any algebraic number field corresponding to that of the exponential function in the field of rational numbers and of the elliptic modular functions in the imaginary quadratic number field.*

Passing to algebra, I shall mention a problem from the theory of equations and one to which the theory of algebraic invariants has led me.

13. Impossibility of the Solution of the General Equation of the 7th Degree by Means of Functions of Only Two Arguments

Nomography[28] deals with the problem: to solve equations by means of drawings of families of curves depending on an arbitrary parameter. It is seen at once that every root of an equation whose coefficients depend upon only two parameters, that is, every function of two independent variables, can be represented in manifold ways according to the principle lying at the foundation of nomography. Further, a large class of functions of three or more variables can evidently be represented by this principle alone without the use of variable elements, namely all those which can be generated by forming first a function of two arguments, then equating each of these arguments to a function of two arguments, next replacing each of those arguments in their turn by a function of two arguments, and so on, regarding as admissible any finite number of insertions of functions of two arguments. So, for example, every rational function of any number of arguments belongs to this class of functions

[27] Cf. Hilbert, "Ueber die Theorie der relativ-Abelschen Zahlkörper", *Gött. Nachrichten*, 1898.

[28] d'Ocagne, Traité de Nomographie, Paris, 1899.

constructed by nomographic tables; for it can be generated by the processes of addition, subtraction, multiplication and division and each of these processes produces a function of only two arguments. One sees easily that the roots of all equations which are solvable by radicals in the natural realm of rationality belong to this class of functions; for here the extraction of roots is adjoined to the four arithmetical operations and this, indeed, presents a function of one argument only. Likewise the general equations of the 5th and 6th degrees are solvable by suitable nomographic tables; for, by means of Tschirnhausen transformations, which require only extraction of roots, they can be reduced to a form where the coefficients depend upon two parameters only.

Now it is probable that the root of the equation of the seventh degree is a function of its coefficients which does not belong to this class of functions capable of nomographic construction, *i.e.*, that it cannot be constructed by a finite number of insertions of functions of two arguments. In order to prove this, the proof would be necessary *that the equation of the seventh degree* $f^7 + xf^3 + yf^2 + zf + 1 = 0$ *is not solvable with the help of any continuous functions of only two arguments*. I may be allowed to add that I have satisfied myself by a rigorous process that there exist analytical functions of three arguments x, y, z which cannot be obtained by a finite chain of functions of only two arguments.

By employing auxiliary movable elements, nomography succeeds in constructing functions of more than two arguments, as d'Ocagne has recently proved in the case of the equation of the 7th degree.[29]

14. Proof of the Finiteness of Certain Complete Systems of Functions

In the theory of algebraic invariants, questions as to the finiteness of complete systems of forms deserve, as it seems to me, particular interest. L. Maurer[30] has lately succeeded in extending the theorems on finiteness in invariant theory proved by P. Gordan and myself, to the case where, instead of the general projective group, any subgroup is chosen as the basis for the definition of invariants.

An important step in this direction had been taken already by A. Hurwitz,[31] who, by an ingenious process, succeeded in effecting the proof, in its entire generality, of the finiteness of the system of orthogonal invariants of an arbitrary ground form.

The study of the question as to the finiteness of invariants has led me to a simple problem which includes that question as a particular case and whose solution probably requires a decidedly more minutely detailed study of the theory of elimination and of Kronecker's algebraic modular systems than has yet been made.

[29] "Sur la résolution nomographique de l'équation du septième degré", *Comptes rendus*, Paris, 1900.
[30] Cf. *Sitzungsber. d. K. Acad. d. Wiss. zu München*, 1899, and an article about to appear in the *Math. Annalen*.
[31] "Ueber die Erzeugung der Invarianten durch Integration", *Nachrichten d. K. Gesellschaft d. Wiss. zu Göttingen*, 1897.

Let a number m of integral rational functions $X_1, X_2, \dots, X_m$ of the n variables $x_1, x_2, \dots, x_n$ be given,

$$\begin{aligned} X_1 &= f_1(x_1, \dots, x_n), \\ X_2 &= f_2(x_1, \dots, x_n), \\ &\dots\dots\dots\dots \\ X_m &= f_m(x_1, \dots, x_n). \end{aligned} \tag{S}$$

Every rational integral combination of $X_1, \dots, X_m$ must evidently always become, after substitution of the above expressions, a rational integral function of $x_1, \dots, x_n$. Nevertheless, there may well be rational fractional functions of $X_1, \dots, X_m$ which, by the operation of the substitution S, become integral functions in $x_1, \dots, x_m$, which becomes integral in $x_1, \dots, x_n$ after the application of the substitution S, I propose to call a *relatively integral* function of $X_1, \dots, X_m$. Every such integral function of $X_1, \dots, X_m$ is evidently also relatively integral; further the sum, difference and product of relative integral functions are themselves relatively integral.

The resulting problem is now to decide whether it is always possible *to find a finite system of relatively integral functions* $X_1, \dots, X_m$ *by which every other relatively integral function of* $X_1, \dots, X_m$ *may be expressed rationally and integrally.*

We can formulate the problem still more simply if we introduce the idea of a finite field of integrality. By a finite field of integrality I mean a system of functions from which a finite number of functions can be chosen, in terms of which all other functions of the system are rationally and integrally expressible. Our problem amounts, then, to this: to show that all relatively integral functions of any given domain of rationality always constitute a finite field of integrality.

It naturally occurs to us also to refine the problem by restrictions drawn from number theory, by assuming the coefficients of the given functions $f_1, \dots, f_m$ to be integers and including among the relatively integral functions of $X_1, \dots, X_m$ only such rational functions of these arguments as become, by the application of the substitutions S, rational integral functions of $x_1, \dots, x_n$ with rational integral coefficients.

The following is a simple particular case of this refined problem: Let m integral rational functions $X_1, \dots, X_m$ of one variable x with integral rational coefficients, and a prime number p be given. Consider the system of those integral rational functions of x which can be expressed in the form

$$\frac{G(X_1, \dots, X_m)}{p^h},$$

where G is a rational integral function of the arguments $X_1, \dots, X_m$ and p^h is any power of the prime number p. Earlier investigations of mine[32] show immediately that all such expressions for a fixed exponent h form a finite domain of integral-

[32] Math. Annalen, vol. 36 (1890), p. 485.

ity. But the question here is whether the same is true for all exponents h, *i.e.*, whether a finite number of such expressions can be chosen by means of which for every exponent h every other expression of that form is integrally and rationally expressible.

From the boundary region between algebra and geometry, I will mention two problems. The one concerns enumerative geometry and the other the topology of algebraic curves and surfaces.

15. Rigorous Foundation of Schubert's Enumerative Calculus

The problem consists in this: *To establish rigorously and with an exact determination of the limits of their validity those geometrical numbers which Schubert*[33] *especially has determined on the basis of the so-called principle of special position, or conservation of number, by means of the enumerative calculus developed by him.*

Although the algebra of to-day guarantees, in principle, the possibility of carrying out the processes of elimination, yet for the proof of the theorems of enumerative geometry decidedly more is requisite, namely, the actual carrying out of the process of elimination in the case of equations of special form in such a way that the degree of the final equations and the multiplicity of their solutions may be foreseen.

16. Problem of the Topology of Algebraic Curves and Surfaces

The maximum number of closed and separate branches which a plane algebraic curve of the nth order can have has been determined by Harnack.[34] There arises the further question as to the relative position of the branches in the plane. As to curves of the 6th order, I have satisfied myself—by a complicated process, it is true—that of the eleven branches which they can have according to Harnack, by no means all can lie external to one another, but that one branch must exist in whose interior one branch and in whose exterior nine branches lie, or inversely. *A thorough investigation of the relative position of the separate branches when their number is the maximum seems to me to be of very great interest, and not less so the corresponding investigation as to the number, form, and position of the sheets of an algebraic surface in space.* Till now, indeed, it is not even known what is the maximum number of sheets which a surface of the 4th order in three dimensional space can really have.[35]

In connection with this purely algebraic problem, I wish to bring forward a question which, it seems to me, may be attacked by the same method of continuous variation of coefficients, and whose answer is of corresponding value for the topol-

[33] Kalkül der abzählenden Geometrie, Leipzig, 1879.

[34] *Math. Annalen*, vol. 10.

[35] Cf. Rohn, "Fläachen vierter Ordnung", Preisschriften der Fürstlich Jablonowskischen Gesellschaft, Leipzig, 1886.

ogy of families of curves defined by differential equations. This is the question as to the maximum number and position of Poincaré's boundary cycles (cycles limites) for a differential equation of the first order and degree of the form

$$\frac{dy}{dx} = \frac{Y}{X},$$

where X and Y are rational integral functions of the nth degree in x and y. Written homogeneously, this is

$$X\left(y\frac{dz}{dt} - z\frac{dy}{dt}\right) + Y\left(z\frac{dx}{dt} - x\frac{dz}{dt}\right) + Z\left(x\frac{dy}{dt} - y\frac{dx}{dt}\right) = 0,$$

where X, Y, and Z are rational integral homogeneous functions of the nth degree in x, y, z, and the latter are to be determined as functions of the parameter t.

17. Expression of Definite Forms by Squares

A rational integral function or form in any number of variables with real coefficients such that it becomes negative for no real values of these variables, is said to be *definite*. The system of all definite forms is invariant with respect to the operations of addition and multiplication, but the quotient of two definite forms—in case it should be an integral function of the variables—is also a definite form. The square of any form is evidently always a definite form. But since, as I have shown,[36] not every definite form can be compounded by addition from squares of forms, the question arises—which I have answered affirmatively for ternary forms[37]—whether every definite form may not be expressed as a quotient of sums of squares of forms. At the same time it is desirable, for certain questions as to the possibility of certain geometrical constructions, to know whether the coefficients of the forms to be used in the expression may always be taken from the realm of rationality given by the coefficients of the form represented.[38]

I mention one more geometrical problem:

18. Building Up of Space From Congruent Polyhedra

If we enquire for those groups of motions in the plane for which a fundamental region exists, we obtain various answers, according as the plane considered is Riemann's (elliptic), Euclid's, or Lobachevsky's (hyperbolic). In the case of the elliptic plane there is a finite number of essentially different kinds of fundamental regions, and a finite number of congruent regions suffices for a complete covering of the whole plane; the group consists indeed of a finite number of motions only. In

[36] *Math. Annalen*, vol. 32.

[37] *Acta Mathematica*, vol. 17.

[38] Cf. Hilbert: Grundlagen der Geometrie, Leipzig, 1899, Chap. 7 and in particular §38.

the case of the hyperbolic plane there is an infinite number of essentially different kinds of fundamental regions, namely, the well-known Poincaré polygons. For the complete covering of the plane an infinite number of congruent regions is necessary. The case of Euclid's plane stands between these; for in this case there is only a finite number of essentially different kinds of groups of motions with fundamental regions, but for a complete covering of the whole plane an infinite number of congruent regions is necessary.

Exactly the corresponding facts are found in space of three dimensions. The fact of the finiteness of the groups of motions in elliptic space is an immediate consequence of a fundamental theorem of C. Jordan,[39] whereby the number of essentially different kinds of finite groups of linear substitutions in n variables does not surpass a certain finite limit dependent upon n. The groups of motions with fundamental regions in hyperbolic space have been investigated by Fricke and Klein in the lectures on the theory of automorphic functions,[40] and finally Fedorov,[41] Schoenflies[42] and lately Rohn[43] have given the proof that there are, in euclidean space, only a finite number of essentially different kinds of groups of motions with a fundamental region. Now, while the results and methods of proof applicable to elliptic and hyperbolic space hold directly for n-dimensional space also, the generalization of the theorem for euclidean space seems to offer decided difficulties. The investigation of the following question is therefore desirable: *Is there in n-dimensional euclidean space also only a finite number of essentially different kinds of groups of motions with a fundamental region?*

A fundamental region of each group of motions, together with the congruent regions arising from the group, evidently fills up space completely. The question arises: *Whether polyhedra also exist which do not appear as fundamental regions of groups of motions, by means of which nevertheless by a suitable juxtaposition of congruent copies a complete filling up of all space is possible.* I point out the following question, related to the preceding one, and important to number theory and perhaps sometimes useful to physics and chemistry: How can one arrange most densely in space an infinite number of equal solids of given form, *e.g.*, spheres with given radii or regular tetrahedra with given edges (or in prescribed position), that is, how can one so fit them together that the ratio of the filled to the unfilled space may be as great as possible?

If we look over the development of the theory of functions in the last century, we notice above all the fundamental importance of that class of functions which we

[39] *Crelle's Journal*, vol. 84 (1878), and *Atti d. Reale Acad. di Napoli*, 1880.
[40] Leipzig, 1897. Cf. especially Abschnitt I, Chapters 2 and 3.
[41] Symmetrie der regelmässigen Systeme von Figuren, 1890.
[42] Krystallsysteme und Krystallstruktur, Leipzig, 1891.
[43] *Math. Annalen*, vol. 53.

now designate as analytic functions—a class of functions which will probably stand permanently in the center of mathematical interest.

There are many different standpoints from which we might choose, out of the totality of all conceivable functions, extensive classes worthy of a particularly thorough investigation. Consider, for example, *the class of functions characterized by ordinary or partial algebraic differential equations*. It should be observed that this class does not contain the functions that arise in number theory and whose investigation is of the greatest importance. For example, the before-mentioned function $\zeta(s)$ satisfies no algebraic differential equation, as is easily seen with the help of the well-known relation between $\zeta(s)$ and $\zeta(1-s)$, if one refers to the theorem proved by Hölder,[44] that the function $\Gamma(x)$ satisfies no algebraic differential equation. Again, the function of the two variables s and x defined by the infinite series

$$\zeta(s, x) = x + \frac{x^2}{2^s} + \frac{x^3}{3^s} + \frac{x^4}{4^s} + \cdots,$$

which stands in close relation with the function $\zeta(s)$, probably satisfies no algebraic partial differential equation. In the investigation of this question the functional equation

$$x\frac{\partial\zeta(s, x)}{\partial x} = \zeta(s-1, x)$$

will have to be used.

If, on the other hand, we are led by arithmetical or geometrical reasons to consider the class of all those functions which are continuous and indefinitely differentiable, we should be obliged in its investigation to dispense with that pliant instrument, the power series, and with the circumstance that the function is fully determined by the assignment of values in any region, however small. While, therefore, the former limitation of the field of functions was too narrow, the latter seems to me too wide.

The idea of the analytic function on the other hand includes the whole wealth of functions most important to science, whether they have their origin in number theory, in the theory of differential equations or of algebraic functional equations, whether they arise in geometry or in mathematical physics; and, therefore, in the entire realm of functions, the analytic function justly holds undisputed supremacy.

19. Are the Solutions of Regular Problems in the Calculus of Variations Always Necessarily Analytic?

One of the most remarkable facts in the elements of the theory of analytic functions appears to me to be this: That there exist partial differential equations whose integrals are all of necessity analytic functions of the independent variables, that is, in

[44] *Math. Annalen*, vol. 28.

short, equations susceptible of none but analytic solutions. The best known partial differential equations of this kind are the potential equation

$$\frac{\partial^2 f}{\partial x^2} + \frac{\partial^2 f}{\partial y^2} = 0$$

and certain linear differential equations investigated by Picard;[45] also the equation

$$\frac{\partial^2 f}{\partial x^2} + \frac{\partial^2 f}{\partial y^2} = e^f,$$

the partial differential equation of minimal surfaces, and others. Most of these partial differential equations have the common characteristic of being the lagrangian differential equations of certain problems of variation, viz., of such problems of variation

$$\int\int F(p, q, z; x, y)\, dx\, dy = \text{minimum}$$
$$\left[p = \frac{\partial z}{\partial x}, q = \frac{\partial z}{\partial y}\right],$$

as satisfy, for all values of the arguments which fall within the range of discussion, the inequality

$$\frac{\partial^2 F}{\partial p^2} \cdot \frac{\partial^2 F}{\partial q^2} - \left(\frac{\partial^2 F}{\partial p \partial q}\right)^2 > 0,$$

F itself being an analytic function. We shall call this sort of problem a *regular* variation problem. It is chiefly the regular variation problems that play a rôle in geometry, in mechanics, and in mathematical physics; and the question naturally arises, whether all solutions of regular variation problems must necessarily be analytic functions. In other words, *does every lagrangian partial differential equation of a regular variation problem have the property of admitting analytic integrals exclusively?* And is this the case even when the function is constrained to assume, as, *e.g.*, in Dirichlet's problem on the potential function, boundary values which are continuous, but not analytic?

I may add that there exist surfaces of constant *negative* gaussian curvature which are representable by functions that are continuous and possess indeed all the derivatives, and yet are not analytic; while on the other hand it is probable that every surface whose gaussian curvature is constant and *positive* is necessarily an analytic surface. And we know that the surfaces of positive constant curvature are

[45] *Jour. de l'Ecole Polytech.*, 1890.

most closely related to this regular variation problem: To pass through a closed curve in space a surface of minimal area which shall inclose, in connection with a fixed surface through the same closed curve, a volume of given magnitude.

20. The General Problem of Boundary Values

An important problem closely connected with the foregoing is the question concerning the existence of solutions of partial differential equations when the values on the boundary of the region are prescribed. This problem is solved in the main by the keen methods of H. A. Schwarz, C. Neumann, and Poincaré for the differential equation of the potential. These methods, however, seem to be generally not capable of direct extension to the case where along the boundary there are prescribed either the differential coefficients or any relations between these and the values of the function. Nor can they be extended immediately to the case where the inquiry is not for potential surfaces but, say, for surfaces of least area, or surfaces of constant positive gaussian curvature, which are to pass through a prescribed twisted curve or to stretch over a given ring surface. It is my conviction that it will be possible to prove these existence theorems by means of a general principle whose nature is indicated by Dirichlet's principle. This general principle will then perhaps enable us to approach the question: *Has not every regular variation problem a solution, provided certain assumptions regarding the given boundary conditions are satisfied* (say that the functions concerned in these boundary conditions are continuous and have in sections one or more derivatives), *and provided also if need be that the notion of a solution shall be suitably extended?*[46]

21. Proof of the Existence of Linear Differential Equations Having a Prescribed Monodromic Group

In the theory of linear differential equations with one independent variable z, I wish to indicate an important problem, one which very likely Riemann himself may have had in mind. This problem is as follows: *To show that there always exists a linear differential equation of the Fuchsian class, with given singular points and monodromic group.* The problem requires the production of n functions of the variable z, regular throughout the complex z plane except at the given singular points; at these points the functions may become infinite of only finite order, and when z describes circuits about these points the functions shall undergo the prescribed linear substitutions. The existence of such differential equations has been shown to be probable by counting the constants, but the rigorous proof has been obtained up to this time only in the particular case where the fundamental equations of the given substitutions have roots all of absolute magnitude unity. L. Schlesinger has given this proof,[47] based upon Poincaré's theory

[46] Cf. my lecture on Dirichlet's principle in the *Jahresber. d. Deutschen Math.-Vereinigung*, vol. 8 (1900), p. 184.

[47] Handbuch der Theorie der linearen Differentialgleichungen, vol. 2, part 2, No. 366.

of the Fuchsian ζ-functions. The theory of linear differential equations would evidently have a more finished appearance if the problem here sketched could be disposed of by some perfectly general method.

22. Uniformization of Analytic Relations by Means of Automorphic Functions

As Poincaré was the first to prove, it is always possible to reduce any algebraic relation between two variables to uniformity by the use of automorphic functions of one variable. That is, if any algebraic equation in two variables be given, there can always be found for these variables two such single valued automorphic functions of a single variable that their substitution renders the given algebraic equation an identity. The generalization of this fundamental theorem to any analytic non-algebraic relations whatever between two variables has likewise been attempted with success by Poincaré,[48] though by a way entirely different from that which served him in the special problem first mentioned. From Poincaré's proof of the possibility of reducing to uniformity an arbitrary analytic relation between two variables, however, it does not become apparent whether the resolving functions can be determined to meet certain additional conditions. Namely, it is not shown whether the two single valued functions of the one new variable can be so chosen that, while this variable traverses the *regular* domain of those functions, the totality of all regular points of the given analytic field are actually reached and represented. On the contrary it seems to be the case, from Poincaré's investigations, that there are beside the branch points certain others, in general infinitely many other discrete exceptional points of the analytic field, that can be reached only by making the new variable approach certain limiting points of the functions. *In view of the fundamental importance of Poincaré's formulation of the question it seems to me that an elucidation and resolution of this difficulty is extremely desirable.*

In conjunction with this problem comes up the problem of reducing to uniformity an algebraic or any other analytic relation among three or more complex variables—a problem which is known to be solvable in many particular cases. Toward the solution of this the recent investigations of Picard on algebraic functions of two variables are to be regarded as welcome and important preliminary studies.

23. Further Development of the Methods of the Calculus of Variations

So far, I have generally mentioned problems as definite and special as possible, in the opinion that it is just such definite and special problems that attract us the most and from which the most lasting influence is often exerted upon science. Nevertheless, I should like to close with a general problem, namely with the indication of a branch of mathematics repeatedly mentioned in this lecture—which, in spite

[48] *Bull. de la Soc. Math. de France*, vol. 11 (1883).

of the considerable advancement lately given it by Weierstrass, does not receive the general appreciation which, in my opinion, is its due—I mean the calculus of variations.[49]

The lack of interest in this is perhaps due in part to the need of reliable modern text-books. So much the more praiseworthy is it that A. Kneser in a very recently published work has treated the calculus of variations from the modern points of view and with regard to the modern demand for rigor.[50]

The calculus of variations is, in the widest sense, the theory of the variation of functions, and as such appears as a necessary extension of the differential and integral calculus. In this sense, Poincaré's investigations on the problem of three bodies, for example, form a chapter in the calculus of variations, in so far as Poincaré derives from known orbits by the principle of variation new orbits of similar character.

I add here a short justification of the general remarks upon the calculus of variations made at the beginning of my lecture.

The simplest problem in the calculus of variations proper is known to consist in finding a function y of a variable x such that the definite integral

$$J = \int_a^b F(y_x, y; x)\, dx, \quad y_x = \frac{dy}{dx}$$

assumes a minimum value as compared with the values it takes when y is replaced by other functions of x with the same initial and final values.

The vanishing of the first variation in the usual sense

$$\delta J = 0$$

gives for the desired function y the well-known differential equation

$$\frac{dF_{y_x}}{dx} - F_y = 0, \tag{1}$$

$$\left[F_{y_x} = \frac{\partial F}{\partial y_x}, \quad F_y = \frac{\partial F}{\partial y}\right].$$

In order to investigate more closely the necessary and sufficient criteria for the occurrence of the required minimum, we consider the integral

[49] Text-books: Moigno-Lindelöf, Leçons du calcul des variations, Paris, 1861, and A. Kneser, Lehrbuch der Variations-rechnung, Braunschweig, 1900.

[50] As an indication of the contents of this work, it may here be noted that for the simplest problems Kneser derives sufficient conditions of the extreme even for the case that one limit of integration is variable, and employs the envelope of a family of curves satisfying the differential equations of the problem to prove the necessity of Jacobi's conditions of the extreme. Moreover, it should be noticed that Kneser applies Weierstrass's theory also to the inquiry for the extreme of such quantities as are defined by differential equations.

$$J^* = \int_a^b \{F + (y_x - p)F_p\}\, dx,$$

$$\left[F = F(p, y; x), F_p = \frac{\partial F(p, y; x)}{\partial p}\right].$$

Now we inquire how p is to be chosen as function of x, y in order that the value of this integral J^* shall be independent of the path of integration, i.e., of the choice of the function y of the variable x. The integral J^* has the form

$$J^* = \int_a^b \{Ay_x - B\}\, dx,$$

where A and B do not contain y_x and the vanishing of the first variation

$$\delta J^* = 0$$

in the sense which the new question requires gives the equation

$$\frac{\partial A}{\partial x} + \frac{\partial B}{\partial y} = 0,$$

i.e., we obtain for the function p of the two variables x, y the partial differential equation of the first order

$$\frac{\partial F_p}{\partial x} + \frac{\partial (pF_p - F)}{\partial y} = 0. \tag{1*}$$

The ordinary differential equation of the second order (1) and the partial differential equation (1^*) stand in the closest relation to each other. This relation becomes immediately clear to us by the following simple transformation

$$\begin{aligned}
\delta J^* &= \int_a^b \{F_y \delta y + F_p \delta p + (\delta y_x - \delta p)F_y + (y_x - p)\delta F_p\}\, dx \\
&= \int_a^b \{F_y \delta y + \delta y_x F_p + (y_x - p)\delta F_p\}\, dx \\
&= \delta J + \int_a^b (y_x - p)\delta F_p\, dx.
\end{aligned}$$

We derive from this, namely, the following facts: If we construct any *simple* family of integral curves of the ordinary differential equation (1) of the second order and then form an ordinary differential equation of the first order

$$y_x = p(x, y) \tag{2}$$

which also admits these integral curves as solutions, then the function $p(x, y)$ is always an integral of the partial differential equation (1^*) of the first order; and conversely, if $p(x, y)$ denotes any solution of the partial differential equation (1^*) of the first order, all the non-singular integrals of the ordinary differential equation (2) of the first order are at the same time integrals of the differential equation (1) of the second order, or in short if $y_x = p(x, y)$ is an integral equation of the first order of the differential equation (1) of the second order, $p(x, y)$ represents an integral of the partial differential equation (1^*) and conversely; the integral curves of the ordinary differential equation of the second order are therefore, at the same time, the characteristics of the partial differential equation (1^*) of the first order.

In the present case we may find the same result by means of a simple calculation; for this gives us the differential equations (1) and (1^*) in question in the form

$$y_{xx}F_{y_x y_x} + y_x F_{y_x y} + F_{y_x x} - F_y = 0, \tag{1}$$

$$(p_x + pp_y)F_{pp} + pF_{py} + F_{px} - F_y = 0, \tag{1*}$$

where the lower indices indicate the partial derivatives with respect to x, y, p, y_x. The correctness of the affirmed relation is clear from this.

The close relation derived before and just proved between the ordinary differential equation (1) of the second order and the partial differential equation (1^*) of the first order, is, as it seems to me, of fundamental significance for the calculus of variations. For, from the fact that the integral J^* is independent of the path of integration it follows that

$$\int_a^b \{F(p) + (y_x - p)F_p(p)\}\, dx = \int_a^b F(\overline{y}_x)\, dx, \tag{3}$$

if we think of the left hand integral as taken along any path y and the right hand integral along an integral curve $\overline{y}$ of the differential equation

$$\overline{y}_x = p(x, \overline{y}).$$

With the help of equation (3) we arrive at Weierstrass's formula

$$\int_a^b F(y_x)\, dx - \int_a^b F(\overline{y}_x)\, dx = \int_a^b E(y_x, p)\, dx, \tag{4}$$

where E designates Weierstrass's expression, depending upon y_x, p, y, x,

$$E(y_x, p) = F(y_x) - F(p) - (y_x - p)F_p(p).$$

Since, therefore, the solution depends only on finding an integral $p(x, y)$ which is single valued and continuous in a certain neighborhood of the integral curve $\overline{y}$, which we are considering, the developments just indicated lead immediately—without the introduction of the second variation, but only by the application of the polar process to the differential equation (1)—to the expression of Jacobi's condition and to the answer to the question: How far this condition of Jacobi's in conjunction with Weierstrass's condition $E > 0$ is necessary and sufficient for the occurrence of a minimum.

The developments indicated may be transferred without necessitating further calculation to the case of two or more required functions, and also to the case of a double or a multiple integral. So, for example, in the case of a double integral

$$J = \int F(z_x, z_y, z; x, y)\, d\omega, \quad \left[z_x = \frac{\partial z}{\partial x}, \quad z_y = \frac{\partial z}{\partial y}\right]$$

to be extended over a given region ω, the vanishing of the first variation (to be understood in the usual sense)

$$\delta J = 0$$

gives the well-known differential equation of the second order

$$\frac{dF_z}{dx} + \frac{dF_{z_y}}{dy} - F_z = 0, \tag{I}$$

$$\left[F_{z_x} = \frac{\partial F}{\partial z}, F_z = \frac{\partial F}{\partial z_y}, F_z - \frac{\partial F}{\partial z}\right],$$

for the required function z of x and y.

On the other hand we consider the integral

$$J^* = \int \{F + (z_x - p)F_p + (z_y - q)F_q\}\, d\omega,$$

$$\left[F = F(p, q, z; x, y), F_p = \frac{\partial F(p, q, z; x, y)}{\partial p}, \quad F_q = \frac{\partial F(p, q, z; x, y)}{\partial q}\right],$$

and inquire, how p and q are to be taken as functions of x, y and z in order that the value of this integral may be independent of the choice of the surface passing through the given closed twisted curve, i.e., of the choice of the function z of the variables x and y.

The integral J^* has the form

$$J^* = \int \{Az_x + Bz_y - C\}\, d\omega$$

and the vanishing of the first variation

$$\delta J^* = 0,$$

in the sense which the new formulation of the question demands, gives the equation

$$\frac{\partial A}{\partial x} + \frac{\partial B}{\partial y} + \frac{\partial C}{\partial z} = 0,$$

i.e., we find for the functions p and q of the three variables x, y and z the differential equation of the first order

$$\frac{\partial F_p}{\partial x} + \frac{\partial F_q}{\partial y} + \frac{\partial (pF_p + qF_q - F)}{\partial x} = 0.$$

If we add to this differential equation the partial differential equation

$$p_y + qp_z = q_x + pq_z, \tag{I*}$$

resulting from the equations

$$z_x = p(x, y, z), \quad z_y = q(x, y, z),$$

the partial differential equation (I) for the function z of the two variables x and y and the simultaneous system of the two partial differential equations of the first order (I^*) for the two functions p and q of the three variables x, y, and z stand toward one another in a relation exactly analogous to that in which the differential equations (1) and (1^*) stood in the case of the simple integral.

It follows from the fact that the integral J^* is independent of the choice of the surface of integration z that

$$\int \{F(p, q) + (z_x - p)F_p(p, q) + (z_y - q)F_q(p, q)\}\, d\omega$$
$$= \int F(\overline{z}_x, \overline{y})\, d\omega,$$

if we think of the right hand integral as taken over an integral surface $\overline{z}$ of the partial differential equations

$$\overline{z}_x = p(x, y, \overline{z}),\ \overline{z}_y = q(x, y, \overline{z});$$

and with the help of this formula we arrive at once at the formula

$$\int F(z_x, z_y) d\omega - \int F(\overline{z}_x, \overline{z}_y)\, d\omega = \int E(z_x, z_y, p, q)\, d\omega, \qquad \text{(IV)}$$

$$[E(z_x, z_y, p, q) = F(z_x, z_y) - F(p, q) - (z_x - p)F_p(p, q) \\ -(z_y - q)F_q(p, q)],$$

which plays the same rôle for the variation of double integrals as the previously given formula (4) for simple integrals. With the help of this formula we can now answer the question how far Jacobi's condition in conjunction with Weierstrass's condition $E > 0$ is necessary and sufficient for the occurrence of a minimum.

Connected with these developments is the modified form in which A. Kneser,[51] beginning from other points of view, has presented Weierstrass's theory. While Weierstrass employed to derive sufficient conditions for the extreme those integral curves of equation (1) which pass through a fixed point, Kneser on the other hand makes use of any simple family of such curves and constructs for every such family a solution, characteristic for that family, of that partial differential equation which is to be considered as a generalization of the Jacobi-Hamilton equation.

The problems mentioned are merely samples of problems, yet they will suffice to show how rich, how manifold and how extensive the mathematical science of to-day is, and the question is urged upon us whether mathematics is doomed to the fate of those other sciences that have split up into separate branches, whose representatives scarcely understand one another and whose connection becomes ever more loose. I do not believe this nor wish it. Mathematical science is in my opinion an indivisible whole, an organism whose vitality is conditioned upon the connection of its parts. For with all the variety of mathematical knowledge, we are still clearly conscious of the similarity of the logical devices, the *relationship* of the *ideas* in mathematics as a whole and the numerous analogies in its different departments. We also notice that, the farther a mathematical theory is developed, the more harmoniously and uniformly does its construction proceed, and unsuspected relations are disclosed between hitherto separate branches of the science. So it happens that, with the extension of mathematics, its organic character is not lost but only manifests itself the more clearly.

But, we ask, with the extension of mathematical knowledge will it not finally become impossible for the single investigator to embrace all departments of this

[51] Cf. his above-mentioned textbook, §§14, 15, 19 and 20.

knowledge? In answer let me point out how thoroughly it is ingrained in mathematical science that every real advance goes hand in hand with the invention of sharper tools and simpler methods which at the same time assist in understanding earlier theories and cast aside older more complicated developments. It is therefore possible for the individual investigator, when he makes these sharper tools and simpler methods his own, to find his way more easily in the various branches of mathematics than is possible in any other science.

The organic unity of mathematics is inherent in the nature of this science, for mathematics is the foundation of all exact knowledge of natural phenomena. That it may completely fulfil this high mission, may the new century bring it gifted masters and many zealous and enthusiastic disciples.

Notes

Almost all of the primary references are noted in the text. As much as possible I tried to leave signposts in the text for a knowledgeable reader about what sources I was using. Here are a few additional notes and sources for quotes. I have tried to keep these notes relatively brief. If an author appears more than once in the bibliography that follows, I include the year of the relevant entry in my note, if it isn't obvious. When I communicated with sources, I did so in phone conversations, e-mail and conventional mail. Usually, I sent what I wrote to the source and asked if it was correct. Usually, a given detail in the text involved more than one interaction. Therefore I often say "private communication" to refer to the process.

1 "That I have been able to accomplish anything in mathematics" [Reid p. 168]

The Origin of the Coordinates

This mostly came from Constance Reid's Biography of Hilbert and Weyl's essay about Hilbert's mathematical career.

1 "such a famous man" [Reid p. 12]

2 "His wise and gay eyes testified as to his spirit" and "to the apple tree" and "On unending walks we engrossed ourselves" [Reid p. 14]

3 "That the objections to Kant's theory of the *a priori* nature" [Reid p. 17]

4 "Over this gloomy November day" [Reid p. 19]

5 "Content and full of joy" [Reid p. 27]

6 "a systematic exploration" [Reid p. 28]

7 "*Das ist nicht Mathematik*" and "*unheimlich*" ("uncomfortable, sinister, weird") [Reid p. 34]

8 In 1892, using this new result and his existence theorem as a pushing-off point [Weyl 1944 p. 622-623]

9 "She was a full human being in her own right, strong and clear" [Reid p. 40]

10 "another great discovery" [Minkowski in Reid p. 40]

11 "Away from Göttingen there is no life" [Reid p. 47]

12 "the first of a long line of terriers" [Reid p. 53]

13 "The Greeks had conceived of geometry as a deductive science" [Weyl 1944 p. 635]

14 "tables, chairs, and beer mugs" [Reid p. 57]

15 "It is one thing to build up geometry on sure foundations" [Weyl 1944 p. 636]

16 "In the construction of his models Hilbert displays" [Weyl 1944 p. 636]

17 "Doesn't it seem to you, *Herr Geheimrat*" [Reid p. 96]

18 "He spoke his regrets upon his fate" and "The doctors themselves stood around his bed" [Reid p. 115]

19 "Hilbert thought the war was stupid, and said so" [Reid p. 137]

20 "And I thought we would get swine!" [Reid p. 144]

21 "If students without the gymnasium diploma" [Reid p. 143]

22 "Every boy in the streets of Göttingen" [Reid p. 142]

23 "the passionate scientific life" [Weyl in Reid p. 216]

24 "When I was young, I resolved never to repeat" and "Mathematics in Göttingen?" [Reid p. 205]

25 "Memory only confuses thought" [Reid p. 209]

26 "That I have been able to accomplish anything in mathematics" [Reid p. 168]

Set Theory, Anyone?

I relied most on Dauben and Grattan-Guinness for Cantor's biography.

1 "Cube one-third the coefficient of x; add to" [Dunham p. 142]

2 Material on notation [*Britannica* 1910–1911—in "Algebra" vol. 1 p. 619]
My point here is that someone introduced each symbol. A reader interested in the detail might look at Florian Cajori's two volume *A History of Mathematical Notations*.

3 "The algebraic numbers are spotted over the" [Bell p. 569]

4 The quote from Cantor: "that the greatest achievement of a genius (like Newton) despite the subjective religiosity of its author, when it is not united with a true philosophical and historical spirit, leads to effects (and, I assert necessarily does so) by which it seems highly questionable whether . . ." is slightly different than in Dauben—"when it is not united" instead of "when it is now united." When I asked him, he told me that his quote was in error. [Dauben p. 295]

5 "No one will drive us out of this paradise" [Reid p. 177]

6 "That was a winter cold and wild" [Dauben p. 284]

I Am Lying (Mathematics Is Consistent)

Dawson, Feferman, Rudolf Gödel, Kreisel, Taussky-Todd, and Wang were the most important biographical sources. I wrote the first draft of this section before Dawson's excellent biography appeared. When it appeared I checked what I had written against it and made a few additions.

1 "If contradictory attributes be assigned to a concept" [Hilbert second problem]

2 "Science does not remain confined to serving industry" [Brouwer p. 4]

3 "In Brouwer's intuitionism, mathematics consists" [Troelstra]

4 "At the beginning of this century a self-destructive democratic principle" [Arnold 1995 p. 7–8]

5 "The antinomies of set theory are usually regarded" [Weyl in Van Dalen p. 147]

6 Further, he said Ackermann had proved the consistency of [Wang p. 54]

7 The whole paragraph that begins, "In 1913, when Kurt was seven, the family built a well-situated villa with a beautiful garden" and the following paragraph are based on information from Rudolf Gödel's article.

8 "eccentric beauty" [Kreisel 1980 p. 153]

9 "Another singular aspect of those lectures (which Gödel did not mention)" [Kreisel 1980 p. 153]

10 "a band of positivist philosophers" [Kreisel 1980 p. 153]

11 "did not like the Tractatus" [Wang p. 77]

12 "There is no doubt about the fact that Gödel" [Taussky-Todd p. 32]

13 "We and my brother often" and "At that time my brother was very interested in the theater" [Rudolf Gödel p. 20]

14 Hilbert had stated four specific problems [see Wang p. 54–59]

15 "*Let S be any consistent set of statements*" [Cohen p. 13]

16 "If A is not derivable from S" [Cohen p. 16]

17 "Incompleteness is a truism for them" [Davis private communication]

18 "as an early example" [Dawson 1997 p. 108]

19 "Our task here will be to explain" and "We shall prove that all the finitary conclusions" [van Heijenoort p. 416]

20 "We cannot agree that mathematical objects are" [Kolmogorov 1991 p. 454]

21 "it was the antinomies of set theory" [Gödel, vol. 3 p. 377]

22 he delivered the first of his lectures entirely facing the blackboard [Moore in *Dictionary* article on Gödel]

23 In his memoir, Kreisel admits to sharing this affliction [Kreisel 1980 p. 158]

24 "Is his illness a consequence of proving" [Taussky-Todd p. 39]

25 "more than twenty years later he still spoke of the frustrations" [Kreisel 1980 p. 154]

26 "I was mustered and found fit for garrison duty" [Dawson 1984 p. 13]

27 "on the superiority of the austere life" [Kreisel 1980 p. 152]

28 "Kurt had a friendly attitude towards people of the Jewish faith" [Taussky-Todd p. 33]

29 "The fact is that he was bitterly frustrated" [Kreisel 1980 p. 156]

30 Stanislaw Ulam and Freeman Dyson [Dawson 1984 p. 15]

31 "Some felt that Gödel would not welcome" and "Princeton society proved unreceptive" [Dawson 1984 p. 15]

32 "so much more friendly" [Wang p. 100, in his words]

33 "Einstein was enchanted by Gödel's" [Kreisel 1980 p. 157]

34 "by far Einstein's best friend" [Straus p. 422]

35 "Einstein has often told me" [Wang p. 31]

36 "a soft spot for new sects in the New World" [Kreisel 1980 p. 152]

37 "Gödel had an interesting axiom" [Straus quoted in Wang p. 32]

38 "Morgenstern had many stories to tell about Gödel" [Feferman p. 12]

39 "[It] soon became too much for me to watch" [Kreisel p. 160]

40 "malnutrition and inanition" and "personality disturbance" [Dawson 1997 p. 253]

41 Hao Wang heard that Gödel weighed sixty-five pounds at the end [Wang p. 133]

42 "Gödel's main published papers from 1930 to 1940" [Feferman p. 32]

The Perfect Spy—How Many Real Numbers Are There?

I spoke with Cohen on the phone on a number of occasions. Most information was obtained in 1996. We had a fairly long conversation discussing what I had written in 2001.

1 "had no objection to Metaphysics" [Rice p. 42]

2 "It is probably the great advantage" [Lewis p. 51]

3 "Frege's notation, it must be admitted, is against him" [Lewis p. 115]

4 "There is just one point where" [Russell in van Heijenoort p. 124]

5 "observe zodiacal light" [*Dict.* on Skolem]

6 “It would in any case be of much” [van Heijenoort p. 299]

7 “What does it mean for a set to exist” [Skolem in van Heijenoort p. 300]

8 “The most important result above is” [Skolem in van Heijenoort p. 300–301]

9 he had requested it, but it hadn’t yet arrived [Wang p. 271]

10 As early as 1930 Gödel was thinking about the continuum hypothesis [Wang p. 108]

11 “fictitious set” [van Heijenoort p. 202]

12 “We may think of the axiom of choice as asserting the possibility” [Royden p. 18]

13 Peano was one of the first to explicitly comment [Moore 1982 p. 76]

14 “During the summer of 1942, while on vacation” [Moore *Dict.* Gödel]

15 “strange kind of combativeness” [MMP p. 43]

16 “fairly successful grocery jobber” [MMP p. 44]

17 “We were in terrible shape” [phone conversation 2001]

18 “that there was a world that was permanently beyond me” [MMP p. 45]

19 “However, math especially appealed to me” [MMP p. 45]

20 “verbal” [MMP p. 46]

21 “about two years older” [MMP p. 47]

22 “in the best high school in the greatest city in the world” [MMP p. 48]

23 “Strangely enough—I don’t know why” [MMP p. 48]

24 “I always thought I could have done better” [private communication]

25 “to devote myself to mathematics as much as I was able” [Weil p. 126]

26 “There were several times in my life” [MMP p. 128–129]

27 “I was rather depressed when I realized Gödel was right” [phone conversation 2001]

28 He briefly thought he had a consistency proof for analysis [Moore 1988 p. 155]

29 As early as 1959 he had asked Feferman [Moore 1988 p. 155]

30 “I don’t know who said it to me” [MMP p. 52]

31 “Solovay knew more logic” [Moore 1988 p. 161]

32 “As a matter of fact, I was told by many people” [MMP p. 54]

33 “Again it was one of those problems” [MMP p. 55–56]

34 Davis has a handwritten note from Cohen saying that Cohen [private communication]

35 “Cohen recalls that both Kreisel and [Dana] Scott” [Moore 1988 p. 155]

36 “I suddenly felt that I had the general idea” [MMP p. 53]

37 On the contrary, on May 3, 1963, thirty minutes before [Moore 1988 p. 157]

38 "concluded that all the interesting questions" [Moore 1988 p. 161]

39 "I hope you are not under some nervous strain" [Moore 1988 p. 157–158]

40 "Cohen was telling Gödel to change anything" [Moore 1988 p. 159]

41 "For every element *x* in M" [Cohen p. 105 Paraphrased with my italics]

42 "All our intuition comes from our belief" [Cohen p. 107]

43 "rather tedious" [Cohen p. 148]

44 (Scott is actually responsible for this simple form.)[Moore 1988 p. 160 or Cohen p. 117]

45 "They have a new method and they'll get everything" [Moore 1988 p. 164]

46 "ridiculous assumption" [Moore 1988 p. 159]

47 Cohen says he took this approach because [Cohen p. 148]

48 "Once upon a time, not so very long ago" [Baumgartner p. 462]

49 "One might say in a humorous way" [MMP p. 58]

50 "Paul and I were with a group of people waiting" [private communication]

Can't We Do This with a Computer?

I had a fairly large correspondence with Matiyasevich and Davis and a smaller one with Putnam. I spoke with Davis and Putnam.

Harvard University was able to send me useful information on Putnam from a file that they had in the publicity department—clippings, press releases, etc.

1 "Church published a pair of substantial papers" [Davis 1982 p. 4]

2 "and was circulating among the logicians at Princeton" [Kleene in Davis 1982 p. 9]

3 "Computing is normally done by writing certain symbols on paper" [Turing in Davis 1965 p. 135]

4 "an overwhelming confusion of classes of cases" and "led to a reversal of our entire program" [Post in Davis 1965 p. 373]

5 Davis says he believes Post lectured [Davis 1982 p. 22]

6 He would work on one for two weeks and then [Davis 1993 p. xxiv]

7 "two way infinite sequence" of "boxes" and a "problem solver or worker" [Post in Davis 1965 p. 289]

8 This was the first breakout for unsolvability theory [Boone et al. p. x]

9 "My mother as a teenage girl frequented the lending library" [private communication]

10 "begs for an unsolvability proof" and "my lifelong obsession with the problem" [Davis 1993 p. xiii]

11 "sure to yield results" [Davis 1993 p. xiv]

12 "I couldn't stop myself from thinking about" [Davis 1993 p. xiv]

13 "She listened attentively and amended or deleted" [Preface *Julia*]

14 "Kafka-like" and "I made many stupid and embarrassing mistakes" [*Julia* p. 11]

15 "would marry a professor" [*Julia* p. 19]

16 "conceived an absolute passion" [*Julia* p. 27]

17 "really more interested in shopping for furniture" [*Julia* p. 43]

18 "I was deeply depressed by the fact" [*Julia* p. 45]

19 "he ranks with Gödel as a logician" [*Julia* p. 45]

20 At the same conference Davis gave a ten-minute talk [Davis, private communication]

21 "Afterwards Martin told me" [Smorynski p. 78]

22 "Sometimes I have escaped from some boring task" [Quine p. 45]

23 "almost all American logicians" and "He and I began collaborating" [Davis 1993 p. xiv]

24 "We had a wonderful time" and "regularly stayed up" [Davis 1993 p. xiv–xv]

25 "I am very pleased, and surprised" [Davis 1993 p. xvi]

26 "drastic simplification of the proof" [Davis 1993 p. xvi]

27 "almost unable to function as a philosopher" [Sanoff]

28 "I had my reply ready" [Davis 1993 p. xiii]

29 "I felt that I couldn't bear" [*Julia* p. 69]

30 "everyone should be in equal conditions" [Matiyasevich, private communication]

31 "In spring 1964 I was rather tired" [Matiyasevich, private communication]

32 Kreisel wrote the following about the Davis, Putnam, Robinson paper for *Mathematical Reviews*: "These results are superficially related to Hilbert's Tenth Problem on (ordinary, i.e., non-exponential) Diophantine equations. The proof of the authors' result, though very elegant, does not use recondite facts in the theory of numbers nor in the theory of r.e. [recursively enumerable] sets, and so it is likely that the present result is not closely connected with Hilbert's Tenth Problem. Also it is not altogether plausible that all (ordinary) Diophantine problems are uniformly reducible to those in a fixed number of variables of fixed degree, which would be the case if all r.e. sets were Diophantine."

33 "I was spending almost all my free time" [Matiyasevich 1992 p. 40]

34 Chudnovsky and his brother building a "supercomputer" [Preston]

35 "This is a remarkable achievement" [Kaplansky p. 71]

In the Original

I spoke with/interviewed and corresponded with Dehn's three children, Helmut, Maria Dehn Peters, and Eva Dehn in 2000.

I spoke and corresponded with Trueman MacHenry in 2000.

I interviewed Peter Nemenyi twice on the phone in 2000.

I communicated with Paul Chessin through e-mail.

I received many photocopied documents from the Black Mountain Archive at the Department of Cultural Resources in Raleigh, North Carolina.

1 "I have met two men in my life who make me think of Socrates" [Weil p. 52]

2 "was one of the secularized" [Helmut Dehn interview]

3 "intuitive geometer" [Fisher in *Dict.*]

4 Dehn proved this result for polygons with great economy [Magnus p. 132]

5 he solved the third Hilbert problem before [Sah, preface]

6 "Dehn's own exposition was hard to understand" and "In 1903 Kagan published a paper" [Boltianskii p. 100]

7 He did not realize who the passing dignitary was [Helmut Dehn. Helmut recounts having had a similar experience with a passing general in the U.S. armed forces.]

8 "Founded early in the eighth century by Charlemagne" [Maria Dehn Peters from a letter]

9 However, a serious error was found in the proof [Stillwell p. 971]

10 "The fundamental idea of these seminar sessions" [Siegel 1965 p. 10]

11 "already a legend" and "drawers full of inspired manuscripts" and "the theory that was circulating" [Weil p. 52–53]

12 "It's a bourgeois, who still does algebra!" [Weil p. 53]

13 "The progress of our discipline depends" [Dehn p. 25]

14 "Instead of trucks rumbling through the streets" [Maria D. Peters in Stillwell 1999 p. 967]

15 The material about life in Frankfurt comes from the three children and specific sources are indicated in the text.

16 "He would climb up the trellis" [Maier p. 33]

17 "In retrospect, these hours together" [Siegel 1965 p. 10]

18 "As he said, he wanted to save the German bureaucrats" [Siegel 1965 p. 10]

19 "He couldn't believe they were that bad" [Eva Dehn interview]

20 "The actual large-scale terror in Germany" [Siegel 1965 p. 14]

21 "a somewhat retiring life" [Siegel 1965 p. 14]

22 "He wanted to see just how far" [Siegel 1965 p. 15]

23 "Isn't it lovely here?" and "Epstein acted as wisely as he could" [Siegel 1965 p. 17]

24 "he then spent several days riding all over the railroads" [Maria Dehn Peters in Stillwell 1999 p. 976]

25 "sly Aryan businessmen" *"schlauen arischen Händlern"* [Siegel 1965 p. 18]

26 "Hello, Professor. We're the dumb ones" [Siegel 1965 p. 24]

27 "The whole thing seemed like a malicious parody" [Siegel 1965 p. 21]

28 "It was a very welcoming wilderness, but there were a certain amount of snakes" [Fuller in transcript of interview for Martin Duberman]

29 "He was a little wisp of a fellow" [Corkran in transcript of interview for Martin Duberman]

30 Dahlberg "delighted" in talking [Duberman p. 305]

31 "The table was round and made of a heavy wood" [Chessin e-mail]

32 "We are building a cathedral" [Nemenyi interview]

33 "stopped out" and "explosions" [MacHenry private communication]

34 "very much like a mother to me" [MacHenry private communication]

35 "I asked [Dehn] if it wasn't frustrating" [Mildred Harding p. 86]

36 "Time's metric" [MacHenry e-mail]

37 he received $2,750 from Wisconsin [From documents supplied by the University of Wisconsin]

38 "For Max Dehn, the incident assumed" [Duberman p. 366]

Distance

I conducted an interview in 1999 with Pogorelov in several parts using e-mail.

1 "Hilbert asks for the construction" [Busemann in *Proceedings* p. 131]

2 Busemann "wasted" [Reid p. 330]

3 Busemann himself begged [Reid p. 377]

4 "chockablock with dozens of large canvases" [Dembart]

5 but his 1966 talk . . . sparked the imagination of A. V. Pogorelov [Pogorelov p. 2]

6 "talented, self-educated" [Pogorelov e-mail]

7 "I was lucky enough to solve many of these problems" [Pogorelov e-mail]

8 “In my article published in 1973 I have admitted” [Pogorelov e-mail]

9 “This book is devoted to Hilbert’s fourth problem” [Pogorelov p. 7]

10 “A lot of people forget that Hilbert also asked” [Alvarez e-mail]

Something for Nothing

Gleason

I interviewed Gleason on the phone and corresponded with him 2000–2001.

Harvard University was able to supply some useful information that was in a publicity file—clippings, press releases, etc.

1 “Even among those of Hilbert, there are still several” and “Beyond this, he [the American mathematician] runs” [Weil 1950 p. 295]

2 “shows that you can’t have a little bit of orderliness” [Harvard Gazette]

3 “From this point on, the reasoning becomes lengthy and quite technical” [Kaplansky p. 19]

4 “Since the proof of Theorem 1 [Hilbert’s problem] is very complicated” [C. T. Yang in *Proceedings* p. 143]

5 “I don’t know how others may have been thinking about the fifth problem” [private communication]

6 “an exploring trip” [MMP p. 81]

7 “had a peripatetic youth” [private communication]

8 “How the World War Was Fought” and “leanings” [MMP p. 82]

9 Since then I have not been so fierce” [MMP p. 82]

10 “She was an absolute stickler for points” [MMP p. 83]

11 “He received me cordially and asked what [calculus] book” [private communication]

12 “It was almost the same as being in graduate school” [private communication]

13 “bootleg operation” [MMP p. 88]

14 “metrically straight” and “theorem by von Neumann” [Gleason p. 194]

15 “putting the whole problem” [private communication]

16 In this nomenclature Gleason was “G-squared” [Nasar p. 146]

17 “proofs really aren’t there to convince you that something is true” [MMP p. 86]

Montgomery

Much of the material on Montgomery comes from the little booklet from the Institute for Advanced Study: *Deane Montgomery 1909–1992*. Mary Jane Hayes at the Institute kindly sent me a copy of this. I corresponded using e-mail with Jim Stasheff who was a the University of North Carolina at the end of Montgomery's life. I was able to fill in some gaps from county records in Minnesota. I spoke with Mostow briefly on the phone.

18 "about a hundred people today" [I called the post office.]

19 "John Montgomery came to Minnesota" [Mostow p. 12]

20 "several" [Mostow p. 12]

21 Deane's mother did not think it "safe for Methodists" [Mostow p. 12]

22 "From his widowed mother Deane inherited a rigorous egalitarianism" [Mostow p. 13]

23 "My first impression was that he was a remarkably handsome man" [Selberg p. 4]

24 "This was the climax of a major effort" [Borel p. 685]

25 "If we don't show appreciation, then who will?" [Mostow p. 13]

26 Raoul Bott recalls one party ending with him [Bott p. 22]

27 "the meanest sonofabitch he ever met" [James Stasheff private com.]

28 "In my mind, Professor Deane Montgomery was much more than a friend" [Yang p. 15]

Zippin

I interviewed on the phone and exchanged e-mail with Zippin's two daughters, Nina Baym and Virginia Narehood.

The material about Central High School came from Dr. Sanders, a retired teacher and historian for the school.

Mary Jane Hayes at the Institute for Advanced Study sent me a photocopy of a CV and some other information in Zippin's file. I received additional information from Professor Diamond at Queens College.

29 "He had a manly, square face and twinkly blue eyes" [Narehood e-mail]

30 "I am a little tired and I had better lie down to rest" [Zippin in memorial booklet]

Physics and Math—The Steamy Details

1 "The maze of experimental facts" [Weyl 1944 p. 653]

2 "Since physics has gone through two revolutions" [Wightman p. 147]

3 "No. I was repelled by it" [Lewy in MMP p. 184]

4 "The marriage between mathematics and physics" [Dyson p. 635—This figure of speech came from his friend, Res Jost]

5 "It should be noted that most of the paper deals" [Preface in *The Arnold-Gelfand Mathematical Seminars*]

6 "The problem now is that there is an impending divorce" [Dyson e-mail 1999]

First State the Tune

1 "anything resembling a regular job" [Kac p. 92]

2 "His visit to Madison became the beginning" [Ulam p. 134–5]

3 "Stan, I am so glad to see you are alive" [Ulam p. 174]

4 "Look Stan! What a nice epsilon" [p. 179]

5 "the number of steps it takes any mathematician to connect with Erdös" [Ulam p. 135]

6 458 mathematicians have this number [Pach p. 42]

7 Gauss himself published six different proofs [Stewart p. 232]

8 "Almost as useless as number theory" [Hardy p. 131]

9 Some, including Weyl, believe that Hilbert's own greatest work [Weyl 1944 p. 634]

10 300,000 primes [Dunham p. 83]

11 formula [from Ramanujan's letter to Hardy]

Transcending Local Conditions

1 "All methods of proof of the transcendence of a number" [Gelfond p. 1]

2 Our first oddball number is approximated very well compared to the error by $\frac{1}{10}$ better still by $\frac{1}{10^{10}}$ remarkably well by $\frac{1}{10} + \frac{1}{10^{10}} + \frac{1}{10^{10^{10}}}$ and so on. Therefore we can see it is not algebraic.

Siegel

I talked about Siegel with Natascha Brunswick and with Karin, Tom, and Michael Artin.

3 His father was a mailman—specifically a "money-order carrier" [Braun p. 19]

4 This led to lower scores than he thought fair [Hlawka p. 509]

5 ranged from average or mediocre to even "malign" *mittelmässigen oder sogar bösartigen* [Siegel 1968 p. 64]

6 "I loved drawing and showed early talent in it, as well" [Hlawka p. 509]

7 "how it really was" and "how it happened by chance" "*wie es wirklich war*" and "*wie es sich zufällig ereignet hat*" [Braun p. 18]

8 "No doubt, C. L. Siegel was the unpopular model pupil" [Braun p. 19]

9 "Although I had no understanding of the politics" [Siegel 1968 p. 64]

10 "He obviously had a great deal of fun rattling off" [Siegel 1968 p. 64]

11 "When I attempted to read this" [Thue p. xxx]

12 "fit for military service" and "before the military tried to abuse me" [Siegel in 1968 p. 64]

13 "this war is not my war" and "He took off, disappearing into thin air" [Weil p. 127]

14 "Then they made plans for Siegel" [Braun p. 20]

15 "He was released after five, for him, very hard weeks" [Hlawka p. 374]

16 In summer of 1919 Siegel moved to Göttingen [Schneider p. 148]

17 "I do not know which of the three was more decided" [Braun p. 21]

18 "I had the impression that Schaffeld was the only authority figure" [Braun p. 20]

19 "There are theorems that only Siegel could prove" [Grauert p. 218]

20 "cold, hungry and unhappy" [Reid p. 310]

21 "he worried constantly" [Reid p. 165]

22 "sinking" "drowning" "evil" [Braun p. 21]

23 "He made poor Hermann go into the woods" and "Meet me at the restaurant near the palm trees" [Natascha Brunswick interview]

24 from burnout (*uberanstrengung*) [Hlawka p. 510]

25 "probably his deepest and most original" [Dieudonné in *Dict.* Siegel p. 827]

26 "I will not allow myself, nor am I able" [Schneider p. 52]

27 Siegel wrote the final version of the paper in Pontresina [Schneider p. 152]

28 "Most" cubic equations are nonsingular [Silverman, Tate p. 6]

29 "Swedish healing exercises" [Braun p. 21–22]

30 Dehn's daughter Eva described Betty as "well-built" [Eva Dehn interview]

31 "I have two girls in my class who can do the work" [Braun p. 7]

32 "What are you doing here?" and "Why don't you make some coffee, will you?" [Braun p. 35–6]

33 "Ah, so you can't make coffee either, huh?" [Braun p. 35–6]

34 "This was probably typical of an evening he usually had" and the rest of the crab dinner story [Braun p. 40–41]

35 "The work of Siegel in this domain may be considered" [Dieudonné *Dict.* Siegel p. 829]

36 "day of national solidarity" and "Not in the mood, no time, no money" [H. Moser]

37 "earlier in his life Siegel expected a change of scene" [Braun p. 44]

38 "take a wrong path and the DMV might follow it too" [Davenport. p. 77]

39 "With Hasse and Siegel Göttingen became a worthwhile place" [Grauert p. 218]

40 Earlier, in the '20s, Siegel had said in conversation with Hasse [Davenport p. 79]

41 "I can't stand to be in the same room with Hasse" [Braun p. 46–7]

42 *Naturschwärmerei* [Braun p. 50]

43 "how to really live" [Braun p. 61]

44 Nine displaced scholars were supported [Dawson 1997 p. 154]

45 studying notes of lectures Siegel gave during the war [Piatetski-Shapiro p. 204]

46 winter of 1946-47 Siegel returned [Hlawka p. 511]

47 She recounts being sick with the flu and Siegel [Brunswick interview 1995]

48 "one crazy man on the faculty" [Dawson 1997 p. 194]

49 "The natural route is by Swiss postal bus" [Davenport p. 79]

50 "It is entirely clear to me what circumstances have led" [Siegel quote in Grauert p. 218–19]

51 "Only such great mathematicians as Siegel could afford to ignore" [Freudenthal p. 169]

52 Brunswick visited Siegel at his home late in his life [interview]

53 "Increasingly isolated but still spiritually vital" [Schneider p. 150]

Gelfond

In an effort to get a little more information than was available in the *Dictionary* and obituary articles, I contacted Ilya Piatetski-Shapiro, who in turn contacted Andrei Shidlovski in Moscow. Shidlovski wrote back to Piatetski-Shapiro responding to questions I had asked. Then Piatetski-Shapiro forwarded the responses to me. Much of the biographical detail in this section comes from Shidlovski, either through Piatetski-Shapiro or from published sources.

54 "His daughter Julia keeps his birth certificate" [Shidlovski e-mail]

55 This book of Lenin's became required reading for all students [Piatetski-Shapiro p. 202]

56 "His role in the events of 1917 was quite outstanding" [Deutscher]

57 "sabotage" [Demidov p. 45]

58 "red" [Demidov p. 42]

59 "A. O. Gelfond studied mathematics hard when he was young" [Shidlovski e-mail]

60 He traveled to Germany for four months in 1930. One six-page narrative about his trip to Germany in 1930 was published in 1977, almost 10 years after his death. Gelfond originally gave it as a speech in East Berlin in 1965. It is a strange document coming out of the Soviet state, printed on poor paper, careful politically, and limited mainly to an account of mathematical contacts and influences—and I don't claim to have understood it. A Russian friend tried to translate it for me as we sat and talked and was himself confused. The most personal piece of information is that Gelfond discussed chess with Hilbert.

61 "The proof of Gelfond, though more advanced" [Hille p. 654]

62 "one of the most sinister personalities of the science vs. ideology" [Demidov p. 44]

63 "Proletarian students 'declared war'" [Demidov p. 44]

64 "It is the enforcement of a standard outlook" [Demidov p. 45]

65 "The tone of the ideological accusations against Luzin" [Demidov p. 49]

66 For example, Gelfond along with others [according to Demidov p. 46]

67 "The campaign was abruptly stopped" [Demidov p. 49]

68 "black angel" [Demidov p. 50]

69 "inherited" [Piatetski-Shapiro p. 202]

70 "He was a very warm person, very humane and sensitive" [Piatetski-Shapiro p. 202]

71 "I was present in the hospital" [Piatetski-Shapiro p. 201]

72 "His mathematical ability was esteemed" [Evrafov et al. p. 177]

73 "When one studies Gelfond's arithmetic works" [Piatetski-Shapiro and Shidlovski p. 238]

Schneider

The main, relatively comprehensive sources were the obituaries of Bundschuh et al. and Kappe et al.

74 "After a few months I gave him a work of six pages" [Bundschuh p. 131]

75 "You are the eleventh who started writing a dissertation" [Bundschuh 130]

76 "For, as of autumn 1933 success depended" [Kappe p. 113–114]

77 This done, Schneider went to Oberwolfach [Kappe p. 115]

78 "Saying goodbye to the active university life" [Kappe p. 118]

79 "In his last years of life" and "a hurting cut" [Kappe p. 118]

The Inordinate Allure of the Prime Numbers

1 "the density of prime numbers appears" [Edwards p. 2]

2 "When Riemann makes an assertion" [Edwards p. x]

3 "lost proof" Therefore God couldn't sink the boat [Reid p. 163]

4 "The difficulty of Siegel's undertaking" [Edwards p. 136]

5 His mother put food in his mouth as he worked [Kanigel p. 13]

6 2 formulas [Hardy p. xviii]

7 "Very recently I came across a tract" [Hardy p. xv]

8 "This may be said to have been Ramanujan's one great failure" [Hardy p. xvii]

Castles of Air

Edwards (*Fermat's Last Theorem*) and Iyanaga were main sources for the first section of this chapter.

1 "Archimedes will be remembered when Aeschylus is forgotten" [Hardy p. 81]

2 "utter" [Honda p. 141]

3 "As a child, Germain had been fascinated" [Dunham p. 241]

4 "Legendre's proof of the second case was rather artificial" [Edwards p. 70]

5 "practically said that Lamé's brainchild" [Edwards p. 77]

6 "a curiosity of number theory rather than a major item" [Edwards p. 80]

7 This idea first appeared (under a different name) in the work of Kronecker in 1882 [Frei p. 425–426 or Hasse p. 266]

Takagi

8 "an unusually vast farm" [Honda p. 142]

9 "an extremely austere and hardworking man" [Honda p. 143]

10 "in another person's home" [Honda p. 143]

11 "Chinese literature, calligraphy, Chinese painting" [Honda p. 145]

12 "succinct, with a lofty tone" [Honda p. 145]

13 "Since he was such a lovely little boy, I wished to touch him" [Honda p. 145]

14 "He is indeed a genius" [Honda p. 145]

15 "strong-minded and romped as if she were a boy" [Honda p. 145–46]

16 "Great Earthquake of Nobi" [Honda p. 147]

17 The timing of Takagi's schooling is somewhat confusing. We are told that he skipped three years of primary school, yet he entered college at the age of nineteen, which would not seem to be early. Be that as it may, his path had led him to the national university,

18 "one of the most famous parlors in Tokyo" and "One day, after the Second World War, when Takagi once happened" [Honda p. 148]

19 "Buddha" [Honda p. 148]

20 "You are ordered to go to Germany" [Honda p. 150]

21 "Perhaps it must have been Prussian taste" [Honda p. 151]

22 "To the eyes of Japanese young men of that time" [Honda p. 152]

23 "In order to study science in Germany" [Honda p. 152]

24 "It is just one year today since I arrived in Berlin" [Takagi in Honda p. 152]

25 "I was much astonished by the striking contrast in the atmospheres" [Takagi in Honda p. 153–154]

26 "That was quite enough. Such attendance for six weeks" [Honda p. 154]

27 "That's fine" [Honda p. 155]

28 "This book played the role of the Bible in Takagi's scientific life" [Honda p. 154]

29 "that dandy gentleman" [Honda p. 156]

30 "a cheerful, sociable woman" [Honda p. 157]

31 "When I am talking with Teiji, I feel quite stiff" [Honda p. 157]

32 "Generally speaking, I am a man who" [Honda p. 158]

33 "utter (scientific) solitude" [Honda p. 141]

34 "the magnificent edifice of class field theory" [Honda p. 141]

35 "There is nothing in any other science that, in subtlety and complexity" [Weyl p. 523]

36 "Its proofs were when it appeared" [Iyanaga p. 496]

37 "the long and difficult proofs of Takagi's theorems" [Honda p. 162]

38 as does Richard Brauer [Brauer p. 31]

39 "Since it was then quite an unexpected result" [Takagi in Honda p. 159]

40 "I hear that Japanese fellow will talk on number theory" [Honda p. 160]

41 "I felt strong admiration for it" [Artin in Honda p. 161]

42 The language he used to frame the reciprocity results helped suggest to Artin [Honda p. 161]

43 "It seems that over the years, everything that has been" [Lang p. iii]

44 "You should be warned that acquaintance with only one" [Lang p. 176]

45 "If the war had been" [Hasse in Honda p. 166]

46 "The dinner party seems to have been the best time" [Honda p. 166]

47 "Observing my old master" [Takagi in Honda p. 165]

48 "the unique gay period in the long quiet life of Takagi" [Honda p. 158]

Artin

I interviewed Natascha Brunswick in 1995 and also Michael Artin in the same year. I later interviewed/spoke with, exchanged correspondence with Karin Artin, Tom Artin and Michael Artin (again).

49 Emil was number 43 [Natascha Brunswick interview 1995]

50 Emil made a telescope, and the two boys also rigged [Michael Artin]

51 telegraph between their houses [Michael Artin]

52 "There was something of the nineteenth century" [Charlotte John in Reid p. 351]

53 "discloses a very great skill with computations" [Brauer p. 35]

54 Gian-Carlo Rota said that Artin loved calculation [Rota p. 13]

55 MANIAC [Silverman and Tate p. 119]

56 "Artin, whose music was as pure and rigorous" [Reid p. 311]

57 inspired by one of Takagi's observations [Iyanaga p. 500]

58 "In 1923, no possibility of proof for this new theorem was in sight" [Brauer p. 32]

59 "living like a student, earning money" [Stevenhagen and Lenstra p. 27]

60 "I devised my best result while carrying water" [Stevenhagen and Lenstra p. 27]

61 "a badly paid and ill-defined position" [Stevenhagen and Lenstra p. 27]

62 "I was very touched by Artin's meticulousness" [Stevenhagen and Lenstra p. 35]

63 "was perhaps the first triumph" [Brauer p. 34]

64 "Artin developed a strong aversion" [Brauer p. 36]

65 "There is the saying of G. B. Shaw" [Brauer p. 39]

66 His students and friends took to calling him "Ma" [Brauer p. 27]

67 "The appeal to the sense of motion" [Zassenhaus p. 3]

68 "We all believe that mathematics is an art" [Artin p. 475]

69 "In concluding I wish again to emphasize" [Artin p. 479]

70 "It is widely felt that it works for no good reason" [Stevenhagen and Lenstra p. 35]

71 According to Reid, Artin was afraid that the fact [Reid p. 427]

72 Hasse suggested he could get Artin's children declared Aryans [Michael Artin private communication]

73 "The intellectual atmosphere of German universities" [Brauer p. 27]

74 "What is applied mathematics, actually?" [Tom Artin]

75 "In my view, Emil never came to feel at home in the U.S." [Tom Artin]

76 "I saw Artin for the last time" [Brauer p. 28–29]

Hasse

77 "emergency school leaving examination" [Frei p. 55]

78 "The 11th problem is simply this" [Kaplansky p. 84]

79 "even today" [Kaplansky p. 84]

80 "high point of the local–global principle" [Edwards in *Dict.* Hasse]

81 "You may worry about your geometric intuition" [Silverman and Tate p. 230]

82 They followed this with the boxed saying [Silverman and Tate p. 230]

83 "I have been harder hit by the turn of events" [Reid p. 367]

84 "It will not be long before the German people find Hitler out" [Reid p. 364]

85 *untragbar*—"unbearable" [Reid p. 370]

86 "If the beast of tyranny takes over his country" [Reid p. 381]

87 "notoriously crazy" [Courant, quoted by Reid 380]

88 Remarkably, he seems later to have admitted this [Segal p. 52]

89 "The situation in Göttingen is really extremely absurd" [Harald Bohr in Segal p. 51]

90 "such an unpleasant demonstration by pro-Nazi mathematics students" [Reid p. 386]

91 "In the night of the famous 30, June, Tornier wrote" [Bohr in his own English from Segal p. 53]

92 "appears like a last resort for outspoken Nazis among mathematicians" [Mehrtens p. 158]

93 "There he embarrassed the mathematics faculty by" [Reid p. 402]

94 but there is some evidence that he later reapplied and was admitted [Segal p. 54]

95 "After the November pogrom [*Kristallnacht*], when I returned" [Siegel quoted in Segal p. 54]

96 "My political feelings have never been National–Socialistic" [Reid p. 474–475]

97 "Hasse never seemed to have realized" [Segal p. 55]

Twelfth

98 Hilbert also proved the result in 1896, thinking he was only offering an alternate method [Schappacher p. 253]

99 "the class field theory alluded to above gives" [Silverman and Tate p. 185]

100 "Whether justly or not, the twelfth problem" [Langlands p. 408]

101 "Whereas physics in its development" [Weyl 1951 p. 523]

102 "When the great German logician David Hilbert" [Ribet and Singh p. 69]

What Is Algebra?

1 To make his new algebra of quaternions work Hamilton had to jump to four bases instead of the desired three, it turned out, so the resulting quaternion is written $a + bi + cj + dk$. There was a deep belief that all "multiplication" was commutative, in other words $ab = ba$. However, in Hamilton's algebra $jk = -kj$. This was a tremendous break with tradition. The general question then became, what do we get if we don't assume the axiom that members of an algebraic structure commute under multiplication?

2 "an energetic and very nearsighted washerwoman" and "loud and disagreeable voice" and "The graces did not preside at her cradle" [Reid p. 143]

3 "Noether-guard uniform" [Reid p. 352]

4 “was overly optimistic” [Mumford p. 431]

5 The quotes from Nagata come from letters he wrote to me in response to my questions 1997–1998. Corrections added 2000.

Schubert's Variety Show

Steven Kleiman was very helpful for this chapter.

1 In 1876, Schubert was appointed *Oberlehrer* at the *Johanneum* [Kleiman p. 446]

2 “How many figures of a certain sort fulfill” [Coolidge p. 180]

3 “grand machine” [Steven Kleiman e-mail]

4 “twisted cubic space curves tangent to 12 given quadric surfaces” [Kleiman p. 445]

5 “furiously attacked” [Kleiman]

6 “there has not been a more contentious topic” [Coolidge p. 181]

7 “Today's interpretation of Hilbert's 15th problem” [Kaplansky p. 133]

8 “Our results include all that is required” [Weil p. viii] However, not all were satisfied, and Kleiman points out that Weil himself notes in the 1962 revision on page 331 that he hadn't closed all the gaps.

9 “Schubert's ideas were slowly interpreted” [Kleiman p. 322]

10 “In an actual situation calculating a number is only the first step” [Kleiman e-mail]

11 The enumerative theory of conics and the greater theory of plane correlations [Kleiman p. 360]

12 “Tremendous progress has been made over the last decade” [Kleiman p. 327]

13 “It is a tribute to Hilbert's vision that he could foresee” [Kleiman p. 327]

Graph That Curve

See articles by Degtyarev and Kharlamov, Gudkov, and Viro. I communicated with Viro and Kharlamov by e-mail. This is a fairly technical chapter, and it is better now than it would have been without these communications.

1 “Pitifully, it seems that the book will never appear” [Degtyarev and Kharlamov p. 4]

How Many Kinds of Crystals Are There, and Does the Grocer Know How to Stack Oranges?

Main sources on Bieberbach were Mehrtens, Freudenthal, Grunsky, and Shields.

1 “In his youth he used to look at the most difficult problems” [Reid p. 146]

2 “unfit for military service” [Mehrtens p. 197]

3 “the ‘*siegerfüllten*’ (filled with victory) spring of 1918” [Mehrtens p. 203]

4 “incredible displays of *ad hoc* synthetic geometry” [Wolf p. 105]

5 “The tendency towards formalism must not” [Mehrtens p. 202]

6 Gödel checked out Bieberbach's book on functional analysis both in Brunn and in Vienna. [Dawson 1997 p. 53]

7 Bieberbach's books, translated into Russian, were important [Yushkevich p. 7]

8 "This notion touches on visual perception" [Mehrtens p. 205]

9 "edifice of symbols which we pile up to infinity" [Mehrtens p. 206]

10 "he was both physically and intellectually a very lively man" [Mehrtens p. 199]

11 "Herr Bieberbach's love and devotion for himself" [Mehrtens p. 200]

12 "excursion to the 'liberated area' of Southern Tirol" [Mehrtens p. 214]

13 "At ours in Berlin Bieberbach was known" [Freudenthal p. 168]

14 "It was impossible to face him" [Freudenthal p. 168]

15 "The postcard showed a small German town" [Mehrtens p. 217]

16 "Thus, to have a current example, the manly rejection (*mannhafte Ablehnung*)" [Shields p. 9]

17 "The majority of the German intelligentsia was conservative" [Freudenthal p. 170]

18 "student followers" and "have himself installed as '*Führer*' of the DMV" [Mehrtens p. 221–222]

19 "was earning his living" [Mehrtens p. 233]

20 After the war he had made his way to Göttingen [Gautschi p. 32–34]

21 "One cannot forget all the suffering" [Reid p. 501–502]

22 "When I visited Bieberbach shortly before his death" [Mehrtens p. 233]

23 patented designs for use in a quilted toilet paper [*Los Angeles Times* 4/17/97]

24 "scandalous situation" [Milnor *Proc.*]

25 "many mathematicians believe, and all physicists know" [Conway and Sloane p. 3]

26 "At one point while working on this book" [Conway and Sloane p. v]

27 "a lot of" Sun computers [Hales private communication]

28 "We try very hard to understand how nature operates" [Hsiang telephone conversation October 1998]

29 "Proofs really aren't there to convince you" [MMP p. 86]

Programs (Analysis Takes At Least Seven Years)

How Famous Can a Function Theorist Be?

Koebe

1 "When on the first day of my study I was inscribed" [Freudenthal 1984 p. 77]

2 "ran a renowned fire engine plant" [Freudenthal 1984 p. 77]

3 "conceited and disagreeable" [Reid p. 257]

4 "younger people, when introduced to Koebe" [Freudenthal 1984 p. 77]

5 "Koebe's mathematical style is prolix" [Freudenthal in *Dict.* on Koebe]

Poincaré

Darboux was by far the best source I found for details on the life of Poincaré.

6 "To give an adequate idea of this immense labor" [Bell p. 538]

7 "To give any complete account of his work" [Baker p. vii]

8 "*lorrains tous les deux*" [Darboux p. lxxxii]

9 "solid, massive and without ornament" [Darboux p. xxxii]

10 "*très original*" [Darboux p. lxxxiv]

11 "*très bonne, très active et très intelligente*" [Darboux p. lxxxiv]

12 Gaston Julia writes that Poincaré's maternal grandmother [*Livre* p. 165]

13 "His mother watched over him with a solicitude full of intelligence" [Darboux p. lxxxiv]

14 "little masterpiece" [Darboux p. lxxxv]

15 "passive resistance" [Darboux p. lxxxvi]

16 "Henri wanted to see everything, understand everything" [Darboux p. lxxxvi]

17 "He had the gaiety and the expansiveness of a child, the reason of a man" [Darboux p. lxxxvii]

18 "From this moment this vocation did nothing but grow" [Darboux p. lxxxviii]

19 Kolmogorov had a theory, offered as an explanation for the childish behavior [Ruelle p. 8]

20 "in glacial cold, through burnt villages, empty of inhabitants" [Darboux p. lxxxix]

21 "particularly obscure" and "without thinking a minute" [Darboux p. xci]

22 "From the beginning of term his intelligence" [Appell in Darboux p. xci]

23 "I desire that you not leave elementary geometry" [Darboux p. xcii]

24 "This is a horse" [Darboux p. xcii]

25 "*sang-froid*" [Darboux p. xcvi]

26 "Poincaré begins in the wilds of differential equations" [Stillwell p. 16]

27 "For fifteen days I strove to prove that there could not be any" [Poincaré 1913 p. 388]

28 "Poincaré's ignorance of the mathematical literature" [Dieudonné *Dict.* p. 53]

29 "A Kleinian group action on the plane turns out" [Stillwell p. 26]

30 "The most extraordinary production" [Dieudonné in *Dict.* p. 56]

31 Neither Hill nor Poincaré actually proved [Siegel 1970 p. 132]

32 "Obviously, no revolution is actually fully a revolution" [Galgani p. 246]

33 "Every problem I had attacked led me to *Analysis situs*" [Dieudonné in *Dict.* p. 59]

34 "Imagine any sort of model and a copy" [Monastyrsky 1987 p. 76]

35 "It has been rightly said that until the discovery" [Dieudonné in *Dict.* p. 60]

36 "a chore for which he does not seem to have shown great reluctance" [Dieudonné in *Dict.* p. 60]

37 He received at least nine honorary doctorates [*Livre*]

38 "was the first paper on what we now call 'algebraic geometry over the field of rationals.' " [Dieudonné *Dict.* p. 55]

39 Arithmetic algebraic geometry: When Andrew Wiles proved Fermat's last theorem, the key step was to prove that all (or at least enough) functions of the type $y^2 = x^3 + ax + b$ for a and b integers (which are called elliptic curves) can be parameterized using only modular functions—a narrow class of automorphic function. This is what is some times called the Taniyama–Shimura conjecture. It was known already that if Fermat's last equation had a solution, A, B, and C, such that $A^n + B^n = C^n$ $n \geq 3$, then the equation $y^2 = x(x - A^n)(x + B^n)$, which is an elliptic curve, though written in a different form, could not be parameterized using modular functions. Therefore Fermat's last equation cannot have a solution, since if it did there would be a contradiction.

40 "It seems that Poincaré was completely unaware of the work" [Bashmakova p. 78]

41 "I saw him often between the years" [Dantzig p. 2]

42 "the first example of an existence proof in analysis based on algebraic topology" [Dieudonné p. 60]

43 "He thought in the street" [Boutroux *Livre* p. 171]

44 "energy" and "the art of giving the same name to different things" [Poincaré 1908 p. 171–172]

45 "I think for my own account, and I am not alone, that the important point is to not introduce things that one cannot define completely in a finite number of words" from "*Je pense pour mon compte, et je ne suis pas seul, que l'important c'est de ne jamais introduire que des êtres que l'on puisse définir complètement en un nombre fini de mots*" [Poincaré 1908 p. 182]

46 "David Hilbert ... formulated twenty-three" [Weyl 1951 p. 525]

Schools Amid Turbulence

Kolmogorov

All of *Russian Math. Surveys* 43:6 (1988) and *The Annals of Probability* 17:3 (1989) are devoted to the life and work of A. N. Kolmogorov as well as p. 31–100 of *Bulletin London Math. Soc.* 22:1 (1990)

1 “I have formulated the thirteenth Hilbert problem” [Shiryaev p. 915]

2 “On 20 October 1987 there ended the life” [Tikhomirov p. 1]

3 “Kolmogorov, Poincaré, Gauss, Euler, Newton” [Arnold 1988 p. 43]

4 “were independent women with high social ideals” [Shiryaev p. 867]

5 “against the wishes of her parents” [Tikhomirov. p. 21]

6 “perished” [Kendall p. 301]

7 “disappeared without a trace” [Tikhomirov p. 2]

8 “These sheets, all speckled over” [Kovalevskaya p. 122–3]

9 “Photographs still exist of the grand old house” [Tikhomirov p. 2–3]

10 “ ‘They took’ her to the haymaking, where alongside the country women” [Tikhomirov p. 3]

11 *Vesennie Lastochki*, “The Swallows of Spring” [Tikhomirov p. 4]

12 “The gymnasium rooms were small” [Kolmogorov in Shiryaev p. 868]

13 “mischief, daring, bravery, dexterity [and] sportiveness” [Tikhomirov p. 5]

14 “organizing a state—a commune on a desert island” [Tikhomirov p. 6]

15 “Together with other senior students” [Shiryaev p. 868]

16 “one man on one horse ploughs an *obzha*” [Yanin p. 190]

17 “Not you, young man!” [Tikhomirov p. 7]

18 “I have decided to go into science” [Yanin p. 184]

19 “Had the work of Andrei Nikolaevich” [Yanin p. 191]

20 “are regarded with awe by specialists in the field” [Kendall p. 303]

21 “We shall prove that all the finitary conclusions” [Kolmogorov in van Heijenoort p. 416]

22 “All my joint work in probability theory with Khinchin” [Shiryaev p. 873]

23 “Pavel Sergeevich Aleksandrov and I date our friendship” [Tikhomirov p. 10]

24 “The typical scenery on the banks of a big Central Russian river” [Kolmogorov 1986 p. 229]

25 “crowds of unemployed people, often very badly dressed” [Kolmogorov 1986 p. 234]

26 “a little inn by the small lake” and “From there we did” [Kolmogorov 1986 p. 237]

27 “Some more time passed” [Aleksandrov 1980 p. 319]

28 “in those years he was so full of ideas” [Tikhomirov p. 11]

29 “Do not fail to buy yourself all these things” and “almost entirely according to their nutritive value” [Aleksandrov in Kolmogorov 1986 p. 242–243]

30 "Some days ago I went for a long four-hour walk" [Aleksandrov in Kolmogorov p. 244]

31 "Often I was woken in the morning" [Aleksandrov 1980 p. 324]

32 "Our work began" [Kac p. 48–49]

33 "I had long had a peculiar sort of contact" [Weiner p. 145]

34 "His work on the foundations of probability theory" [Arnold p. 150]

35 "an unlikely person at the conference" [Whitney p. 110]

36 "the flow of liquid over a manifold" [Tikhomirov p. 25]

37 "Nevertheless, all my really deep ideas" [Weiner p. 261]

38 "The average (mean) square of the difference" [Shiryaev p. 905]

39 "inspired considerable bewilderment" [Gleick p. 76]

40 "After Kiev I went back to Moscow" [Smale 1980 p. 151]

41 Both Kolmogorov and Khinchin, at considerable risk to themselves [Vere-Jones p. 122]

42 "Well, old man, you sure started something" [Yushkevich p. 25]

43 "Sometime I will explain everything to you" [Arnold 1993 p. 134]

44 "I didn't think that at all. The main thing was that in 1953 there appeared some hope" [Arnold 1993 p. 130]

45 In a 1997 interview with H. S. Lui, Arnold advanced a simple, more elegant, and certainly funnier theory about where KAM theory came from than the one he questioned Kolmogorov about. Kolmogorov's work on this problem came from a required homework problem he assigned to his second year undergraduates. He asked them to find a Hamiltonian system of a certain kind that was irregular in its motion. His own intuition and common wisdom suggested that it should be easy. However, the students couldn't solve this problem. Kolmogorov couldn't solve the problem. This was not a state of affairs that could be tolerated.

46 "the quasiperiodical motions of Hamiltonian systems under small perturbations" [Shiryaev p. 911]

47 "imply the stability of the motion of a planetoid" [Arnold 1989 p. 150]

48 "In conservative systems, asymptotically stable motions are impossible" [Kolmogorov p. 356]

49 "It would be important to obtain also" [Siegel 1941 p. 808]

50 "It is strange . . . but he has done it" [*Arnoldfest* p. 3]]

51 "Of course careful note must be taken of the fact that continuous functions" [Kendall p. 311]

52 Kolmogorov said that this proof required the longest period of concentration [Tikhomirov 1993 p. 117]

53 "intricate" [Kaplansky p. 105]

54 "His conjectures differ from the modern 'strange attractors' program only in terminology" [Arnold 1989 p. 50]

55 "I last saw him on the Courland Spit (Kursu Neringa) in Lithuania" [Rosenfeld p. 80] I think it likely that the accident—as opposed to recuperation—happened in the Black Sea.

56 "There was talk about my allegedly inexhaustible youth" [Kolmogorov in Shiryaev p. 935]

Arnold

57 "helped our family to survive when my father died" [Arnold 1999 p. 13]

58 "He never explained anything, just posed problems" [Zdravkovska p. 29]

59 "wanted to choose something completely orthogonal" [Arnold 1999 p. 7]

60 "It is almost impossible for me to read contemporary mathematicians" [Zdravkovska p. 30]

61 "criminal bourbakizers" [Arnold 1995 p. 8]

62 "some criminal algebraist" [Arnold 1996 p. 3]

63 "administrators in most countries, like pigs under an oak tree" [Arnold 1996 p. 3]

64 "By the way, the 200-year interval from Huygens and Newton" [Zdravkovska p. 30]

65 "Bizarre questions like Fermat's problem" [Arnold 1995 p. 8]

66 "almost daily marathon hiking, biking, swimming or skiing trips" [Zdravkovska p. 29]

67 "After some time I saw that he finally was turning back" [Marina Ratner e-mail 2000]

68 "Terrible, yesterday I failed to cross Golden Gate" [Yakov Eliashberg e-mail 2000]

69 Richard Montgomery says that Arnold told him [e-mail 2000]

70 "When I cannot prove something, I put on my skis" [Zdravkovska p. 32]

71 "Research Interests: Dynamical Systems, Differential Equations, Hydrodynamics" http://elib.zib.de/IMU/EC/ArnoldVI.html

72 "After the tiring Congress, I spent a day" [Arnold 1995 p. 10]

Past Chernaya Rechka, to 61 Savushkina Street

I communicated with Bolibruch via e-mail.

1 "This is a distinctly formulated problem" [Anosov and Bolibruch p. 7]

2 Plemelj was appointed by imperial decree first [Dobrovolskii p. 224]

3 "When I was 6 years old my father decided" [private communication]

4 "fishing with disks" [private communication]

5 "This kind of fishing is not usual" [private communication]

6 "I spent a lot of time in such camps" [private communication]

7 "In those days the boarding school was located" [Bolibruch 1996 p. 1]

8 "Every other student of our school wrote poetry" [Bolibruch 1996 p. 3]

9 "the peak of this poetic activity was reached in winter of 1966" [Bolibruch 1996 p. 3]

10 "Now I see what is really going on here" [Bolibruch 1996 p. 4]

11 "At the time we were still in the classrooms, but 'good news spreads quickly'" [Bolibruch 1996 p. 4]

12 "Today our boarding school is housed" [Bolibruch 1996 p. 5]

13 "It was maybe the best place to study math" [Bolibruch private communication]

14 "Of course, misunderstandings are common" [Anosov and Bolibruch p. 8]

15 "Traditionally, it was viewed as a problem in 'function theory'" [Katz p. 541–542]

16 "another system with the same monodromy" [Bolibruch 1990 p. 6]

17 "After the publication of Plemelj's result, the subject matter" [Bolibruch 1990 p. 6]

18 "Hilbert's twenty-first problem on the existence" [Katz p. 527] [my italics]

19 "The final solution of this Hilbert Problem" [Holzapfel p. 45]

20 "I cannot estimate the number of wrong works" [Bolibruch private communication]

21 "are a good possibility to do mathematics" [Bolibruch private communication]

Work on It

1 "He proved and generalized the theorem" [Aleksandrov et al. p. 170]

2 "This result was a complete solution of Hilbert's nineteenth problem" [Aleksandrov, Oleinik p. 3]

3 "Unlike many of Luzin's students" [Landis p. 70]

4 "He knew a great deal about painting" [Landis p. 71]

5 "[When speaking] in modern terms" [Marcellini p. 330]

6 the first time anyone called a partial differential equation elliptic [Kline vol. 2 p. 700]

7 "S. Bernstein (1904) has furnished the first response" [Marcellini p. 329]

8 "With contributions from many authors" [Marcellini p. 330]

9 "Garabedian looked De Giorgi up on his way back through Rome" [Nasar p. 220]

10 "I see this as a journey throughout which, until the end" [De Giorgi in Interview with Emmer p. 1098]

11 "abstract powers from outer space . . . were communicating" [Nasar p. 241]

12 "A few days later, at teatime, Nash approached him again" [Nasar p. 349]

13 "Among the prophetic problems in Hilbert's famous list" [Serrin p. 507]

14 "brilliant underlying idea" [Serrin p. 509]

We Come to Our Census

See this book.

Disclaimer

At the present time the spelling of Russian names in English, is an undecidable question. Piatetski-Shapiro spells his own name at least four ways in his collected works. Bolibruch one-upped him, spelling his own name four ways in one short e-mail letter to me. Here is a quote from Smilka Zdravkovska in her preface to *The Golden Years of Moscow Mathematics*: "The revolutionary changes in the past few years have affected this volume in several ways. I will mention just a minor consequence: whereas originally we were going to use the Mathematical Reviews transliteration of Russian names, this became impossible, with many of the mathematicians traveling abroad and using Western versions of their names and some insisting on using specific spellings for various people mentioned in their articles. So we had to resign ourselves to inconsistent spellings of names." Zdravkovska, I think, was overly optimistic that *Mathematical Reviews* could supply consistent spellings. It spells both Bolibruch and Piatetski-Shapiro many different ways.

Selected Bibliography

This bibliography contains the main sources I used for the biographical passages of the book and only major sources for the mathematical passages. If I recommended a book in the text or used it for a note, it is here.

General Biographical Sources

Dictionary of Scientific Biography. Edited by Gillispie, Charles Coulston. New York: Scribner (1970–90). Referred to as *Dictionary*.

http://www-groups.dcs.st-and.ac.uk/~history/
Mactutor History of Mathematics page: Biographies.
This Web-site is a work in progress. I didn't feel I could trust it as much as the *Dictionary of Scientific Biography*. I found some material on it that was incorrect, and the articles change over time—usually with new material added. That said, I found the bibliographies particularly useful and went to it often. If there was something new to me I generally tried to track down the Mactutor sources.

http://www.agnesscott.edu/lriddle/women/women.htm
A site for "Biographies of Women Mathematicians." (Same comment as for Mactutor.)

More Mathematical People. Edited by Albers, Donald J.; Alexanderson, Gerald L; Reid, Constance. Boston, San Diego, New York: Harcourt Brace Jovanovich (1990). Referred to as MMP.

Sources for the Mathematics of Hilbert's Problems

Someone looking into the mathematics of Hilbert's problems should start with *Mathematical Developments Arising From Hilbert Problems*, edited by Felix E. Browder, and Kaplansky's book. These books generally contain good mathematical bibliographies.

Mathematical Developments Arising From Hilbert Problems. Edited by Browder, Felix E. Volume 28 of *Proceedings of Symposia in Pure Mathematics*. Providence, Rhode Island: American Mathematical Society (1976). Referred to as *Proceedings of AMS Conference* or simply *Proceedings*.

Individual problems were treated by:

1 Martin, Donald A.
2 Kreisel, G.
4 Busemann, Herbert
5 Yang, C. T.
6 Wightman, A. S.
7 Tijdeman, R.
8 Bombieri, E.; Katz, Nicholas M.; Montgomery, Hugh L.
9 Tate, J.
10 Davis, Martin; Robinson, Julia; Matiyasevich, Yuri
11 O'Meara, O. T.
12 Langlands, R. P.
13 Lorentz, G. G.
14 Mumford, David
15 Kleiman, Steven L.
17 Pfister, Albrecht
18 Milnor, J.
19 Bombieri, Enrico
20 Serrin, James

Problems 19 and 20 are not the most clearly worded of Hilbert's problems and Bombieri and Serrin both write about both.

21 Katz, Nicholas M.
22 Bers, Lipman
23 Stampacchia, Guido

Kaplansky, Irving. *Hilbert's Problems*. Lecture notes, Department of Mathematics, University of Chicago (1977).

There are also articles and books on individual problems—see specific sources below. These generally contain efficient presentations and accounts of the solutions or partial solutions to the problems. Look to Cohen on the first (with a treatment of second), Boltianskii on the third, Pogorelov on the fourth, Montgomery and Zippin on the fifth, Hille on the seventh, Edwards for background for the eighth. Junusc has a very concise treatment (though still exceeding 200 pages) of class field theory and Artin's reciprocity theorem that is also relevant to the eleventh and twelfth. Continuing with sources for individual problems, there is Matiyasevich on the tenth, Schappacher on the twelfth, Nagata for the fourteenth, Kleiman and Thorup for the fifteenth, Viro for the algebraic half of the sixteenth, Ilyashenko and Yakoshenko for the second half of the sixteenth and Auslander for the first part of the eighteenth, Hales for the third part of the eighteenth, Marcellini for the nineteenth and twentieth, Anosov and Bolibruch for the twenty-first, and Lehner for background to the twenty-second.

Specific Sources

Abraham, Ralph and Marsden, Jerrold E. *Foundations of Mechanics, A Mathematical Exposition of Classical Mechanics.* London: Benjamin Cummings Publishing Company (1978).

Aleksandrov, A. D.; Marchenko, V. A.; Novikov, S. P.; Reshetnyak, Yu.G. "Aleksei Vasil'evich Pogorelov (On his Seventieth Birthday)." *Russian Math. Surveys* 44:4 (1989), 217–223.

Aleksandrov (Alexandrov), P. S., editor. *Hilbertschen Probleme.* Translated from Russian into German by Bernhardt, Hannelore. Leipzig: Akademische Verlags (1971).

Aleksandrov, P. S. "Pages From an Autobiography, Part One." *Russian Math. Surveys* 34:6 (1979), 267–302.

Aleksandrov, P. S. "Pages From an Autobiography, Part Two." *Russian Math. Surveys* 35:3 (1980), 315–358.

Aleksandrov. P. S.; Akhiezer, N. I.; Gnedenko. B. V.; Kolmogorov, A. N. "Sergei Natanovich Bernstein." *Russian Math. Surveys* 24 (1969), 169–176.

Aleksandrov. P. S.; Oleinik, O. A. "On the Eighteenth Anniversary of the Birth of Ivan Geogrevich Petrovskii." *Russian Math. Surveys* 36:1 (1981), 1–8.

Alvarez, J. C.; Gelfand, I. M.; Smirnov, M. "Crofton Densities, Symplectic Geometry and Hilbert's Fourth Problem." *The Arnold–Gelfand Mathematical Seminars.* Edited by Arnold, V. I. et al. Boston: Birkhäuser (1994), 77–92.

Anosov, D. B. et al. (21). "Vladimir Igorevich Arnold (on his sixtieth birthday)." *Russian Math. Surveys* 52:3 (1996), 1117–1139.

Anosov, D. V. and Bolibruch, A. A. *The Riemann–Hilbert Problem.* Braunschweig-Wiesbaden: Vieweg (1994).

Appell, Paul. *Henri Poincaré.* Paris: Plon-Nourrit et cie. (1925).

Apostol, Tom. "A Centennial History of the Prime Number Theorem." *Engineering and Science* (Caltech) 59:4 (1996), 18–28.

Arnold, V. I. "Distribution of Ovals of the Real Plane of Algebraic Curves of Involutions of Four-dimensional Manifolds and the Arithmetic of Integer-valued Quadratic Forms." *Functional Analysis Appl.* 5 (1971), 169–176.

Arnold, V. I. *Mathematical Methods of Classical Mechanics.* Translated by Vogtmann, K. and Weinstein, A. New York: Springer-Verlag (1978).

Arnold, V. I. "A Few Words on Andrei Nikolaevich Kolmogorov." *Russian Math. Surveys* 43:6 (1988), 43–44.

Arnold, V. I. "A. N. Kolmogorov." *Physics Today* (October 1989), 148–150.

Arnold, V. I. "On A. N. Kolmogorov." In *Golden Years of Moscow Mathematics.* Edited by Zdravkovska, Smilka and Duren, Peter. Providence: AMS (1993), 129–153.

Arnold, V. I. "Will Mathematics Survive? Report on the Zurich Conference." *Mathematical Intelligencer* 17:3 (1995), 6–10.

Arnold, V. I. "Topological Problems of the Theory of Wave Propagation." *Russian Math. Surveys* 51:1 (1996), 1– 47.

Arnold, V. I. "From Hilbert's Superposition Problem to Dynamical Systems." In *The Arnoldfest* (1999), 1–18.

Arnold, V. I. "On A.N. Kolmogorov." In *Kolmogorov in Perspective.* (2000), 89–108.

Arnold, V. I. et al. "Vladimir Abramovich Rokhlin." *Russian Math. Surveys* 41:3 (1986), 189–195.

Arnold, V.; Atiyah, M.; Lax, P.; Mazur, B., eds. *Mathematics: Frontiers and Perspectives.* International Mathematical Union and AMS (2000).

The Arnoldfest: Proceedings of a Conference in Honour of V. I. Arnold for his Sixtieth Birthday. Edited by Bierstone, Edward; Khesin, Boris; Khovanskii, Askold; Marsden, Jerrold E. Providence, Rhode Island: American Mathematical Society (1999).

Artin, Emil. *The Collected Papers of Emil Artin.* Edited by Lang, Serge and Tate, John T. Reading, Mass.: Addison-Wesley (1965).

Artin, Emil. "Review of Bourbaki's Algebra." *Bulletin AMS* 59 (1953), 474–479.

Auslander, Louis. "An Account of the Theory of Crystallographic Groups." *Proc. AMS* 16 (1965), 1230–36.

Baker, H. F. "Jules Henri Poincaré. 1854–1912." *Phil. Trans. Royal Society London A* (1914), vi-xvi.

Barrow-Green, June. *Poincaré and the Three Body Problem. History of Mathematics Volume 11.* AMS and London Mathematical Society (1997).

Bashmakova, Isabella Grigoryevna. *Diophantus and Diophantine Equations.* Updated by Silverman, Joseph. Translated by Schenitzer, Abe. *Dolciani Mathematica Expositions: Number 20.* Mathematical Association of America (1997).

Bashmakova, Isabella and Smirnova, Galina. *The Beginnings and Evolution of Algebra.* Mathematical Association of America (2000).

Baumgartner, James E. A review of *Set Theory. An introduction to Independence Proofs* by Kenneth Kunen in *Journal of Symbolic Logic* 51:2 (1986), 462–464.

Bell, E. T. *Men of Mathematics.* New York: Simon and Schuster (1986) (originally published 1937).

Beaulieu, Liliane. "A Parisian Café and Ten Proto-Bourbaki Meetings (1934–1935)." *Mathematical Intelligencer* 15:1 (1993), 27–35.

Bers, Lipman. "Finite Dimensional Teichmüller Spaces and Generalizations." In Browder, Felix, ed. *The Mathematical Heritage of Henri Poincaré.* (1983).

Bliss, Gilbert Ames. *Calculus of Variations*. Chicago: Published for the Mathematical Association of America by the Open Court Pub. Co. (1925).

Blum, Lenore and Smale, Steve. "The Gödel Incompleteness Theorem and Decidability over a Ring." In *From Topology to Computation: Proceedings of the Smalefest*. Edited by Hirsch, M. W.; Marsden, J. E.; Shub, M. New York etc.: Springer-Verlag (1993), 321–339.

Boas, R. P. "Bourbaki and Me." *Mathematical Intelligencer* 8:4 (1986), 84.

Bolibruch, A. A. "The Riemann–Hilbert Problem." *Russian Math. Surveys* 45:2 (1990), 1–47.

Bolibruch, A. A. "Some memories of Boarding School #45." *Translations AMS* 174 (1996), 1–5.

Boltianskii, Vladimir G. *Hilbert's Third Problem*. Translated by Silverman, Richard A. Washington D.C.: Winston and Sons (1978).

Bombieri, Enrico. "Problems of the Millennium: The Riemann Hypothesis." *http://www.claymath.org/prizeproblems/riemann.htm*

Boone, W. W.; Cannonito F. B.; Lyndon, R. C. *Word Problems: Decision Problems and the Burnside Problem in Group Theory*. Amsterdam: North-Holland Publishing Company (1973).

Borel, Armand. "Deane Montgomery 1909–1992." *Notices AMS* 39:7 (1992), 684–686.

Brauer, R. "Emil Artin." *Bulletin AMS* 73 (1967), 27–43.

Braun, Hel. *Eine Frau und die Mathematik 1933–1949—Der Beginn einer wissenschaftlichen Laufbahn.* Berlin: Springer-Verlag (1980).

Brezis, Haïm and Browder, Felix. "Partial Differential Equations in the 20th Century." *Advances in Mathematics* 135 (1998), 76–144.

Brouwer, L. E. J. *Collected Works: Philosophy and Foundations of Mathematics*. Vol. 1. Edited by Heyting, A. Amsterdam: North-Holland Publishing Co. and New York: American Elsevier Publishing Co. Inc. (1975).

Browder, Felix E., ed. *The Mathematical Heritage of Henri Poincaré*. 2 vols. Volume 39 of *Proceedings of Symposia in Pure Mathematics*. Providence, RI: AMS (1983).

Bundschuh, Peter and Zassenhaus, Hans. "Nachruf: Theodor Schneider (1911–1988)." *Journal of Number Theory* 39 (1991), 129–143.

Cajori, Florian. *A History of Mathematical Notations*. 2 Vols. Chicago: Open Court Publishing Company (1928–29).

Cartan, H. "Emil Artin." *Abhandlungen aus dem Mathematischen Seminar der Universität Hamburg* 28 (1965), 1–5.

Chern, Shiing-shen. "Remarks on Hilbert's 23rd Problem." *Mathematical Intelligencer* 18:4 (1996), 7–8.

Chevalley, C. "Emil Artin, [1898–1962]." *Bulletin de la Société mathématique de France* 92 (1964), 1–10.

Cipra, Barry. "A Prime Case of Chaos." *http://www.ams.org/new-in-math/cover/prime-chaos.pdf*

Cohen, Paul J. "A Minimal Model for Set Theory." *Bulletin AMS* 69 (1963), 537–540.

Cohen, Paul J. "The Independence of the Continuum Hypothesis, Parts I and II." *Proc. Nat. Acad. Sci. U.S.A.* 50 (1963), 1143–1148 and 51 (1964), 105–110.

Cohen, Paul J. *Set Theory and the Continuum Hypothesis*. Reading, Mass: W. A. Benjamin (1966).

Conway, J. H.; Hales, T. C.; Muder, D. J.; Sloane, N. J. A. "On the Kepler Conjecture." *Mathematical Intelligencer* 16:2 (1994), 5.

Conway, J. H., and Sloane, N. J. A. *Sphere Packings, Lattices and Groups*. New York: Springer-Verlag (1988).

Coolidge, Julian Lowell. *A History of Geometrical Methods*. Oxford: Clarendon Press (1940).

Dantzig, Tobias. *Henri Poincaré. Critic of Crisis: Reflections on his Universe of Discourse*. New York: Scribner (1954).

Darboux, Gaston. "Éloge historique d'Henri Poincaré." *Mémoires de l'Académie des Sciences* 52 Paris: Gauthier-Villars (1914), lxxxi–cxlviii.

Dauben, Joseph Warren. *Georg Cantor: His Mathematics and Philosophy of the Infinite*. Princeton: Princeton University Press (1979).

Dauben, Joseph Warren. "Peirce's Place in Mathematics." *Historia Mathematica* 9 (1982), 311–325.

Davenport, Harold. "Reminiscences of Conversations with Carl Ludwig Siegel." Edited by Davenport, Mrs. Harold. *Mathematical Intelligencer* 7:2 (1985), 76–79.

Davis, Martin. "Arithmetical Problems and Recursively Enumerable Predicates." *Journal of Symbolic Logic* 18:1 (1953), 33–41.

Davis, Martin. *Computability and Unsolvability*. New York: McGraw Hill (1958).

Davis, Martin, editor. *The Undecidable: Basic Papers on Undecidable Propositions, Unsolvable Problems and Computable Functions*. Hewlitt, New York: Raven Press (1965).

Davis, Martin. "Why Gödel Didn't Have Church's Thesis." *Information and Control* 54 (1982), 3–24.

Davis, Martin. "Foreword." In Matiyasevich, Yuri. *Hilbert's Tenth Problem*. Cambridge, Mass: MIT Press (1993), xiii–xvii.

Davis, Martin. "Emil L. Post: His Life and Work." Introduction in *Solvability, Provability, Definability: The Collected Works of Emil L. Post.* (1994), xi–xxviii.

Davis, Martin. "American Logic in the 1920s." *Bulletin of Symbolic Logic* 1:3 (1995), 273–278.

Davis, Martin; Matiyasevich, Yuri; Robinson, Julia. "Hilbert's Tenth Problem. Diophantine Equations: Positive Aspects of a Negative Solution." *Proceedings of AMS Conference.*

Davis, Martin; Putnam, Hilary; Robinson, Julia. "The Decision Problem for Exponential Diophantine Equations." *Annals of Mathematics* 74:3 (1961), 425–436.

Dawson, John W. Jr. "Kurt Gödel in Sharper Focus." *Mathematical Intelligencer* 6:4 (1984), 9–17.

Dawson, John W. Jr. *Logical Dilemmas: The Life and Work of Kurt Gödel.* Natick, Mass.: A K Peters (1997).

Deane Montgomery 1909–1992. Princeton: Institute for Advanced Study (1992).

Degtyarev, A. and Kharlamov, V. "Topological Properties of Real Algebraic Varieties: Du Côte de Chez Rokhlin." Preprint (2000).

Dehn, Max. "The Mentality of the Mathematician. A Characterization." *Mathematical Intelligencer* 5:2 (1983), 18–26.

Dehn, Max. *Papers on Group Theory and Topology.* Translated and Introduction by Stillwell, John. New York: Springer-Verlag (1987).

Dembart, Lee. "An Unsung Geometer Keeps His Own Place." *Los Angeles Times* July 14 (1985), Opinion 3.

Demidov S. S. "The Moscow School of the Theory of Functions in the 1930s." In *Golden Years of Moscow Mathematics.* Edited by Zdravkovska, Smilka and Duren, Peter. Providence: AMS (1993), 35–54.

Deuring, Max. "Carl Ludwig Siegel, 31.12.1896–4.4.1981." *Acta Arithmetica* 45 (1985), 92–107.

Deutscher, Isaac. "Introduction." In Lunacharsky, Anatoly Vasilievich. *Revolutionary Silhouettes.* Translated and edited by Michael Glenny. New York: Hill and Wang (1968).

Diacu, Florin and Holmes, Philip. *Celestial Encounters: the Origins of Chaos and Stability.* Princeton, New Jersey: Princeton University Press (1996).

Dieudonné, Jean. "Poincaré, Jules Henri." In *Dictionary.*

Dieudonné, Jean. "Siegel, Carl Ludwig." In *Dictionary.*

Di Francesco, P. and Itzakson, C. "Quantum Intersection Rings." In *The Moduli Space of Curves.* Edited by Dijkgraaf, Robert et al. Boston: Birkhäuser (1995).

Dobrovol'skii, V. A. "Josip Plemelj (on the centenary of his birth)." *Russian Math. Surveys* 28:6 (1973), 223–226.

Duberman, Martin. *Black Mountain: An Exploration in Community.* New York: E. P. Dutton (1972).

Dunham, William. *Journey Through Genius: The Great Theorems of Mathematics*. New York: John Wiley and Sons (1990).

Dyson, Freeman. "Missed Opportunity." *Bulletin AMS* 78:5 (1972), 635–652.

Edwards, H. M. *Fermat's Last Theorem: A Genetic Introduction to Algebraic Number Theory*. New York: Springer-Verlag (1977).

Edwards, H. M. *Riemann's Zeta Function*. New York: Academic Press (1974).

Edwards, H. M. "Dedekind's Invention of Ideals." *Bulletin of the London Mathematical Society* 15 (1983), 8–17.

Emmer, Michele. "Interview with Ennio De Giorgi." *Notices of the AMS* 44:9 (1997), 1097–1101.

Encylopædia Britannica, 11th edition. Edited by Chisholm, Hugh. New York: Encyclopædia Britannica Company (1910–1911).

Enderton , H. B. "In Memoriam: Alonzo Church: 1903–1995." *Bulletin of Symbolic Logic* 1:4 (1995), 486–488.

Evgrafov, M. A.; Korobov, N. M.; Linnik, Yu. V.; Pyatetskii-Shapiro, I. I.; Fel'dman, N. I. "Aleksandr Osipovich Gel'fond." *Russian Math. Surveys* 24:3 (1969), 177–178.

Feferman, Anita Burdman. *Politics, Logic, and Love*. Wellesley, Mass: A K Peters (1993).

Feferman, Solomon. "Gödel's Life and Works." In *Gödel's Collected Works*. Vol. 1.

Feferman, Solomon. *Julia Bowman Robinson: 1919–-1985*. Washington, D.C.: National Academy Press (1994).

Figuier, Louis. *La Terre avant le Déluge*, fifth edition. Paris: Librarie de L. Hachette et cie. (1866).

Fisher, C. S. "Max Dehn" In *Dictionary.*

Folland, Gerald B. *Introduction to Partial Differential Equations,* second edition. Princeton: Princeton University Press (1995).

Frei, Günther. "Helmut Hasse (1898–1979)." *Expositiones Math.* 3 (1985), 55–69.

Frei, Günther. "Heinrich Weber and the Emergence of Class Field Theory." In Rowe, David E. and McCleary, John. *The History of Modern Mathematics, Volume I: Ideas and Their Reception*. Boston: Academic Press: Harcourt Brace and Jovanovich (1989).

Freudenthal, Hans. "A Bit of Gossip: Koebe." *Mathematical Intelligencer* 6:2 (1984), 77.

Freudenthal, Hans. "Commentary." In Mehrtens (1989), 167–70.

Freudenthal, Hans. Notes on Koebe incident. In Bouwer's *Collected Works*. (1975), 572–587.

Freudenthal, Hans. "David Hilbert." In *Dictionary*.

Fuchs, D. B. "On Soviet Mathematics of the 1950s and 1960s." *Golden Years of Moscow Mathematics.* Edited by Zdravkovska, Smilka and Duren, Peter. Providence: AMS (1993), 213–222.

Galgani, Luigi. "Ordered and Chaotic Motions in Hamiltonian Systems and the Problem of Energy Partition." In *Chaos in Astrophysics*. Edited by Buchler, J. R. et al. Dordrecht: D. Reidel Publishing Company (1985), 245–257.

Gardner, Richard J. *Geometric Tomography*. In the series *Encyclopedia of Mathematics and its Applications*. Cambridge, England: Cambridge University Press (1995).

Gautschi, Walter. "Ostrowski and the Ostrowski Prize." *Mathematical Intelligencer* 20:3 (1998), 32–34.

Gelbart, Stephen. "An Elementary Introduction to the Langlands Program." *Bulletin AMS* 10:2 (1984), 177–219.

Gelfond, A. O. *Transcendental and Algebraic Numbers*. Translated by Boron, Leo I. New York: Dover Publications Inc. (1960).

Gelfond, A. O. "Some Impressions of a Scientific Visit to Germany in 1930." *Istor. Mat. Issled. 22* (1977), 246–251.

Gleason, Andrew. "Groups Without Small Denominators." *Annals of Mathematics* 56:2 (1952), 193–212.

Gilbarg, D. and Trudinger, N. S. *Elliptic Partial Differential Equations of the Second Order*. *Grundl. der Math. Wiss*. 224. Berlin: Springer-Verlag (1977) c. 1957.

Gleick, James. *Chaos: Making a New Science*. New York: Viking Penguin (1987).

Gnedenko, B. V. and Kolmogorov, A. N. "Aleksandr Yakovlevich Khinchin (1894–1959) Obituary." *Russian Math. Surveys* 15:4 (1960), 93–106.

Gödel, Kurt. *Collected Works*. 3 Vols. Edited by Feferman, Solomon et. al. New York, Oxford: Oxford University Press (1986–1995).

Gödel, Rudolf. "History of the Gödel Family." In *Gödel Remembered*. Edited by Weingartner, P. and Schmetterer, I. Napoli: Bibliopolis (1987).

Golden Years of Moscow Mathematics. Edited by Zdravkovska, Smilka and Duren, Peter. Providence: AMS (1993).

Goldstine, Herman H. *The Computer: From Pascal to von Neumann*. Princeton: Princeton University Press (1972).

Golomb, Solomon W. "Tiling Rectangles With Polynminoes." *Mathematical Intelligencer* 18:2 (1996), 38–47.

Grattan-Guinness, I. "Towards a Biography of Georg Cantor." *Annals of Science* 27 (1971), 345–391.

Grauert, Hans. "Gauss und die Göttinger Mathematik." *Naturwissenschaftliche Rundschau* 47:6 (1994), 211–219.

Gray, J. J. "Algebraic Geometry in the Late Nineteenth Century." In *The History of Modern Mathematics, Volume I: Ideas and Their Reception*. Edited by Rowe, David E. and McCleary, John. Boston: Academic Press: Harcourt Brace and Jovanovich (1989).

Grunsky, H. "Ludwig Bieberbach zum Gedächtnis." *Jahresbericht der Deutschen Mathematiker Vereinigung* 88 (1986), 190–205.

Gudkov, D. A. "The Topology of Real Projective Algebraic Varieties." *Russian Math. Surveys* 29:4 (1974), 3–79.

Guedj, Denis. "Nicholas Bourbaki, Collective Mathematician: An Interview with Claude Chevally." Translated by Grey, Jeremy. *Mathematical Intelligencer* 7:2 (1985), 18–22.

Hales, Thomas C. "The Status of the Kepler Conjecture." *Mathematical Intelligencer* 16:3 (1994), 47–48.

Hales, Thomas C. "Cannonballs and Honeycombs." *Notices AMS* 47:4 (2000), 440–449.

Harding, Mildred. "My Black Mountain." *Yale Literary Magazine* 151:1 (1982).

Hardy, G. H. with a foreword by Snow, C. P. *A Mathematicians Apology*. Cambridge: Cambridge University Press (1969).

Hardy, G. H. *Ramanujan: Twelve Lectures on Subjects Suggested by his Life and Work.* New York: Chelsea (1959).

Hardy, G. H. "The J-type and the S-type Among Mathematicians." *Nature* 134 (1934), 250.

Hardy, G. H. "Srinivasa Ramanujan, 1887–1920." *Proc. London Math. Society*: *Series A* 19 (1921), xiii–xxv.

Harvard Gazette. "A Theory for Everything: Andrew Gleason says mathematics embodies a unifying structure." May 8 (1962), 3–4.

Hasse, Helmut. "History of Class field Theory." In Cassels J. W. S. and Frölich, A., editors. *Algebraic Number Theory.* London: Academic Press (1967), 266–279.

Hasse, Helmut. "The Modern Algebraic Method." Translated by Schenitzer, Abe. *Mathematical Intelligencer* 8:2 (1986), 18–25.

Henkin, Leon. "In Memoriam Raphael Mitchel Robinson." *Bulletin of Symbolic Logic* 1:3 (1995), 340–343.

Hilbert, David. "Mathematical Problems." *Bulletin AMS* 8 (1902), 437–479.

Hilbert, David. *The Foundations of Geometry*, second edition. Translated by Townsend, E. J. Chicago: Open Court (1910).

Hilbert, David. "Adolf Hurwitz." *Math. Annalen* 83 (1921), 161–168.

Hilbert, David and Courant, Richard. *Methoden der mathematischen physik.* Berlin: Springer-Verlag, (1924).

Hildebrandt, Stefan and Tromba, Anthony. *The Parsimonious Universe: Shape and Form in the Natural World.* New York: Copernicus Springer-Verlag (1996).

Hille, Einar. "Gelfond's Solution of Hilbert's Seventh Problem." *American Mathematical Monthly* 49 (1942), 654–661.

Hlawka, Edmund. "Carl Ludwig Siegel." *Jour. Number Theory* 20 (1985), 373–404.

Holzapfel, Rolf-Peter. *The Ball and Some Hilbert Problems.* Basel: Birkhäuser (1995).

Honda, Kin-ya. "Teiji Takagi: A Biography." *Commentarii Mathematici Universitatis Sanct. Pauli* 24:2 (1975), 141–167.

Hsiang, Wu-Yi. "A Rejoinder to Hale's Article." *Mathematical Intelligencer* 17:1 (1995), 35–42.

Igoshin, V. I. "A Short Biography of Mikhail Yakovlevich Suslin." *Russian Math. Surveys* 51:3 (1996), 371–383.

Illman, Soren. "Every Proper Smooth Action of a Lie Group is Equivalent to a Real Analytic Action: a Contribution to Hilbert's Fifth Problem." In *Prospects in Topology : Proceedings of a Conference in Honor of William Browder.* Edited by Quinn, Frank. Princeton: Princeton University Press (1995), 189–220.

Ilyashenko, Y. and Yakovenko, S., editors. *Concerning the Hilbert 16th Problem. Translations AMS* 2:165 (1995).

Iyanaga, S. *The Theory of Numbers*. Translated by Iyanaga, K. Amsterdam: North-Holland (1975), 479–518.

Janusz, Gerald J. *Algebraic Number Fields: Second Edition*. Providence: AMS (1996).

Johnson, Dale M. "L. E. J. Brouwer's Coming of Age as a Topologist." In *Studies in the History of Mathematics*. Edited by Phillips, E. Mathematical Association of America *Studies in Mathematics* 26 (1987), 61–97.

Jones, Landon Y. Jr. "Bad Days on Mount Olympus." *Atlantic Monthly* February (1974), 37–46.

Kac, Mark. *Enigmas of Chance: An Autobiography*. New York: Harper and Row (1985).

Kanigel, Robert. *The Man Who Knew Infinity: A Life of the Genius Ramanujan*. New York: Charles Scribner's Sons (1991).

Kantor, Jean-Michel. "Hilbert's Problems and Their Sequels." *Mathematical Intelligencer* 18:1 (1996), 21–30.

Kaplansky, Irving. *Hilbert's Problems*. Lecture notes, Department of Mathematics, University of Chicago (1977).

Kappe L.-Ch.; Schlickewerei H. P.; Schwarz W. "Theodor Schneider zum Gedächtnis." *Jahresberichte der Mathematiker Vereinigung* 92 (1990), 111–129.

Kanimori, Akihiro. "The Mathematical Development of Set Theory From Cantor to Cohen." *Bulletin of Symbolic Logic* 2:1 (1996), 1–71.

Kendall, D. G. "Andrei Nikolaevich Kolmogorov." *Biographical Memoirs of Fellows of the Royal Society of London* 37 (1991), 299–320.

Kennedy, Hubert C. *Peano: Life and Works of Giuseppe Peano*. Dordrecht: D. Reidel (1980).

Kleiman, Steven, with Thorup, Anders, editors. *Enumerative Algebraic Geometry—Proceedings of the 1989 Zeuthen Symposium*. Providence, Rhode Island: American Mathematical Society (1991).

Kleiman, Steven, with Thorup, Anders. "Intersection Theory and Enumerative Geometry: A Decade in Review." *Proceedings of Symposia in Pure Mathematics.* AMS 48 (1987), 321–370.

Klein, Felix, *The Evanston Colloquium: Lectures on Mathematics*. New York: Macmillan and Co. (1894).

Kline, Morris. *Mathematical Thought from Ancient to Modern Times*. New York: Oxford University Press (1972).

Kolmogorov, A. N. "Memories of P.S. Akeksandrov." *Russian Math. Surveys* 41:6 (1986), 225–246.

Kolmogorov, A. N. *Selected Works of A. N. Kolmogorov* three volumes. Edited by Tikhomirov, V. M. et al. Translated by Volosov, V. M. et al. Dordrecht: Kluwer (1991–1993).

Kolmogorov in Perspective. History of Mathematics 20. AMS and London Mathematical Society (2000).

Kovalevskaya, Sofya. *A Russian Childhood.* Translated, edited and introduced by Stillman, Beatrice. New York: Springer-Verlag (1978).

Kreisel, George. "Sums of Squares." *Summer Institute for Symbolic Logic, 1957* 2nd ed. (1960), 313–330.

Kreisel, George. "Mathematical Significance of Consistency Proofs." *Journal of Symbolic Logic* 23:2 (1958), 155–182.

Kreisel, George. "Kurt Gödel." *Biographical Memoirs of Fellows of the Royal Society* 26 (1980), 148–224; corrigenda, 27 (1981), 697.

Kreisel, George. "A3061: Davis, Martin; Putnam, Hilary; Robinson, Julia. The Decision Problem for Exponential Diophantine Equations." *Mathematical Reviews* 24A(6A):573 (1962).

Kreisel, George and Newman, M. H. A. "Luitzen Egbertus Jan Brouwer." In *Biographical Memoirs of Fellows of the Royal Society of London* 15 (1969), 39–68.

Kullman, David E. "Penrose tiling at Miami University." *Mathematical Intelligencer* 18:4 (1996), 66.

Kuznetsov, P. S. "From Autobiographical notes." *Russian Math. Surveys* 43:6 (1988), 193–209.

Landis, E. M. "About Mathematics at Moscow State University in the late 1940s and early 1950s." In *Golden Years of Moscow Mathematics.* Edited by Zdravkovska, Smilka and Duren, Peter. Providence: AMS (1993), 55–74.

Lang, Serge. "Mordell's Review, Siegel's Letter to Mordell, Diophantine Geometry, and 20th Century Mathematics." *Notices AMS* 42:3 (1995), 339–350.

Lang, Serge. *Algebraic Number Theory.* Reading, Mass.: Addison Wesley (1970).

Los Angeles Times. "Mathematician Sues Toilet Paper Maker over Use of a Patented Design." (April 17, 1997), p. B2.

Lehner, Joseph. *A Short Course in Automorphic Functions*. New York: Holt, Rinehart and Winston (1966).

Le Livre du Centenaire de la Naissance de Henri Poincaré 1854–1954. Paris: Gauthier-Villars, (1955).

Leopoldt, H. W. "Obituary: Helmut Hasse." *Journal of Number Theory* 14 (1982), 118–120.

Lewis, C. I. *A Survey of Symbolic Logic*. Berkeley: University of California Press (1918).

Levin, B. V.; Feldman N. I.; Shidlovski, A. B. "Alexander O. Gelfond." *Acta Arithmetica* 17 (1971), 314–336.

Lozinskii, S. M. "On the Hundredth Anniversary of the Birth of S.N. Bernstein." *Russian Math. Surveys* 38:3 (1983), 163–178.

Lui, S. H. "An Interview with Vladimir Arnol'd." *Notices of the AMS* 44:4 (1997), 432–438.

Mac Lane, Saunders. "Mathematics at Göttingen Under the Nazis." *Notices of the AMS* 42:10 (1995), 1134–1138.

Macrae, Norman. *John von Neumann*. New York: Pantheon Books (1992).

Magnus, Wilhelm. "Max Dehn." *Mathematical Intelligencer* 1 (1978–79), 132–43.

Magnus, Wilhelm and Moufang, Ruth. "Max Dehn zum Gedächtnis." *Math. Annalen* 127 (1954), 215–227.

Maistrov, L. E. *Teoriia veroiatnostei*. English: *Probability theory; a historical sketch.* Translated and edited by Samuel Kotz. New York: Academic Press (1974).

Mandelbrot, Benoit B. "Self-Inverse Fractals Osculated by Sigma Disks and the Limit Sets of Inversion Groups." *Mathematical Intelligencer* 5:2 (1983), 9–17.

Marcellini, Paolo. "Alcuni recenti sviluppi nei problemi 19-simo e 20-simo di Hilbert." *Bollettino UMI* 7:11-A (1997), 323–352.

Matiyasevich, Yuri. "Enumerable Sets are Diophantine." *Sov. Math. Dokl.* 11:2 (1970), 354–358.

Matiyasevich, Yuri. "My Collaboration with Julia Robinson." *Mathematical Intelligencer* 14:4 (1992), 38–45.

Matiyasevich, Yuri. *Hilbert's Tenth Problem*. Cambridge, Mass.: MIT Press (1993).

Mehrtens, Herbert. "Mathematics in the Third Reich: Resistance, Adaptation and Collaboration of a Scientific Discipline." In *New Trends in the History of Science*. Edited by Visser, R. P. W. et al. Amsterdam: Rodopi (1989), 151–166.

Mehrtens, Herbert. "Ludwig Bieberbach and 'Deutsche Mathematik.'" In *Studies in the History of Mathematics*. Edited by Phillips, E. Mathematical Association of America *Studies in Mathematics* 26 (1987), 195–241.

Monastyrsky, Michael. *Riemann, Topology, and Physics*. With a forward by Dyson, Freeman J. Translated by King, James and King, Victoria. Edited by Wells, R. O. Jr. Boston: Birkhäuser (1987).

Monastyrsky, Michael. *Modern Mathematics in the Light of the Fields Medals*. Natick, Mass.: A K Peters (1997).

Montgomery, Deane and Zippin, Leo. "Topological Group Foundations of Rigid Space Geometry." *Transactions AMS* 48 (1940), 21–49.

Montgomery, Deane and Zippin, Leo. "Small Subgroups of Finite Dimensional Groups." *Annals of Mathematics* 56:2 (1952), 213–241.

Montgomery, Deane and Zippin, Leo. *Topological Transformation Groups*. New York: Interscience (1955).

Moore, Gregory H. "Kurt Gödel." In *Dictionary*.

Moore, Gregory H. *Zermelo's Axiom of Choice: Its Origins, Developments, and Influence*. New York: Springer-Verlag (1982).

Moore, Gregory H. "The Origins of Forcing*." Logic Colloquium '86*. Edited by Drake, F. R. and Truss, J. K. North-Holland: Elsevier Science Publishers (1988).

Moser, Helmut A. "Das Beispiel de Mathematikers Carl Siegel." *Frankfurter Allgemeine Zitung* (May 8, 1981).

Myshkis, A. D. and Oleinik, O. A. "Vyacheslaus Vasil'vich Stepanov (On the Centenary of His Birth)." *Russian Math. Surveys* 45:6 (1990), 179–182.

Nagata, Masayoshi. *Lectures on the Fourteenth Problem of Hilbert*. Notes by Murthy, M. Pavaman. Bombay: Tata Institute of Fundamental Research (1965).

Nasar, Sylvia. *A Beautiful Mind*. New York: Simon and Schuster (1998).

Nirinberg, Louis. In Mather, John N.; McKean, Henry; Nirenberg, Louis; Rabinowitz, Paul H. "Jürgen K. Moser (1928–1999)." *Notices of the AMS* 47:11 (2000), 1392–1405.

Ono, Takashi. *An Introduction to Algebraic Number Theory*, second edition. New York: Plenum Press (1990).

Pach, Jáno. "Two Places at Once: A Remembrance of Paul Erdös." *Mathematical Intelligencer* 19:2 (1997), 38–48.

Pais, Abraham. *'Subtle is the Lord...': The Science and the Life of Albert Einstein*. Oxford: Oxford University Press (1982).

Parshall, Karen Hunger. "Toward a History of Nineteenth-Century Invariant Theory." In Rowe, David E. and McCleary, John. *The History of Modern Mathematics, Volume I: Ideas and Their Reception*. Boston: Academic Press (1989).

Piatetski-Shapiro. "Étude on Life and Automorphic Forms in the Soviet Union." In *Golden Years of Moscow Mathematics*. Edited by Zdravkovska, Smilka and Duren, Peter. Providence: AMS (1993), 199–212.

[Pitetski-Shapiro] Pyatetskii-Shapiro, I. I. and Shidlovskii, A. B. "Aleksandr Osipovich Gel'fond (On his Sixtieth Birthday)." *Russian Math. Surveys* 22:3 (1967), 234–242.

Plemelj, Josip. *Problems in the Sense of Riemann and Klein*. Edited and Translated by Radok, J. R. M. New York: Interscience Publishers (1964).

Pogorelov, A. V. *Hilbert's Fourth Problem*. Translated by Silverman, Richard A. Washington D.C.: V. H. Winston and Sons, Scripta Mathematica Series; New York: John Wiley and Sons (1979).

Poincaré, Henri. "L'avenir des mathématiques." *Atti de IV Congresso Internazionale dei Matematici*. Vol. 1 (1908), 167–182.

Poincaré, Henri. *The Foundations of Science: Science and Hypothesis, The Value of Science, Science and Method*. Translated by Halsted, George Bruce; with a special preface by Poincaré, and an introduction by Josiah Royce. New York: Science Press (1913).

Poincaré, Henri. *Ouevres de Henri Poincaré*. 11 Vols. Paris: Gauthier Villars et cie. (1916–1954).

Poincaré, Henri. *Mathematics and Science: Last Essays (Dernières Pensées)*. Translated by Bolduc, John W. New York: Dover (1963).

Poincaré, Henri. *Papers on Fuchsian Functions*. Translated by John Stillwell. New York.: Springer-Verlag (1985).

Post Emil L. *Solvability, Provability, Definability: The Collected Works of Emil L. Post*. Edited by Davis, Martin. Boston: Birkhäuser (1994).

Prasad, Ganesh. *Some Great Mathematicians of the Nineteenth Century: Their Lives and Their Works*. Benares: Benares Mathematical Society (1933).

Preston, Richard. "The Mountains of Pi." *New Yorker* 68:2 (March 2, 1992).

Purkert, Walter and Ilgauds, Hans Joachim. *Georg Cantor: 1845–1918*. Basel: Birkhäuser (1987).

Putnam, Hilary. "Peirce the Logician." *Historia Mathematica* 9 (1982), 290–301.

Putnam, Hillary. *Philosophical Papers*. 3 Vols. Cambridge: Cambridge University Press (1975–1983).

Quine, W. V. "Autobiography of W. V. Quine" In *The Philosophy of W. V. Quine*. Edited by Hahn, Lewis and Schilpp, Paul. Series title: *The Library of Living Philosophers*. Vol. 18. La Salle, Ill.: Open Court (1986), 1–46.

Quine, W. V. *The Time of My Life: An Autobiography*. Cambridge, Mass.: MIT Press (1985).

Reid, Constance. *Hilbert—Courant*. New York: Springer-Verlag (1986).

Reid, Constance. *Julia: A Life in Mathematics*. Washington D.C.: Mathematical Association of America (1996).

Rice, Adrian. "Augustus De Morgan (1806–1871)." *Mathematical Intelligencer* 18:3 (1996), 40–43.

Robinson, Abraham. "On Ordered Fields and Definite Functions." *Math. Annalen* 130 (1955), 257–271.

Robinson, Julia. "Existential Definability in Arithmetic." *Transactions of AMS* 72 (1969), 437–449.

Röhrl, Helmut. "Das Riemann–Hilbertsche Problem der Theorie der Linearen Differentialgleichungen." *Math. Annalen* 133 (1957), 1–25.

Rosenfeld, B. A. "Reminiscences of Soviet Mathematicians." In *Golden Years of Moscow Mathematics*. Edited by Zdravkovska, Smilka and Duren, Peter. Providence: AMS (1993), 75–100.

Rota, Gian-Carlo. *Indiscrete Thoughts*. Edited by Palombi, Fabrizio. Boston: Birkhäuser (1997).

Rovnyak, James. "Ernst David Hellinger 1883–1950: Göttingen, Frankfurt Idyll, and the New World." In *Topics in Operator Theory: Ernst D. Hellinger Memorial Volume*. Edited by de Branges, L.; Gohberg, I; Rovnyak, J. Basel: Birkhäuser (1990).

Rowe, David E. "'Jewish Mathematics' at Göttingen in the Era of Felix Klein." *Isis* 77 (1986), 422–49.

Rowe, David E. "Gauss, Dirichlet and the Law of Biquadratic Reciprocity." *Mathematical Intelligencer* 10:2 (1988), 13–25.

Rowe, David E. and McCleary, John. *The History of Modern Mathematics, Volume I: Ideas and Their Reception*. Boston: Academic Press (1989).

Royden, H. L. *Real Analysis*: *Second Edition*. London: Macmillan Company, Collier-Macmillan Limited (1968).

Ruelle, David. *Chance and Chaos*. Princeton: Princeton University Press (1991).

Russell, Bertrand. *The Autobiography of Bertrand Russell: 1972–1914*. Boston: Little, Brown and Company (1967).

Sah, C. H. *Hilbert's Third Problem: Scissors Congruence*. San Francisco: Pitman (1979).

"Samuel Putnam Papers: Collection 59—Biographical Note." *http://www.lib.siu.edu/spcol/SC059.html*

Sanoff, Alvin P. "Bringing Philosophy Back to Life." *U.S. News and World Report* (April 25, 1988) 12.

Schappacher, Norbert. "On the History of Hilbert's Twelfth Problem, A Comedy of Errors." In *Matériaux pour l'histoire des mathématiques au XXe siècle, Actes du colloque à la mémoire de Jean Dieudonné (Nice, 1996), Séminaires et Congrès* (Société Mathématique de France) 3 (1998), 243–273.

Schneider. Theodor. "Das Werk C L Siegels in der Zahlentheorie." *Jahresberichte der Deutschen Mathematiker Vereinigung* 85:4 (1983), 147–157.

Scott, Dana. "Foreword." In Bell, J. L. *Boolean-Valued Models and Independence Proofs in Set Theory*, second edition. Oxford: Oxford University Press (1985).

Segal, S. L. "Helmut Hasse in 1934." *Historia Mathematica* 7 (1980), 46–56.

Serrin, James. "The Problem of Dirichlet for Quasilinear Elliptic Differential Equations With Many Independent Variables." *Phil. Trans. Royal Society London* 264 (1969), 413–496.

Shafarevich, I. R., editor. *Algebraic Geometry I: Algebraic Curves Algebraic Manifolds and Schemes.* Berlin: Springer-Verlag (1994).

Shafarevich, I. R. *Basic Algebraic Geometry.* Translated by Hirsch, K. A. New York: Springer-Verlag (1974).

Shen, A. "Entrance Examinations to the Mekh-mat." *Mathematical Intelligencer* 16:4 (1994), 6–10.

Shields, Allen. "Klein and Bieberbach: Mathematics, Race, and Biology." *Mathematical Intelligencer* 10:3 (1988), 7–11.

Shiryaev, A. N. "Kolmogorov: Life and Creative Activities." *Annals of Probability* 17:3 (1989), 866–944.

Shiryaev, A.N. "Andrei Nikolaevich Kolmogorov (April 25, 1903 to October 20, 1987): A Biographical Sketch of His Life and Creative Paths." In *Kolmogorov in Perspective* (2000).

Siegel, Carl Ludwig. "On the Integrals of Canonical Systems." *Annals of Mathematics* 42:3 (1941), 806–822.

Siegel, Carl Ludwig. *Gesammelte Abhundlungen.* 4 Vols. Berlin: Springer-Verlag (1966–79).

Siegel, Carl Ludwig. *Zur Geschichte des Frankfurter mathematischen Seminars.* Frankfurt/Main: Victorio Klostermann (1965). This is reprinted in *Gesammelte Abhundlungen.* Quotations in my text were translated by Nancy Schrauf. A translation by Kevin Lenzen appears in *Mathematical Intelligencer* 1:4 (1978/79), 223–230.

Siegel, Carl Ludwig. "Erinnerungen an Frobenius." In *Gesammelte Abhundlungen* (piece was published first in 1968), 63–65.

Siegel, Carl Ludwig. "Axel Thue." Preface in *Selected Mathematical Papers of Axel Thue.* Edited by Nagell, Trygve et al. Oslo: Universitetsforlaget (1977).

Siegel, C. L. and Moser, J. K. *Lectures on Celestial Mechanics.* Translated by Kalme, C. I. New York: Springer-Verlag (1971).

Silverman, Joseph H. and Tate, John. *Rational Points on Elliptic Curves*. New York: Springer-Verlag (1992).

Sinai, Ya. G. "Kolmogorov's Work on Ergodic Theory." *Annals of Probability* 17:3 (1989), 833–839.

Sinai, Ya. G. "Remembrances of A. N. Kolmogorov." In *Kolmogorov in Perspective* (2000).

Singh, Simon and Ribet, Kenneth A. "Fermat's Last Stand." *Scientific American* (November 1997), 68–73.

Smale, Stephen. *The Mathematics of Time: Essays on Dynamical Systems, Economic Processes, and Related Topics*. New York: Springer-Verlag (1980).

Smale, Steve. "Mathematical Problems for the Next Century." *Mathematical Intelligencer* 20:2 (1998), 7–15.

Smith, David Eugene and Mikami, Yoshio. *A History of Japanese Mathmatics.* Chicago: Open Court (1914).

Smorynski, C. "Julia Robinson, *In Memoriam*." *Mathematical Intelligencer* 8:2 (1986), 77–79.

Sossinsky, A. B. "In the Other direction." *Golden Years of Moscow Mathematics.* Edited by Zdravkovska, Smilka and Duren, Peter. Providence: AMS (1993), 223–244.

Stevenhagen, P. and Lenstra, H. W. Jr. *Mathematical Intelligencer* 18:2 (1996), 26–37.

Stewart, Ian. "Hilbert's Sixteenth Problem." *Nature* 326:19 (1987), 248.

Stewart, Ian. *Does God Play Dice? The Mathematics of Chaos*. Cambridge Mass. and Oxford, England: Blackwell (1989).

Stewart, Ian and Tall, David. *Algebraic Number Theory*, second edition. London: Chapman and Hall (1987).

Stillwell, John. "Max Dehn." In *History of Topology*. Edited by James, I. M. Amsterdam, New York: Elsevier Science B. V. (1999).

Stone, Marshal H. "Reminiscences of Mathematics at Chicago." *Mathematical Intelligencer* 11:3 (1989), 20–25.

Straus, Ernst G. "Reminiscences." In *Albert Einstein: Historical and Cultural Perspectives.* Edited by Holton, Gerald and Elkana, Yehuda. Princeton: Princeton University Press (1982).

Szabó, Z.I. "Hilbert's Fourth Problem, 1" *Advances in Mathematics* 59 (1986), 185–301.

Taussky-Todd, Olga. "Remembrances of Kurt Gödel." In *Gödel Remembered.* Edited by Weingartner, P. and Schmetterer, I. Napoli: Bibliopolis (1987).

Thurston, William P. "Three Dimensional Manifolds, Kleinian Groups and Hyperbolic Geometry." In Browder, Felix ed. *The Mathematical Heritage of Henri Poincaré.* (1983).

Tikhomirov, V. M. "The Life and Work of Andrei Nikolaevich Kolmogorov." *Russian Math. Surveys* 43:6 (1988), 1–39.

Tikhomirov, V. M. "A. N. Kolmogorov." In *Golden Years of Moscow Mathematics.* Edited by Zdravkovska, Smilka and Duren, Peter. Providence: AMS (1993), 101–128.

Troelstra, A. S. "Arend Heyting" In *Dictionary.*

Ulam, Stanislaw. *Adventures of a Mathematician.* New York: Scribner (1976).

Van Dalen, Dirk. "Hermann Weyl's Intuitionistic Mathematics." *Bulletin of Symbolic Logic* 1:2 (1995), 145–169.

Van der Pooten, Alf. *Notes on Fermat's Last Theorem.* New York: John Wiley and Sons (1996).

van Heijenoort, Jean, editor. *From Frege to Gödel: A Source Book in Mathematical Logic, 1879–1931.* Cambridge, Mass: Harvard University Press (1967).

Vere-Jones, D. "Boris Vladimirovich Gnedenko, 1912–1995. A personal tribute." *Austral. J. Statist.* 39:2 (1997), 121–128.

Vershik, A. "Admission to the Mathematics Faculty in Russia in the 1970s and 1980s." *Mathematical Intelligencer* 16:4 (1994), 4–5.

Viro, O. Ya. "Progress in the Topology of Real Algebraic Varieties Over the Last Six Years." *Russian Math. Surveys* 41:3 (1986), 55–82.

Wang, Hao. *Reflections on Kurt Gödel.* Cambridge, Mass.: MIT Press (1987).

Weil, André. "The Future of Mathematics." *American Mathematical Monthly* 57 (1950), 295–306.

Weil, André. *Foundations of Algebraic Geometry.* Providence: AMS (1962).

Weil, André. *Number Theory: An Approach Through History: From Hammurpi to Legendre.* Boston: Birkhäuser (1984).

Weil, André. *The Apprenticeship of a Mathematician.* Translated by Gage, Jennifer. Basel: Birkhäuser (1992).

Weyl, Hermann. "David Hilbert and His Mathematical Work." *Bulletin AMS* 50 (1944), 612–654.

Weyl, Hermann. "A Half-Century of Mathematics." *American Mathematical Monthly* 56 (October 1952) 523–553.

Whitney, Hassler. "Moscow 1935: Topology Moving Toward America." In *A Century of Mathematics in America.* 3 Vols. Edited by Duren, Peter. Providence, RI: AMS (1988–89).

Wiener, Norbert. *I Am a Mathematician*. Cambridge, Mass.: MIT Press (1964) originally published 1956.

Wolfe, Joseph A. *Spaces of Constant Curvature*, fifth edition. Wilmington, Del.: Publish or Perish, Inc. (1984).

Yanin, V. L. "Kolmogorov as Historian." *Russian Math. Surveys* 43:6 (1988), 183–191.

Yushkevich, A.P. "Encounters with Mathematicians." In *Golden Years of Moscow Mathematics*. Edited by Zdravkovska, Smilka and Duren, Peter. Providence: AMS (1993), 1–34.

Zassenhaus, H. "Emil Artin and His Work." *Notre Dame Journal of Formal Logic* 5:1 (1964), 1–9.

Zdravkovska, Smilka. "Listening to Igor Rostislavovich Shafarevich." *Mathematical Intelligencer* 11:2 (1989), 16–28.

Zdravkovska, Smilka. "Conversations with V. I. Arnold." *Mathematical Intelligencer* 9:4 (1987), 28–32.

Index

A

B

C

D

E

F

G

H

I

J

K

L

M

N

O

P

Q

R

S

W

X

Y

Z